Environmental Sensing

James K. Lein

Environmental Sensing

Analytical Techniques for Earth Observation

James K. Lein
Department of Geography
Ohio University
Athens, OH, USA
lein@ohio.edu

ISBN 978-1-4614-0142-1 e-ISBN 978-1-4614-0143-8
DOI 10.1007/978-1-4614-0143-8
Springer New York Dordrecht Heidelberg London

Library of Congress Control Number: 2011935230

Printed on acid-free paper

Springer is part of Springer Science+Business Media (www.springer.com)

"The sun which is so high and shining is hidden often by a small cloud"

Armenian Proverb

Preface

Remote sensing is neither a difficult or exotic technology. At its most fundamental level, remote sensing is a form of basic field data collection, an activity common to all physical, natural, and social science and motivated by the same overarching goals: to systematically observe phenomena in order to record useful measurements of the variables that define their characteristic properties. Remote sensing simply acquires those measurements in a very different manner. Anyone with a background in subject areas ranging from geography, Earth science, planning, or resource management, to perhaps less obvious fields such as public policy studies, environmental health, and international development can apply remote sensing methods successfully and produce useful information that can address a myriad of problems and issues. To approach this wider audience, I have elected to title this book using the term "environmental sensing," where the verb sensing means "to become aware of"; aware of the technology and of the environmental patterns this technology can illuminate.

My interest in remote sensing science dates to my very first course on the topic as an undergraduate geography major at San Francisco State University and continues to develop as this science continues to evolve. In that first course we marveled at the use of exotic equipment such as the additive color viewer that directed red, green and blue light through gray-scale transparencies of imagery acquired from the Landsat MSS. Today we marvel at the speed by which computer algorithms can process this data in digital form revealing details and complexities of the land surface that have extended remote sensing technology beyond a simple mapping tool. It is here where this book takes focus with the intent of reaching an audience that knows something about remote sensing, but has not attempted to exploit its range of capabilities. I was introduced to this audience a few years ago following an invitation to participate in a workshop aimed at promoting remote sensing to policy makers, government administrators and legal analysts working in the environmental sector. During this experience I was surprised by how a large segment of the environmental community still held numerous misconceptions concerning remote sensing and echoed common frustrations centered around "steep" learning curves of a technology most still regarded as out of reach. My intent in this book is to "reach" this audience and demonstrate that remote sensing science is tamable and can be used with great success when interested parties are equipped with a fundamental appreciation of the theory and methods on which this science is based. Remote sensing no longer requires color additive viewers or other complex and bulky machines, nor does it necessarily require expensive software. Today remote sensing can be practiced on lap top computers with an internet connection and an enterprising user eager to take full advantage of the knowledge contained in the measurements acquired via this technology.

Although it might be convenient to characterize this book as an introductory text on remote sensing, my purpose here is not to replicate what has been done very well elsewhere. Rather, in these pages I have worked to distill the "need to know" information and methods and reassemble them a way that highlights where and how remote sensing can make valuable contributions to the study of our environment. Using a systematic approach, this book explains how remote sensing science can produce pertinent information while at the same time introducing selected methods that are well suited to the situations encountered when contending with the intricacies of human/environmental interaction. Therefore, while not an introductory text, per se, this book is none the less appropriate in a traditional classroom setting in a course that builds on the fundamentals of digital image processing as well as more specialized courses that seek to advance environmental assessment by integrating a remote sensing approach into existing methodologies. Beyond the classroom this book can serve as a resource for those working in the environmental sector that are well grounded in policy matters or the science of the environment, but not well acquainted with remote sensing. To these individuals this book describes a sequenced presentation of the technology and supporting techniques that provide both missing background and the script to follow when developing specific applications where a remote sensing solution is useful.

Whether your focus is the classroom, or professional practice, the paced treatment presented here represents a synopsis of the current approaches to the emerging science of Earth observation. Beginning in Chap. 1 with an overview of what Earth observation means in the context of remote environmental data gathering, the outline of this book continues with the environmental theme by exploring the concept of "environmental sensing" and the areas of concern that direct environmental analysis. From here we explore the tools we can call upon to collect environmental data, describing the expanding array of satellite sensor systems and their products. Chapter 4 introduces the fundamental techniques used to extract information from satellite-based sensors which transitions into a discussion of uncertainty and soft computing (Chap. 5) environmental characterization (Chap. 6), and environmental monitoring (Chap. 7). From there, the book branches out into comparatively new or underutilized applications areas such as thermal analysis and anomaly detection (Chap. 8) followed by an examination of hyperspectral remote sensing in Chaps. 9 and 10. Next we explore object-based image classification (Chap. 11) and introduce an emerging area I have termed forensic remote sensing (Chap. 12). Finally our treatment of remote sensing concludes with a chapter dedicated to the role and integration of remote sensing in the larger field of geomatics and applied geospatial analysis (Chap. 13).

Obviously, none of this work took place in a vacuum, and I am indebted to numerous people for the opportunity to produce this work. First and foremost I would like to thank my wife Christine for the encouragement and guidance as I embarked on this project. It is always challenging to take on the task of writing a book and it is nice to have the support of family when you feel like giving up. I would also like to thank those reviewers whose comments and suggestions greatly improved my initial draft. Finally, I would like to extend a special thank you to Dr. Richard Beck, whose phone call in 1998 made all of this work possible and truly changed how remote sensing technology could be accessed and enjoyed by a broader range of people. It was a simple question, "Would you be interested in helping to improve access satellite data." I had the presence of mind to say yes and the Ohio-View (AmericaView) program was born (http://www.americaview.org). From that day forward "bringing remote sensing down to earth" became an active part of my teaching and research agenda and I hope that idea translates and communicates within the pages of this book.

Athens, OH James K. Lein

Contents

Chapter 1
The Earth Observation Perspective

Environmental sensing and the analytical techniques that transform raw measurements acquired from the instruments deployed in Earth orbit are the focus of this book. The process of rendering these data useful, however, involves more than the application of a method. Integral to this transformation are the innately human abilities to abstract from general principles and to craft a conceptual view of sensing technologies and techniques that place them concretely into the problem-based and decision-oriented roles that they have been designed to support. In this context, environmental sensing activities are neither difficult nor exotic; rather they are often misunderstood and viewed as beyond the reach of those individuals and groups that can benefit from them. Satellite remote sensing can significantly enhance the information available from traditional data sources because it can provide synoptic views of large portions of the Earth. Satellite imagery can also expand the spatial dimensions of limited and sometimes costly field or point-source sampling efforts. Some satellite sensors cover areas that may be physically or politically inaccessible or that are too vast to survey with traditional methods. Remote sensing can also provide consistent repeat coverage at relatively frequent intervals, making the detection and monitoring of change feasible. Satellite-derived data and information are also useful for applications that require fine spatial resolution such as surveys of urban and suburban land use, land cover for agricultural purposes, and natural resource management (Wang et al. 2010). However, regardless of sophistication, satellite remote sensing has obvious disadvantages, such as the inability of some sensors to obtain data and information through cloud cover and other systems where spatial resolutions do not provide sufficient geographic detail of the surface to support meaningful analysis. Similarly, the data gathered by sensor systems need to correct for atmospheric absorption and scattering, and other effects that can make it difficult to obtain desired data and information on particular environmental variables. In practical terms, satellite remote sensing also creates large quantities of data that typically require extensive processing as well as storage and analysis; and in the "real-world" setting of agencies and institutions, the data from satellite remote sensing are often costly hampered by intellectual property restrictions, can limit the dissemination of

J.K. Lein, *Environmental Sensing: Analytical Techniques for Earth Observation*,
DOI 10.1007/978-1-4614-0143-8_1, © Springer Science+Business Media, LLC 2012

the information products derived from these sources. The goal of this chapter is to suggest an alternative perspective that elevates environmental sensing to the status of an accessible and commonplace means of addressing the issues that define human–environmental interaction and examine the fundamental of environmental sensing using satellite-based instruments.

1.1 Earth Observation and Environment

Since the deployment of the first civilian land resource satellite in 1972, the role of satellite-remote sensing as a means to improve our understanding, management, and monitoring of Earth's natural resources has been pivotal as nations of the world make choices that influence not only their futures, but also the future of our planet. With the advent of satellite remote sensing, Earth observation has evolved to become a critical method for gathering information regarding the physical, chemical, biological, and human elements that comprise our environmental system. Although the technologies that support Earth observation have become increasingly varied and sophisticated, their purpose is singular; to facilitate informed decision making at local to global scales.

Connecting the science of Earth observation to the intricacies that define environmental process can be a daunting undertaking. Inferring the environmental parameters from measurements collected remotely requires knowledge of the techniques of remote sensing as well as that of the process that pattern the surface of the Earth. Integrating remote sensing theory and practice into this larger framework focuses on the information that can be gained through the processing of remotely sensed data. Information needs, however, are diverse and the environmental issues confronting society are equally varied and variable. The Earth observation perspective is one approach to better connect the data gathering assets that characterize the current status of satellite-based remote sensing with the informational requirements directing environmental analysis and policy making (Lautenbacher 2006; Goetzs 2007; Boyd 2009).

Earth observation science is based on the premise that continuous and systematic acquisition of information concerning the status of objects and phenomena on the surface can advance our fundamental understanding of the environmental system (Bauer et al. 2006; King and Birk 2004). Programmatic endeavors in the context of societal and national imperatives direct Earth observation science and its associated technologies toward the documentation, modeling, and analysis of practical societal problems. At the center of this science are the satellite systems deployed in a variety of orbits that capture data with contrasting capabilities and an expanding capacity for delivering information products tailored to address specific objectives. The Earth observation perspective exploits this new paradigm of integrated global observation to expedite the assimilation of environmental data as decision support. As a component of decision support, Earth observation data demands timely access to facilitate ease of application. In very simple terms, this emerging model characterizes scenario where for any one location of the surface, numerous sensor systems are actively recording data regarding its status (Thackrah 2006). When an operational need for data arises, connecting the problem to its most appropriate sensor becomes similar to the craftsman's task of selecting the appropriate tool from the array of equipment displayed on the wall of the workshop.

Decision support also recognizes that the environmental system is complex. The problems and situations confronted in the process of decision making are often ill-structured and ill-defined. Incorporating the Earth observation perspective as part of the environmental decision making process reinforces the application of data products derived from Earth observational sensors, where synoptic scale and geographic detail can be matched to the requirements specified by the problem. To the decision maker, the challenge is to identify which sensor instrument to select and how best to maximize the use of the data it collects. As the number and scope of sensor systems increase and technological advancements continue to refine their capabilities, we can conceptual a future where environmental analysis is couple to continuous monitoring and data gathering in a coordinated and policy-relevant manner. In this conceptualization, the Earth observation perspective expands the methods used in the study of the environmental system. However, this model of decision analysis is not a panacea. Success in the design on decision-centric applications develops out of a sensitivity and fundamental awareness of (1) the practical and technological constraints that surround this emerging science and (2) the inherent complexity that envelops the study of human–environmental interaction.

1.2 Confronting Complexity

While it is often convenient to approach a technology only as a technology, concentrating on the how-and-why aspects that describe its functioning, this common predilection creates an artificial separation that segregates those which may benefit from the technology from those charged with its development and management. With its focus on sensor instruments and specialized methods of data processing, environmental sensing is a subject that can easily fall prey to this intellectual trap. However, if the technological system is integrated into the large question of environmental process and change, a synergy can be realized where the complex needs of a decision problem are fitted with a requisite level of environmental information. One useful approach to achieve a synergistic fit between disparate technologies and the environmental system draws from complexity theory.

From its beginnings in the theoretical areas of mathematics and physics near four decades ago, complexity theory has evolved from an equation-based paradigm to an environmental science model that has become increasingly useful when examining how the human–environmental system operates (Manson 2001; Contanza1993; Phelan 1999). The idea of complex systems is an approach to science that explains how relationships between parts give rise to the collective behaviors of a system and how that system interacts to form relationships with its environment. Complexity theory illuminates many interactions between natural and social systems and can offer a new understanding of the general principles at work, which can be used to help address vexing environmental issues. The central feature of this theory is the concept of a system that consists of definable parts that function to produce a result. However, rather than reducing a system to its components, complexity theory recognizes that elements of a system cannot be simplified by grouping them together and instead focuses on how components interact. From this assertion, complexity investigates emergent, dynamic, and self-organizing systems that interact in ways that influence the probabilities of later events; and at a different level of analysis, complexity theory provides a new paradigm that can be employed to formalize how scale and abstraction gives rise to contrasting explanations. By concentrating analysis on complexity, systems are irreducible to elementary laws or simple processes. The equations from which complex system models are developed generally derive from statistical physics, information theory, and non-linear dynamics, and represent organized but unpredictable behaviors of systems of nature that are characterized by the combination of elements in intricate arrangements (i.e., complex).

Within this theory, there are four types of complexity can be observed. The first form of complexity is commonly referred to as static complexity. This is perhaps the simplest definition of complexity and is based on the assumption that what we are interested in is as an image that it does not change over time or space (Kauffman 1995). From the perspective our understanding of static systems, knowing that they are complex, allows us to compare and relate them to other systems to determine the complexity of the system. The second type of complexity is called dynamic complexity and it introduces the dimension of time. Time recognizes set patterns or cycles, but it does not provide a means to organize systems into categories. Therefore, to understand this type of complexity, we rely on being able to test and confirm what might be going on, and to deduce whether a system is either static or some kind of cycle. If we observe Earth environments through the lens of complexity, we see objects that do not change defined as variables. However, complexity theory requires consideration of process or action as well, which introduces the third type of complexity: evolving complexity. Evolving complexity describes the situation where systems over time are able to evolve into different systems (Kauffman 1995). This type of complexity is easily thought of when we consider how systems are able to change and describe new states or conditions. Quickly we recognize that any complex system has many combinations of variables, and many of these variables might not have taken place yet. The final explanation of complexity is the most relevant to the analysis of human–environmental interaction: self-organizing complexity. This type of complexity characterizes the case where we can take a closed system, ecosystem, and combine the constraints observed by the closed system with those of an open system, typically one defined as inherently human (Kauffman 1995). This is called coevolving where we must describe the functions of the system and how they relate to the world. These four types of complexity enable us to do under the complexity theory is predict solutions that will occur from differing constraints and configurations we construct. We provide an environment and let the random systems in it use their variables and come together to form a solution that is useful to us in a sense that we can predict what will happen under certain environments and constraints. Several properties of complex systems can be noted with respect to these definitions:

- Systems influence one another. Since each agent and each system is nested within other systems, all evolving together and interacting, we cannot fully understand any of the agents or systems without reference to the others. This can lead to unexpected actions in response to change and so complicates the prediction.
- Actors defined in a complex adaptive system respond to their environment by using internalized rules that drive behavior. In a biochemical system, the "rules" are chemical reactions. At a human level, the rules can be expressed as instincts, procedural rules, or mental models. These internal rules need not be shared, explicit, or even logical when viewed by another agent.
- Complex systems interact with other complex systems which produces tension and paradox that can never be fully resolved. In complex social systems, the seemingly opposing forces of competition and cooperation often work together in positive ways: fierce competition within an industry can improve the collective performance of all participants.
- Neither the system nor its external environment are, or ever will be, constant.
- Individuals within a system are independent and creative decision makers.
- Uncertainty and paradox are inherent within the system.
- Problems that cannot be solved can nevertheless be "moved forward."
- Effective solutions can emerge from minimum specification.
- Small changes can have big effects.
- Behavior exhibits patterns (that can be termed "attractors").
- Change is more easily adopted when it taps into attractor patterns.
- Complex systems frequently produce fluctuations that are often explicable only at the level of the whole system.

With respect to the Earth observation science, recent understanding of the complexity of natural systems has eliminated the idea that everything is determined. Therefore, knowing that it is impossible to understand every facet of the environmental system, the relationships that govern environmental process observable in the system are directed by the natural interaction of variables and other systems. This understanding of the complexity theory carries important implications practically and philosophically, because it demonstrates that we may need to change our approach to what questions we might be asking in trying to understand the environmental system. The information we gain through the Earth observation perspective should cause us to change the questions we ask as well as our understanding of the environmental system. In this context, the value of complexity thinking rests in its ability to:

- Explain emergent structures (self-organization).
- Measure relative complexity (hierarchical multiparameter).

- Provide control methods for complex systems (steering points).
- Generate effective models (abstractions).
- Give statistical predictors (constraints).
- Solve outstanding problems (breakthroughs).
- Demonstrate possible new applications (novelty).
- Quantify the laws of order and information (if any).

1.3 Remote Sensing Theory and Practice

Over the past four decades, remote sensing has emerged as one of the more important and widely described methods for collecting data on the disposition of Earth surface phenomena (Jensen 2007; Campbell 2002). Today satellite-based remote sensing serves as the foundation for Earth observation and the cornerstone for the future. Indeed. Earth observation from space through various remote sensing instruments has become a rapid and cost-effective means of monitoring land surface dynamics, and remote sensing is one of a suite of tools available to environmental analysts that provides up-to-date, detailed information about condition of the environmental system. Although definitions abound, for our purposes, we can explain remote sensing as the science and art of acquiring information about the Earth's surface without actually being in contact with it. When compared with other methods of environmental data collection, remote sensing focuses on measurements made at a distance, rather than in situ. The process of remote sensing involves the detection and measurement of radiation of different wavelengths reflected or emitted from distant objects or materials, by which they may be identified and categorized by class/type, substance, and spatial distribution. The success of this approach hinges on sensing and recording reflected or emitted energy and processing, analyzing, and applying that information to reveal these patterns. In addition, remote sensing systems, particularly those deployed on satellites, provide a repetitive and consistent view of the Earth which directly supports the conceptual model implied by the Earth observation science (Schowengerdt 2007).

The quantity commonly measured by the current generation of remote sensors is the electromagnetic (EM) energy emanating from the object of interest. The electromagnetic spectrum is the continuous range of electromagnetic radiation, extending from gamma rays (highest frequency and shortest wavelength) to radio waves (lowest frequency and longest wavelength) and including visible light, and each has useful properties that can be exploited for the purposes of remote sensing. Based on the properties of electromagnetic radiation, a set of fundamental principles can be described that direct its use as a measurement tool:

1. Electromagnetic energy is explained by wavelength and arranged to form the electromagnetic spectrum.
2. As electromagnetic energy interacts with the atmosphere and the surface of the Earth, the most important concept to remember is the conservation of energy (i.e., the total energy is constant).

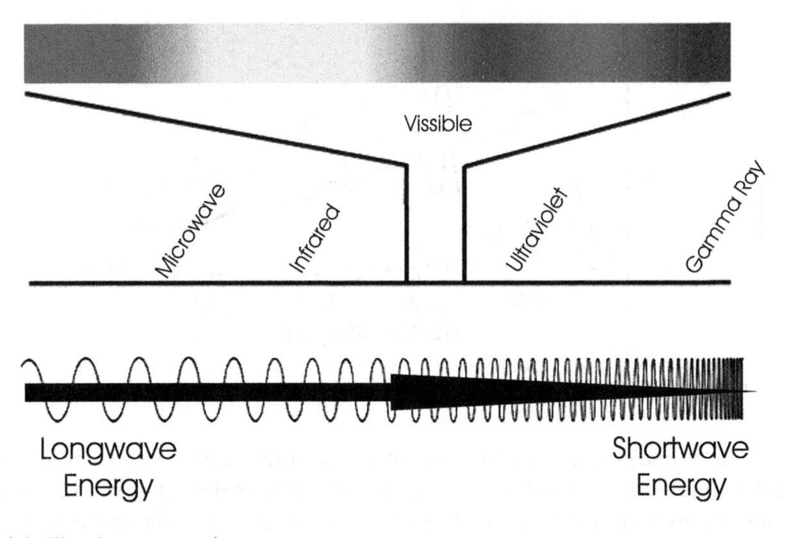

Fig. 1.1 The electromagnetic spectrum

3. As electromagnetic waves travel, they encounter objects (discontinuities in velocity) that reflect some energy such as a mirror and transmit some energy after changing the travel path.
4. The distance (*d*) an electromagnetic wave travels in a certain time (*t*) depends on the velocity of the material (*v*) through which the wave is traveling: $d = vt$.
5. The velocity (*c*), frequency (*f*), and wavelength (*l*) of an electromagnetic wave are related by the equation: $c = fl$.
6. The analogy of a rock dropped into a pond can be drawn as an example to define wave front.
7. It is quite appropriate to look at the amplitude of an electromagnetic wave and think of it as a measure of the energy in that wave.

The EM spectrum can be divided into seven different regions – gamma rays, X-rays, ultraviolet, visible light, infrared, microwaves, and radio waves (Fig. 1.1). Although remote sensing involves the measurement of energy in many parts of the electromagnetic (EM) spectrum, the major regions of interest in satellite-based sensing are visible light, reflected and emitted infrared, and the microwave regions. The measurement of this radiation takes place in what are known as spectral bands. A spectral band is defined as a discrete interval of the EM spectrum. For example, the wavelength range of 0.3–0.4 μm (or 10–6 m) is one spectral band. Satellite sensors have been designed to measure responses within particular spectral bands to enable the discrimination of the major Earth surface materials. In the design of a remote sensing investigation, particular spectral band(s) are selected for data collection depending on what they wish to examine.

The underlying theory that support remote detection and measurement is based on the assumption that particular features of the landscape such as tree cover, agricultural crops, urbanized land, and water reflect (and/or emit) electromagnetic

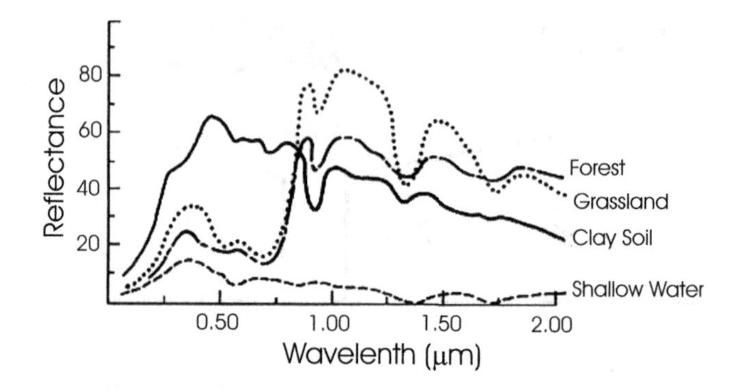

Fig. 1.2 Generalized spectral response pattern of selected land covers

energy differently in specific portions of the electromagnetic spectrum. Detection and discrimination of objects or surface features means detecting and recording of radiant energy reflected or emitted by objects or surface material and transforming these measurements of energy flux to thematic information. Because different objects return different levels of energy in different bands of the electromagnetic spectrum, a pattern of response is produced as a function of the properties of the surface material (structural, chemical, and physical), surface roughness, angle of incidence, intensity, and wavelength of radiant energy. In simple terms, grass on a healthy lawn looks green to the human eye, because it reflects green light and absorbs other visible wavelengths. On digital imagery that pattern can be seen as a peak in the green band in the reflectance spectrum for green grass. This pattern of reflected radiation measured according to wavelength is called the spectral signature of the surface. When solar radiation hits a target surface, it may be transmitted, absorbed, or reflected. Different materials reflect and absorb differently at different wavelengths which can be seen in the reflectance spectrum of a material when displayed as a plot of the fraction of radiation reflected as a function of the incident wavelength (Fig. 1.2). This patterning serves as a unique signature for the material. In principle, a material can be identified from its spectral reflectance signature, if the sensing system has sufficient spectral resolution to distinguish its spectrum from those of other materials. This premise provides the basis for the thematic extraction of information taken from remote measurements. The digital measurements of electromagnetic energy captured by the remote sensing instruments detect these "signature" variations in wavelength intensity and record it as a number typically along a "brightness scale" of values between 0 and 255. Three essential characteristics of our data enable the identification of signature pattern:

1. *Spectral differentiation* – Remote sensing depends upon observed differences in the energy reflected or emitted from features of interest.
2. *Radiometric differentiation* – Examination of any image acquired by remote sensing ultimately depends upon the detection of differences in the brightness of objects and the features.
3. *Spatial differentiation* – Every sensor instrument is limited with respect to the size of the smallest area that can be separately recorded as an entity on an image.

Achieving this level of discrimination in strongly influenced by the interaction between incident radiation and the surface we wish to measure and understand. However, the detection and measurement is based on energy interactions depends upon a number of factors:

- *Source of illumination* (A) – The first requirement for remote sensing is to have an energy source which illuminates or provides electromagnetic energy to the target of interest.
- *Radiation and the atmosphere* (B) – As the energy travels from its source to the target, it will come in contact with and interact with the atmosphere it passes through. This interaction may take place a second time as the energy travels from the target to the sensor.
- *Interaction with the target* (C) – Once the energy makes its way to the target through the atmosphere, it interacts with the target depending on the properties of both the target and the radiation.
- *Recording of energy by the sensor* (D) – After the energy has been scattered by, or emitted from the target, we require a sensor (remote – not in contact with the target) to collect and record the electromagnetic radiation.
- *Transmission, reception, and processing* (E) – The energy recorded by the sensor has to be transmitted, often in electronic form, to a receiving and processing station where the data are processed into an image (hardcopy and/or digital).
- *Interpretation and analysis* (F) – The processed image is interpreted, visually and/or digitally or electronically, to extract information about the target which was illuminated.
- *Application* (G) – The final element of the remote sensing process is achieved when we apply the information we have been able to extract from the imagery about the target to better understand it, reveal some new information, or assist in solving a particular problem.

These components organize to form the remote sensing system (Fig. 1.3). The decisive factor in the successful application of remote sensing data, however, need not be the technical sophistication of the user, but rather the suitability and precise use of the tool to obtain accurate and relevant data. With a general grasp of the technical process that transforms electromagnetic energy into useful information, the appropriate use of these tools can be improved and expanded. The goal of image processing is to detect features and changes in those features over time, and to be sure that what is seen is related to the ground cover rather than to interference caused by the atmosphere. To do this, sequences of images are aligned to each other and to standard map grids (registration and rectification) and are calibrated to remove the effects of atmospheric differences. The utility of remotely sensed data for a given application problem is generally related to its spatial, spectral, radiometric, and temporal resolutions.

- *Spatial resolution* – The size of a pixel that is recorded in a raster image – typically pixels may correspond to square areas ranging in side length from 1 to 1,000 m (3.3–3,300 ft).

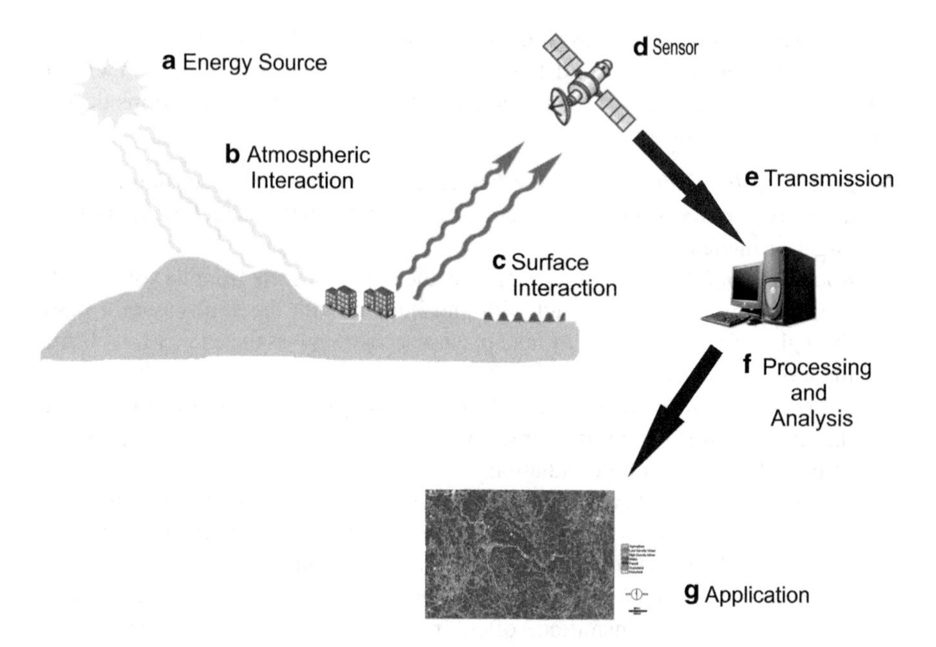

Fig. 1.3 Elements of a remote sensing system

- *Spectral resolution* – The wavelength width of the different frequency bands recorded – usually, this is related to the number of frequency bands recorded by the platform.
- *Radiometric resolution* – The number of different intensities of radiation the sensor is able to distinguish. Typically, this ranges from 8 to 14 bits, corresponding to 256 levels of the gray scale.
- *Temporal resolution* – The frequency or repeat cycle defined by the satellite and its orbital.

As an alternative to field-based data collection, the application potential of remote sensing introduces (1) a practical way of obtaining data from difficult or inaccessible areas, (2) a synoptic view of the site and situational features that define the problem, (3) a method to achieve near-continuous acquisition of data, and (4) a data collection methodology less time-sensitive. Remotely sensed data, however, are not direct samples of the phenomenon of interest but measures that must be calibrated against reality. This data must also be corrected geometrically and geographically references to form a reasonable representation of the surface under investigation. Even when "corrected" issues related to thematic uncertainty and interpretation error remain difficult to resolve. Controlling for the deficiencies inherent to remote sensing data analysis can be accomplished by adopting well thought out procedures that blend the technique into the needs of the problem. In general, the systematic nature of remote sensing analysis is directed by three

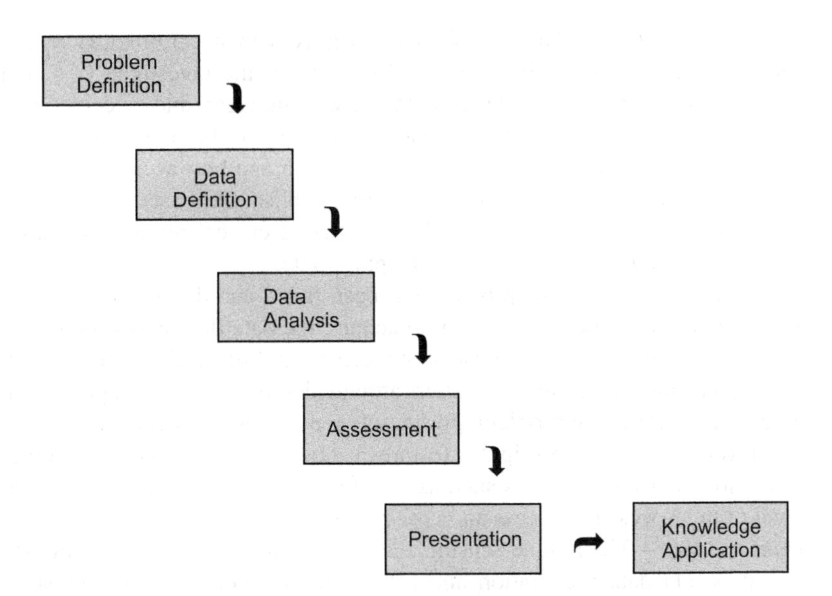

Fig. 1.4 General methodology for conduction a remote sensing investigation

grounding axioms, particularly in modern remote sensing here computer processing is integral to the methodology (Madhok and Landgrebe 2002):

1. Human abilities are different from those of the computer – remote sensing data convey information regarding energy measurements and inter-pixel relationships. The subjective evaluation afforded by the image representation schemes are the interface between the analyst and machine.
2. The machine (in) validates the user's hypothesis – algorithms that process spatially organized data through the optimization of mathematical criteria are often suboptimal in the sense that output may be "noisy" or visually confusing.
3. Every analysis usually requires at least one revision – terminating an analysis depends on the tolerance level for acceptable error, the available resources, and time. Implementing a solution typically requires "experimentation" and the best solution is often a result of several techniques pieced together.

By keeping these "realities" in mind, a workable methodology can be designed that directs choice toward a solution that satisfies the problem. Above all they remind us that remote sensing is as much an "art" as it is "science," where procedures, regardless of their logical structure, demand equal amounts of creative thinking and adaptive problem-solving skills.

As a methodology, remote sensing activities can be organized around a six-phase sequence of steps (Fig. 1.4). Although presented here in linear fashion, for any given investigation image analysis may require significant backtracking and several iterations thought a particular stage before a successful result is produced.

1. *Problem definition* – This initial phase begins with a careful and reasoned statement of the goals, objectives, and purpose or the investigation – simple questions that can be challenging to answer in clear and succinct terms. In addition, there must be some consideration given to the selection or identification of success metrics that communicate when and how well a solution has been achieved. Success metrics also help establish the basis for evaluation, an assessment of the results, criteria to engage a critical review, and suggest where refinements in methods may be appropriate.
2. *Data definition* – Once the problem has been stated and the underlying knowledge defined, it becomes necessary to acquire the data that drives the solution. As with any analysis, data must be collected carefully and in the context of remote sensing, data collection centers around identifying the appropriate source given the nature of the problem. For remote sensing, this directs attention to the senor system and its descriptive features such as spectral, spatial radiometric resolution, as well as factors such as data format, areal coverage and available level of preprocessing, access and cost considerations.
3. *Data analysis* – The remote sensing solution requires two important analytical activities: (1) data preparation and (2) algorithm selection. Data preparation explains the procedure required to remove noise and other artifacts from the data that compromise its fidelity. Through data preparation a comparatively "clean" image is produced. Cleaning might include various corrections to remove defects such as unusable bands, or calibrate the data to enhance contrast and the "correctness" of the radiometric measurements. Algorithm selection focuses on implementing a specific image processing strategy to extract thematic information from the raw imagery. An image processing strategy may include statistically based pattern-recognition techniques, strategies based on approximate reasoning methods, or algorithms that employ methods adapted from artificial intelligence. The specific choice of algorithm depends to a large degree on the nature of the problem and the desired information product to be extracted from the data.
4. *Assessment of results* – This analytical phase concentrates on the correctness (accuracy) and utility of the thematic information derived from the raw imagery. Accuracy assessment is critical undertaking before the derived product can be used to guide decision making. This is also the phase where the success metrics are used to determine both subjectively and quantitatively, if the results of thematic extraction describe reasonable representations of surface conditions.
5. *Presentation* – Once the derived results satisfy the criteria for success, the thematic information must be organized into a format that transmits and communicates the information it contains. A variety of communication formats are available depending on the purpose and intended audience of the information. Commonly used mediums for communication include maps in either printed or digital form, summary statistical reports, graphs, and more recently as inputs to more and more comprehensive spatial database.
6. *Knowledge application* – Recognizing that the product of a remote sensing investigation is more that simply a "pretty picture from space," this final phase

in the information extraction process directs our attention to the actual use of the information in its application environment. At this phase, issues related to both the proper use of the results and the correctness of interpretations, inferences, and conclusions drawn are paramount. Knowledge application moves us beyond the mechanics of accuracy to consider questions pertaining to the answers found in the thematic information, and whether the product delivers knowledge relative to the problem and the product provides support for any conclusions that emanate from its use.

Owing to the technical nature of the data processing operations that are typically involved in a remote sensing application, disseminating the knowledge contained in the results of an analysis is often an afterthought. However, for remote sensing to yield value in relation to environmental decision making, both the technology and the data must be accessible to those who might benefit from it and fit seamlessly into the operational setting through the combination of a reasoned application design executed using a tractable procedural model. Improving access to the environmental community that has an unsatisfied need for timely data and the corresponding means of analysis is addressed in the section to follow.

1.4 Bringing Remote Sensing Down to Earth

Heightened interest in the practical applications of Earth observations from satellite platforms has coincided with renewed concerns related to environmental change and increasing societal pressures of Earth's ecosystems. Over the past four decades, there have been significant improvements in the availability and capabilities of remote sensing data. Yet despite these improvements, critical bottlenecks remain that frustrate the adaptation of remotely sensed data as an active component of institutional decision making. For example, in the USA, satellite remote sensing has been historically until dominated by federal agencies and their private sector contractors and was focused on reconnaissance, scientific and technological innovation, and operational weather monitoring and prediction (Steering Committee on Space Applications and Commercialization and National Research 2001). Satellite-based remote sensing provides sources of information that cannot be easily obtained in other ways, however to achieve the full potential of this technology discovering cost-effective ways to realize the potential useful applications that exist is essential. In general, encouraging a wider use of remote sensing data demands that this often misunderstood technology is made accessible to an audience that may not enjoy the technical expertise to exploit its capabilities.

Earth observation to feed environmental decision making and analysis are activities driven by information rather than by the technical capabilities of the end users. Here, unlike those initial "proof of concept" applications of remote sensing, future application users are likely to have little, if any, knowledge of remote sensing technology or how it is employed to derive information

(Steering Committee on Space Applications and Commercialization and National Research 2001). Rather, their primary focus will likely be centered on issues of accuracy and timeliness of the data as well as its relevance to specific decision problems. The interplay between the intrinsic information content of the raw data produced by a remote sensing instrument, the way the data are processed to produce new information and the operational utility of the applications brought to the technology are central determinants of value gained by adoption an Earth observation perspective. Therefore, extending the benefits of remote sensing to environmental decision makers involves both technology and knowledge transfer and overcoming long-standing barriers than have restricted the adoption of remote sensing and produced:

- Gaps between the raw remote sensing data and the information needs of applications users.
- Gaps in communication and understanding between those with technical experience and training and those who are the potential end users of a remote sensing technology.
- Gaps between the acquisition of remote sensing data and the development of a usable application.

Two important developments have emerged to narrow these gaps and bring remote sensing down to Earth. The first advancement has been the elevation of data distribution archives the repositories and that provide free access to civilian satellite imagery. Perhaps the best example of this revolution in data access is the GLOVIS portal (http://glovis.usgs.gov/). The United States Geological Survery (USGS) Global Visualization Viewer (GLOVIS) is an online search and order tool for selected satellite data (Fig. 1.5). The viewer allows access to all available browse images from the Landsat 7 ETM+, Landsat 4/5 TM, Landsat 1–5 MSS, EO-1 ALI, EO-1 Hyperion, MRLC, and Tri-Decadal data sets, as well as Aster TIR, Aster VNIR, and MODIS browse images from the Land Processes Distributed Active Archive Center (LP DAAC) inventory. Through a graphic map display, the user can select any area of interest and immediately view all available browse images for the specified location. From the browse image viewer page, the user may either navigate to view adjacent scene locations or select a new area of interest. The portal also offers additional features such as cloud cover limits, date limits, user-specified map layer displays, scene list maintenance, and access to metadata. The highly visual nature of the GLOVIS interface offers is a quick and easy method to perform online search for selected satellite data. The viewer allows user-friendly access to browse images from the multiple data holdings and by means of a graphic map display, the user can select any area of interest and immediately view all available browse images within the USGS inventory for the specified location. From the browse image viewer page, the user may either navigate to view adjacent scene locations or select a new area of interest. A downloading interface is also provided for datasets that are available at no charge.

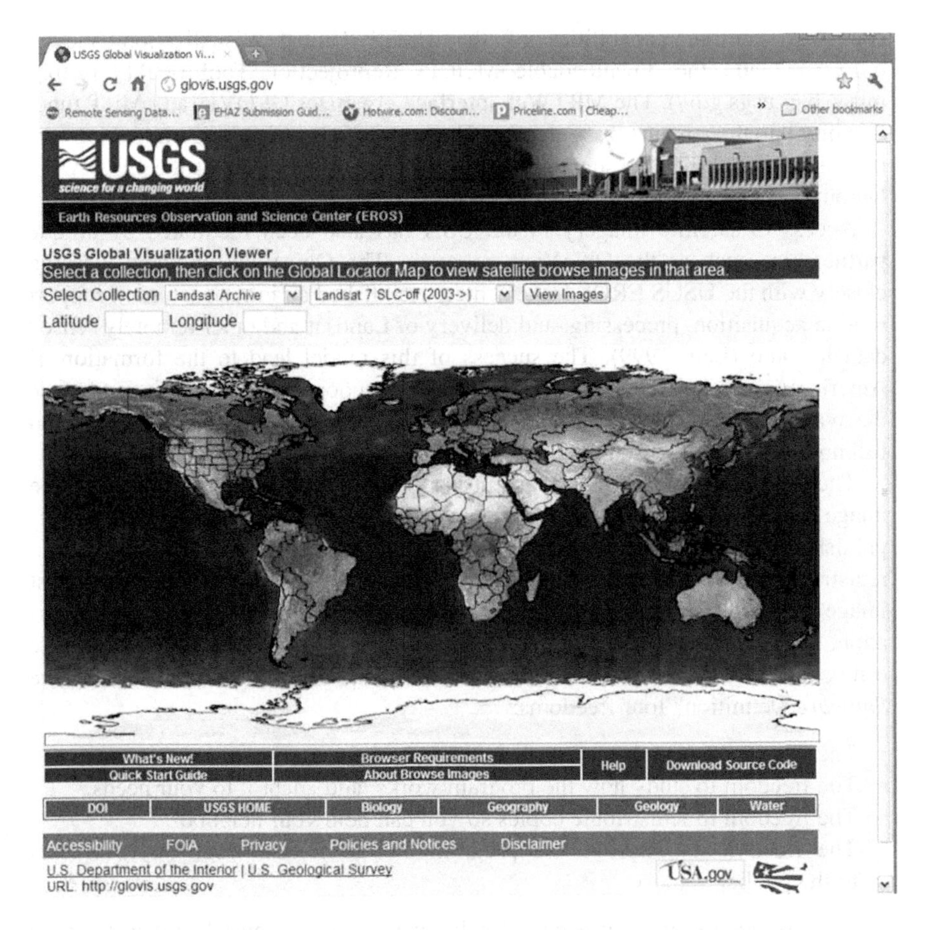

Fig. 1.5 The USGS global visualization server opening Web page

In addition to GLOVIS is a search tool named The Data Pool facilitates access to satellite imagery. The Data Pool is the publicly available portion of the LP DAAC online holdings. Data Pool provides a more direct way to access files by foregoing their retrieval from tape storage devices. All Data Pool holdings are available at no cost to the user. The Data Pool access method allows users to select a collection interest, then define spatial and temporal filters to return a listing of data sets meeting their criteria. Data files returned from this search are exposed to limited on-the-fly reformatting, reprojection, and subsetting services users to navigate directly to the product collection of interest and then to the product period(s) of interest within the online file structure.

Other search tools of the Land Processes Distributed Active Archive Center (LP DAAC) include the MODIS reprojection tool Web (MRTWeb) interface, which was developed to provide enhanced, Web-based discovery and delivery services for MODIS land product tiles archived at the LP DAAC. MRTWeb 2.0

combines familiar capabilities of the USGS Global Visualization Viewer (GLOVIS) and the downloadable MODIS Reprojection Tool (MRT) (http://mrtweb.cr.usgs.gov/). The MRTWeb interface organizes GLOVIS and MRT functionality into three main tabs: Selection, Process, and Download. Submitted jobs are run with MRT 4.0 processing software across multiple servers at the LP DAAC. Output data sets are staged on a job-specific ftp directory for download.

Access to satellite imagery in the USA has also been facilitated by unique partnerships such as the OhioView program. The OhioView partners, working closely with the USGS EROS Data Center, established an infrastructure to support routine acquisition, processing, and delivery of Landsat and other remotely sensed data to Ohio (Lein 1999). The success of this model lead to the formation of AmericaView: a locally controlled and nationally coordinated program to advance the availability, timely distribution, and widespread use of remote sensing data and technology (http://www.americaview.org).

The second significant advancement has been the introduction of open source image processing software. Open source software is software whose source code is published and made available to the public, enabling anyone to copy, modify and redistribute the source code without paying royalties or fees. As opposed to most image commercial systems, open source code evolves through community cooperation. According to the Free Software Foundation (http://www.fsf.org), software can be labeled as *free software if the associated license conditions fulfill the* "*Free Software* Definition" four freedoms:

- The freedom to run the program, for any purpose.
- The freedom to study how the program works, and adapt it to your needs.
- The freedom to redistribute copies so you can help your neighbor.
- The freedom to improve the program and to release your improvements to the public.

Because open source software is freely available and licensed so that it can be freely distributed and modified, it removes several of the bottlenecks that frustrate application development. There are several licensing options for open source software but follow the same set of basic rules:

- Free redistribution
- Source code
- Integrity of the author's source code
- No discrimination against persons or groups
- Distribution of license
- License must not be specific to a product
- License must not restrict other software
- License must be technology-neutral.

In developing a remote sensing application, open source software can be a replacement for or a compliment to propriety software or it may function solely to "fill in the functionality holes" of the propriety software currently in use. Perhaps the most attractive feature of the open source alternative is for those new to remote sensing technology: it is free of charge and something to learn and experiment with. In this context, some of the benefits of open source systems include:

- Cost savings.
- Improved quality assurance.
- Avoiding vendor lock-in.
- Quicker bug fixes.
- Agile deployment changes.
- Decreasing the risk of being stranded by proprietary systems.

Over the last decade, the world of free and open source geospatial software has experienced important advances that have enhanced access to the means of image processing and GIS analysis. For instance, the Web site FreeGIS.org currently list over 325 GIS-related projects that deliver useful GIS functionality. Besides the advent of new software projects and the growth of established projects, a new organization known as the OSGeo Foundation has been established to offer a point of contact for those searching for open source alternatives (http://www.osgeo.org/).

Currently, there are several open source alternatives to commercial products that provide useful image processing and this list is likely to expand as the open source alternative draw greater acceptance. A sample of open source alternatives to commercial remote sensing packages include:

- *OPTICKS* – *Opticks* is an open source, remote sensing application that supports imagery, video (motion imagery), Synthetic Aperture Radar (SAR), multi-spectral, hyper-spectral, and other types of remote sensing data. Opticks is unlike other remote sensing applications because it treats imagery and video alike. Opticks is one of the only remote sensing applications that supports processing remote sensing video. Opticks was initially developed by Ball Aerospace and Technologies Corp. and other organizations for the United States Intelligence Community. Ball Aerospace open sourced Opticks hoping to increase the demand for remote sensing data and broaden the features available in existing remote sensing software. The Opticks software and its plug-ins are developed by over 20 different organizations. Opticks can also be used as a remote sensing software development framework (Fig. 1.6).
- *Multispec* – MultiSpec is a processing system for interactively analyzing Earth observational multispectral image data such as that produced by the Landsat series of Earth satellites and hyperspectral image data from current and future airborne and spaceborne systems such as AVIRIS.
- *OSSIM* – OSSIM provides advanced geo-spatial image processing for remote sensing, photogrammetry, and Geographic Information Systems (Fig. 1.7). Backed by an active open source software development community, OSSIM solutions have been deployed on a number of critical commercial and government systems. OSSIM is a high performance software system for remote sensing, image processing, geographical information systems and photogrammetry. It is an open source software project maintained. at http://www.ossim.org and has been under active development since 1996.

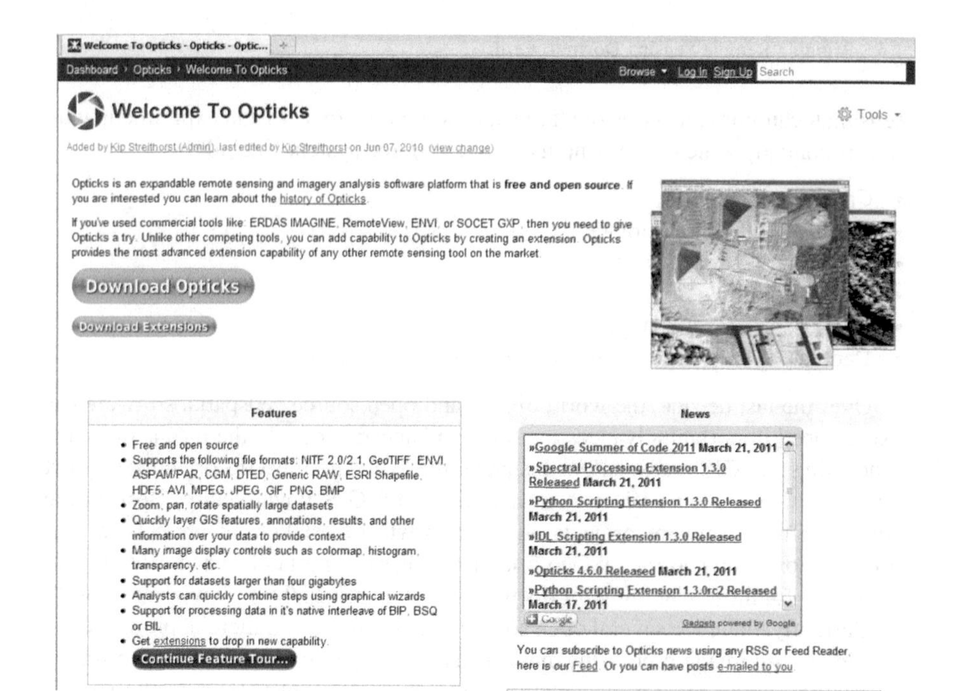

Fig. 1.6 The OPTICKS Web page

Fig. 1.7 The OSSIM Web page

Fig. 1.8 The OpenEV Web page

- *SPRING* – SPRING is a state-of-the-art GIS and remote sensing image processing system with an object-oriented data model which provides for the integration of raster and vector data representations in a single environment. SPRING is a product of Brazil's National Institute for Space Research (Camara et al. 1996). SPRING was developed with the following design objectives (1) operate as a seamless geographical data base, with a large volume of data, without being limited by tiling schemes, scale, and projection, (2) support both raster and vector data geometries and integration of remote sensing data into a GIS, with functions for image processing, digital terrain modeling, spatial analysis and data base query and manipulation, (3) achieve full scalability, that is, be capable of working with full functionality from desktop PCs running Windows or OS/2 to high-performance UNIX workstations, and (4) provide an easy-to-use, yet powerful environment, with a combination of menu driven applications and a spatial algebra language.
- *OpenEV* – The original version of OpenEV was developed by Atlantis Scientific as a prototype viewer for the Canadian Geospatial Data Infrastructure(CGDI). Its development was supported by the Canada Centre for Remote Sensing GeoConnections program and J-2 Geomatics (Canadian Department of National Defence). The goal was to create a free, downloadable advanced satellite imagery viewer that allowed users to work interactively with CGDI data (Fig. 1.8). OpenEV runs on a variety of platforms (Windows, Irix, Solaris, and Linux)

and contains a number of data and image manipulation functions, which support the interaction with, fusion of, and analysis of information. The power of the viewer comes from its design which exploits the cross-platform OpenGL library and the hardware accelerator cards that have become commonplace in Unix workstations and PC computers. The open source nature of OpenEV permitted widespread distribution among members of the remote sensing community and beyond. The business benefits include the ease with which the base library has been leveraged to create custom, proprietary tools, and the contributions of the open source development community back to the project.
- *BILKO* – Bilko is described as a complete system for learning and teaching remote sensing image analysis skills (http://www.noc.soton.ac.uk/bilko/). Bilko routines may be applied to the analysis of any image in an appropriate format and include a wide range of standard image processing functions. The original aim of the project was to facilitate "hands-on" access to remote sensing for those traditionally excluded from such training due to the:
  - High cost of commercial image-processing software.
  - Need for expensive computer equipment to run that software.
  - Difficulty of acquiring remotely sensed images for teaching purposes.
  - Long learning-curves required to master complex commercial software.

1.5 Moving Forward

Earth observation science is an emerging field that couples the study of environmental change with the technologies useful for gathering change-data. This chapter presented the concept of Earth observation and how this paradigm supports environmental analysis and assessment efforts. Understandably, the environment under investigation is inherently complex, and with the framework offered by complexity theory, remote sensing methods can be integrated with a considered view of their functional aspects and technological constraints. The treatment of remote sensing theory presented in this chapter informs the analyst of how this technology works and based on a suite of comparatively straightforward principles demonstrates how useful decision products can be obtained. Perhaps more important than the theoretical basis on which remote sensing science is based are those singular effort to bridge the gap between technical analyst and policy maker. In this chapter, bringing remote sensing down to Earth was explained as a critical step in developing successful and sustainable remote sensing programs. To that end, the traditional bottlenecks surrounding data access and data processing that frustrated wider use of this technology were shown to be breaking with the emergence of internet accessible data archives and open source software; two developments that will greatly facilitate an expanded role of remote sensing in the environmental practice.

References

Bauer, P., Gerard, F., Minster, J. (2006) Observing the Earth: An international endeavor, Comptes Rendus Geosciences, 338, 14–15.

Boyd, D. (2009) Remote Sensing in Physical Geography: a twenty first century perspective, Progress in Physical Geography, 33, 451–458.

Camara G., Souza R., Freitas, U., Garrido, J. (1996) SPRING: Integrating remote sensing and GIS by object-oriented data modeling, Journal of Computers & Graphics, 20, 395–403.

Campbell, J (2002) Introduction to Remote Sensing, Guilford Press, New York, 621p.

Contanza, R. (1993) Modeling Complex Ecological Economic Systems, Bioscience 43, 545–555.

Goetz, S. (2007) Crisis in Earth Observation, Science, 315, 1767.

Jensen, J. (2007) Remote Sensing of the Environment: An Earth Resource Perspective, Prentice Hall, 608p.

Kauffman, S. (1995) At Home in the Universe: the search for the laws of self-organization and complexity, Oxford University Press, 336p.

King, R.L. & Birk, R.J. (2004) Developing Earth System Science Knowledge to Manage the Earth's Natural Resources, Computing in Science & Engineering 6: 1, 45–51.

Lautenbacher, C. (2006) The Global Earth Observation System: science serving society, Space Policy, 22, 8–11.

Lein, J. (1999) Developing a Statewide Environmental Monitoring Program Using Landsat Data: The OhioView Example, Proceedings of the Pecora 14/Land Satellite Information Conference, pp. 303–313.

Madhok, V. & Landgrebe, D.A. (2002) A process model for remote sensing data analysis, IEEE Transactions on Geoscience and Remote Sensing, 40, 680–686.

Manson, S. (2001) Simplifying Complexity: a review of complexity theory, Geoforum, 32, 405–414.

Steering Committee on Space Applications and Commercialization, National Research Council (2001) Transforming Remote Sensing Data into Information and Applications, The National Academies Press.

Phelan, S. (1999) A Note on the Correspondence Between Complexity and Systems Theory, Systems Practice and Action Research, 12, 237–247.

Schowengerdt, R. (2007) Remote Sensing: Models and methods for image processing, Academic Press, 560p.

Thackrah, G. (2006) Using Earth Observation to Monitor Habitats, in Hurford, C. and Schneider, M (eds) Monitoring Nature Conservation is Cultural Habitats, Springer, pp. 325–331.

Wang, K., Franklin, S., Guo, X., Cattel, M. (2010) Remote Sensing of Ecology, Biodiversity and Conservations: A review from the perspective of remote sensing specialist, Sensors, 10, 9647–9667.

Chapter 2
Environmental Sensing

The environment has remained at the forefront of scientific interest for well over four decades, and no other topic will likely captivate our attention in the foreseeable future as we struggle to understand the complexities of this planet, we call home. Understanding is the key, but understanding does not take place in a vacuum. To grasp the significance of our relationship with our environment, we need to comprehend the patterns and processes that characterize its many features; how they interact, how they change, and how they influence behaviors that shape our future. Understanding requires information, which helps to reveal the distinct actors and actions that conspire to define the environment. However, being informed implies not only an improved comprehension of the complexities inherent to the study of the environment, but also a greater sensitivity to the limit of our knowledge, the uncertainties that remain, and the unavoidable realities of our ignorance. In this context, information is intelligence that we not only learn from, but also apply to guide us while we strive to make good environmental decisions. The goal of this chapter is to place the environment into a framework that enables our ability to measure, map, and model its features using remote sensing technology to gain intelligence. Too often remote sensing is discussed from a technological perspective that leaves a gap between the obvious technical aspects of this science and the pragmatic need to obtain relevant data to address a problem. This chapter examines the environment by identifying its descriptive elements that can be explored remotely; characteristics that can not only be measured, but also whose measures communicate essential facts that explain the disposition of the environmental complexity. From this discussion, the notion of environmental sensing is introduced as the conduit between the technology, the myriad of applications it can serve, and our environmental system.

J.K. Lein, *Environmental Sensing: Analytical Techniques for Earth Observation*,
DOI 10.1007/978-1-4614-0143-8_2,

2.1 Sensing the Environment

When used in common language the word "sensing" defines any of the faculties, such as sight, hearing, smell, taste, or touch, by which humans perceive stimuli originating from outside our bodies. Sensing, according to this simple definition, means to detect, perceive, or become aware of some phenomena external to us. Remote sensing technology has long been identified as a means of detecting or perceiving phenomena where the measurements taken at distance from objects and surfaces of interest are transformed into information, in a manner analogous to our brain transforming the perception of touch into concepts such as rough or smooth. The data collected remotely satisfies our desire for knowledge and provides needed information to guide us in a similar way that our hand searches for the light switch in a dark room. It may be argued that our present state of knowledge regarding the environment is not unlike an adventure in a dark room, it can also be argued that our capacity to sense our environment will be integral to becoming aware. How well we sense will determine likely how well we learn and understand.

For the purposes of this discussion, environmental remote sensing may be defined as the measurement and representation of earth surface characteristics that support the information requirements for effective environmental management and decision making. This practical definition suggests that there is an underlying rational that directs the remote collection of data and narrows the scope of the science of remote sensing by focusing on the delivery on information that illuminates the complexities, uncertainties, and dynamic nature of the environmental process. In this regard, environmental remote sensing is an extension of an existing technique that strives to incorporate alternative strategies and sensors that can yield new information and provide new insight into the status of Earth's environments and detect conditions of critical concern.

2.2 The Environmental System

Earth's environments are complex and varied. In simple terms, they can be characterized as biomes; a defining area of ecologically similar geographic and climatic conditions, which support communities of plants, animals, and soils that assume distinctive relationships and patterns (Fig. 2.1). From a remote sensing perspective, these biophysical patterns explain land covers that form as the outcome of abiotic factors and the biomass productivity of the organizing vegetation types that dominate its spatial expanse (Olsen et al. 2001). Land covers also describe human environments where culture has altered patterns of ecosystem process and biodiversity. Such alterations generate distinct surface characteristics that form as the product of sustained and direct human interaction with ecosystems. These anthropogenic biomes emerge as the consequence of human impact range from settlements, croplands, forested areas, and wildlands subject to human modification (Ellis and Ramankutty 2008). Whether biotic or human-induced, this ecosphere is a

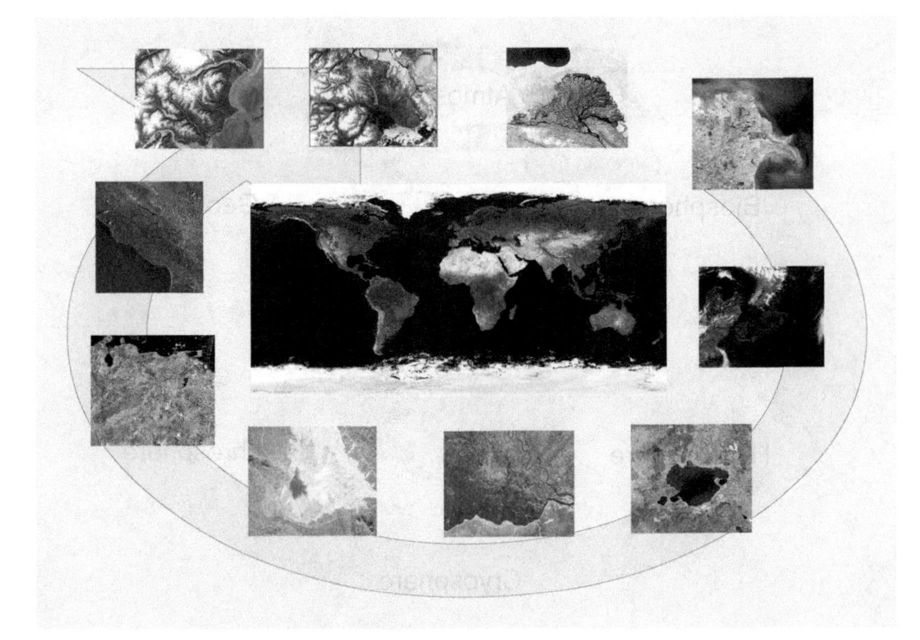

Fig. 2.1 The patterns of Earth's biomes

thin layer of the earth, estimated at less than 14 km, that supports life (MacKenzie 2010). Recognizing that this has taken over four billion years to achieve the present state of the environment, there is both uniqueness and an element of chance-consequence that underlies the conditions we observe.

As a system, Earth's environments explain a collection of interdependent elements (Fig. 2.2). At the most general level, these are commonly referred to as:

- The *lithosphere*, which contains all of the cold, hard, and solid rock of the planet's crust (surface), the hot semi-solid rock that lies underneath the crust, the hot liquid rock near the center of the planet, and the solid iron core (center) of the planet
- The *hydrosphere*, which contains all of the planet's solid, liquid, and gaseous water
- The *biosphere*, which contains all of the planet's living organisms
- The *atmosphere*, which contains all of the planet's air

The elements are closely connected and exhibit cyclic patterns of behavior when materials and energy flow across space and over time. This familiar cascade, characterized in relation to the ecosystem, the flux of solar energy that drives climate, and the process–response progressions that punctuate geomorphic and hydrologic activity, demonstrate how environmental components interact with their surroundings and evolve functional relationships that connect components together to form a definable structure (Figs. 2.3a–c).

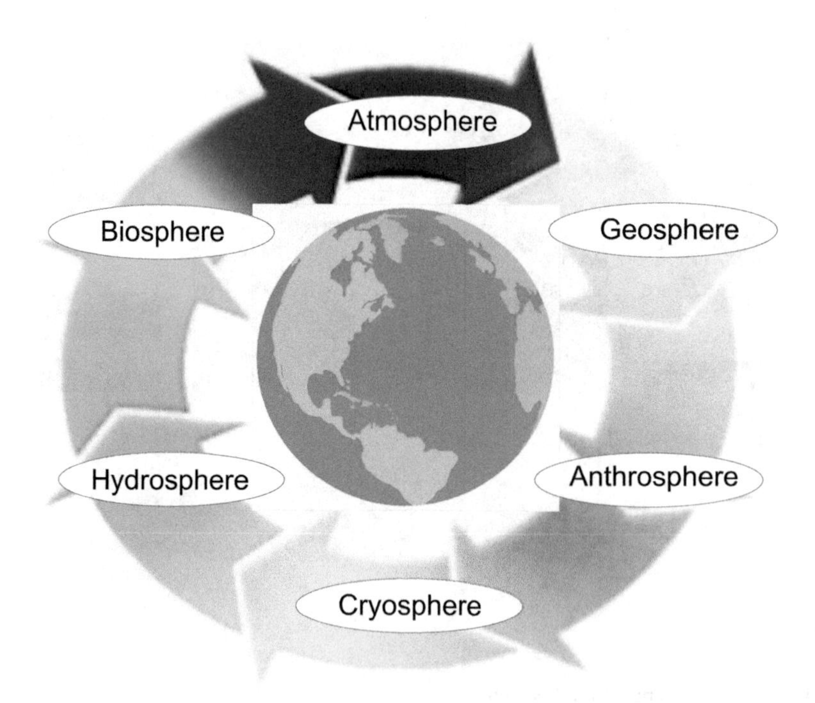

Fig. 2.2 The Earth system

In many respects, the land covers observed via remote sensing serve as evidence of these structures, displaying a morphology that permits inferences to be made regarding their disposition, causation, and variation both spatially and temporally.

Change and the dynamic processes that propel components of the environmental system are basic attributes of variability that produce contrasting patterns over time and space. As an attribute, environmental change varies in form, size, duration, and areal extent and arises not at random, but as a result of basic biological and physical processes operating on the planet (Hidore 1996). The patterns that emerge are the observable consequence of these occurrences. Changes in the environmental system can be described in several ways. At one level, we can recognize a change as short-term – defining cyclic behaviors occur in less than one rhythm of the system. From here we can also identify medium-term changes that explain seasonal rhythms among environmental attributes. Finally, behaviors may characterize long-term patterns, which may not be easily resolved, transitioning from one dynamic equilibrium state to another, or exhibiting stepped fluctuations punctuated by lag times well beyond the common human scales of reference. More purposeful explanations may be offered that give

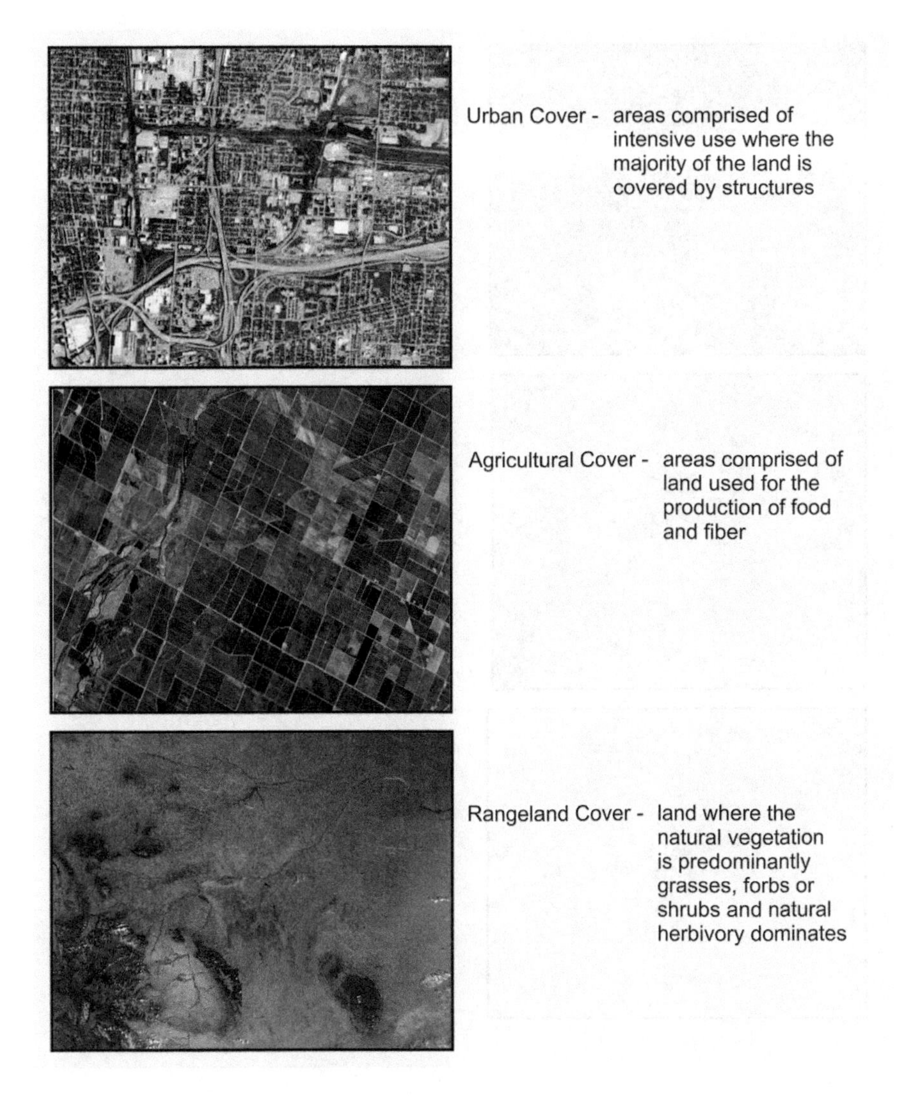

Fig. 2.3 Characteristic land covers and their definition

better insight into the patterns of change and how they channel environmental process over time. Here, environmental change can be categorized according to one of the five different conditions (Hidore 1996):

1. *Persistent change* – unidirectional trajectories typified by slow, steady progressions over time
2. *Rhythmic change* – displaying regular oscillations where periodic fluctuations occur at regular, predictable intervals

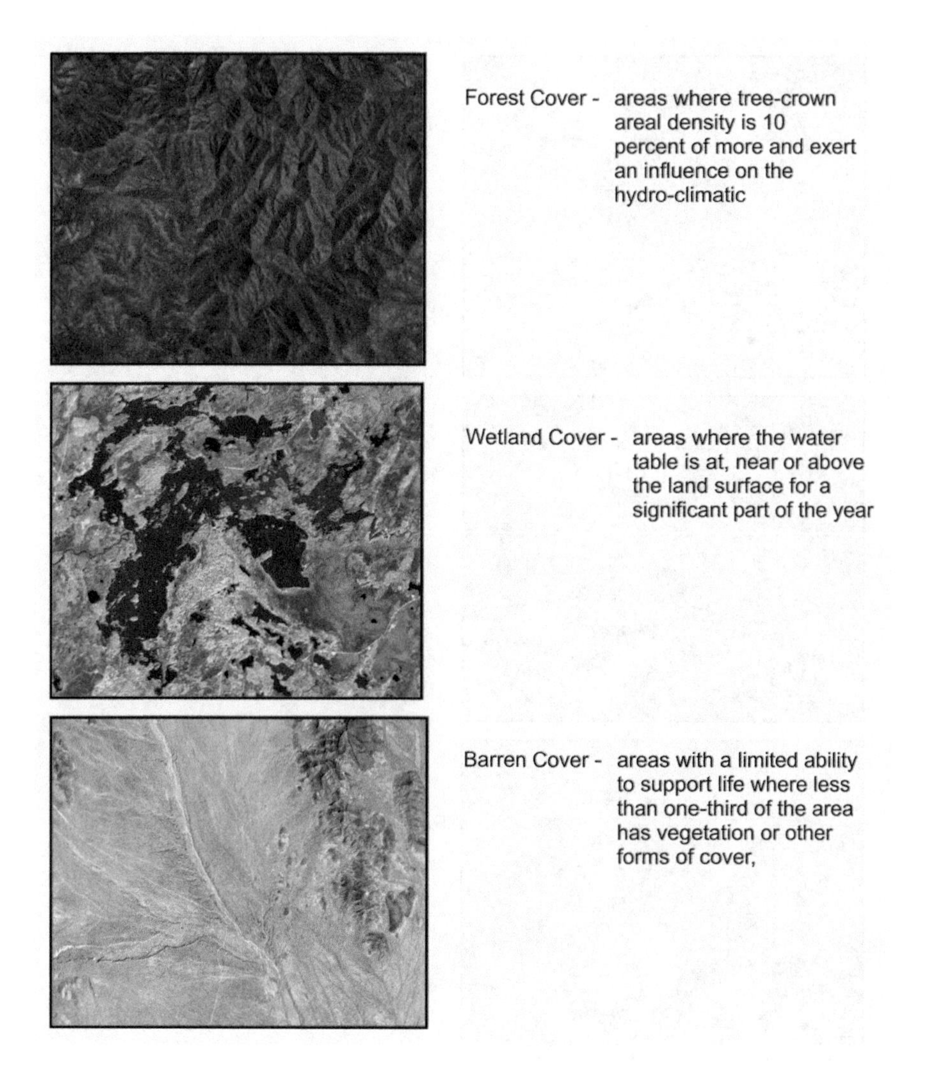

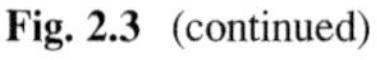
Fig. 2.3 (continued)

3. *Cyclical oscillations* – where change repeats at irregular intervals with varying intensity, but are not periodic
4. *Short-lived events* – explaining sporadic episodes often identifying deviations from average or expected conditions with durations spanning seconds to several days in length
5. *Anthropogenic change* – exemplified by human-induced effects on natural patterns sustained by trajectories of established social and economic drivers acting over time

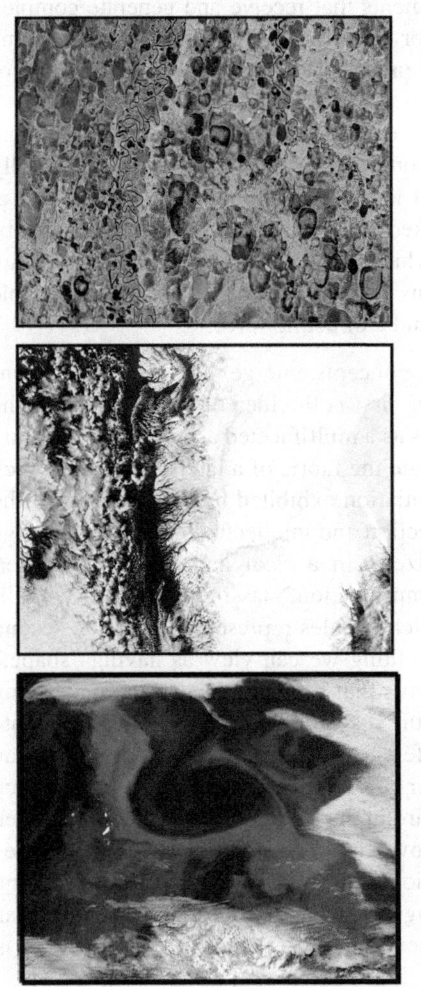

Fig. 2.3 (continued)

As an active force, change contributes to shaping and reshaping of the structure of the environmental system. Since the environment is essentially open to the transfer of matter and energy and generally oscillates between conditions of equilibrium and disequilibrium, an appreciation of the system structure offers a means to observe the environment independent of function or state (Dury 1981). Four general categories of structure, moving from less to greater complexity, can be noted:

1. *Morphological systems* – defined in terms of their internal geometry as expressed by the number, size, shape and linkages displayed by their components, morphological systems have identifiable shapes and patterns such as those descriptive of streams, glacier systems, shorelines, and entire landscapes.

2. *Cascading systems* – explain environments that receive and generate complex inputs and outputs of matter, energy, or both. Cascading systems include some type of regulator and mechanisms that provide storage where the main focus of interest becomes the rate of flow or flux between components.
3. *Process–response systems* – describe environments that alter their internal geometry and/or behavior in response to cascading inputs. Generally process–response systems comprise at least one morphological system and at least one cascading system that are linked and often share common components.
4. *Control systems* – environments in which some aspect of their functions are controlled by intelligence. Such systems vary in scale, but share the inescapable influence of human decision making and a directing force.

From this cursory review, several key concepts emerge that focus an environmental remote sensing investigation. The first is the idea of complexity. As the subject of inquiry, the environment forms as a multifaceted arrangement of living and nonliving elements that blend to create the fabric of a landscape from which our measurements emanate. The differentiation exhibited by the elements of the environment encourage the need for selection and intellectual devices to manage the variety presented to us and organize it in a clear and coherent manner. Through the strategies of abstraction, simplification, classification, and symbolization, complexity is made sensible, which enables representation of the second key concept: structure. Structure is something we can view as having "shape," whether it is the shape of drainage patterns that provide clues to the underlying geologic structure of the environment or the shape of the boundaries that delineate land units that may be indicative of differences in soil type, climate, or human impact. Through structure we can infer arrangement and connectivity which supports a process view of how the environmental system behaves at any given location. More importantly, behavior moves us to consider change, the last key idea that lends itself to remote detection. Through change, the patterns and processes manifest in a dynamic setting that captures the environment in an active and often transient state. The next section explores these concepts in more detail.

2.3 Pattern, Process, and Disturbance

Environmental systems evidence distinctive patterns that develop as the product of energy and material interactions over time. These patterns are identifiable as the communities of living and nonliving elements that not only give rise to a structure, but also define the focus of environmental remote sensing. The agents of the biosphere, atmosphere, hydrosphere, and lithosphere form an interdependence recognizable as the landscape: a land surface of associated habitats that explain an ecology termed as the mosaic (Bissonete and Storch 2004; Huggett 1995; Turner and Gardner 1994). These spatially heterogeneous area characterize a dynamic that can be expressed according the "brash" equations (Huggett 1995). Using this conceptual model, we can represent the environmental system as interacting terrestrial life and life-support components where the biosphere (b), troposphere (r), atmosphere (a), pedosphere (s) and hydrosphere (h) respond over time to each other plus the influence of forcing functions (z) that lie outside the landscape. When these agents are expressed mathematically produce the "brash" set (Huggett 1995) such that:

$$\frac{\mathrm{d}b}{\mathrm{d}t} = f(b, r, a, s, h) + z$$

$$\frac{\mathrm{d}r}{\mathrm{d}t} = f(b, r, a, s, h) + z$$

$$\frac{\mathrm{d}a}{\mathrm{d}t} = f(b, r, a, s, h) + z$$

$$\frac{\mathrm{d}s}{\mathrm{d}t} = f(b, r, a, s, h) + z$$

$$\frac{\mathrm{d}h}{\mathrm{d}t} = f(b, r, a, s, h) + z,$$

which provides an ideal explanation of the landscape that offers an analytical design for exploring pattern and how pattern changes over time.

Pattern is also a function of scale; a dimension that refines the spatial and temporal characterization of the landscape mosaic. As a unit of observation, the structure, function, and dynamics of the landscape are scale-dependent where the processes and resulting patterns at one scale may be insignificant at another. Traditionally, scale has been described using the "cone of resolution" model (Fig. 2.4). This familiar representation depicts the level of detail synonymous with the scale from the macro-scale through the meso-scale and down to the micro-scale and implies a reference to the size (relative or absolute) at which physical or human structures and processes are observable. Scale in this context is also defined in terms of generalization, where scale controls the apparent detail or complexity phenomena that may assume. Using this conceptual model, the connection between the idea of scale, "visibility," and the observational detail a remote sensing device provides can be neatly established, which helps to identify the appropriate level of detail required to address an environmental problem and the capacity of a sensor to deliver that detail. At the macro-level, scale defines pattern and process in its most generalized form. Geographically, this can be visualized as a near-continental perspective that explains a comparatively coarse degree of details. In the language of remote sensing, macro-scale conforms to a level of spatial resolution common to sensor platforms such as the advanced very high resolution radiometer (AVHRR) and MODIS. The meso-scale can be

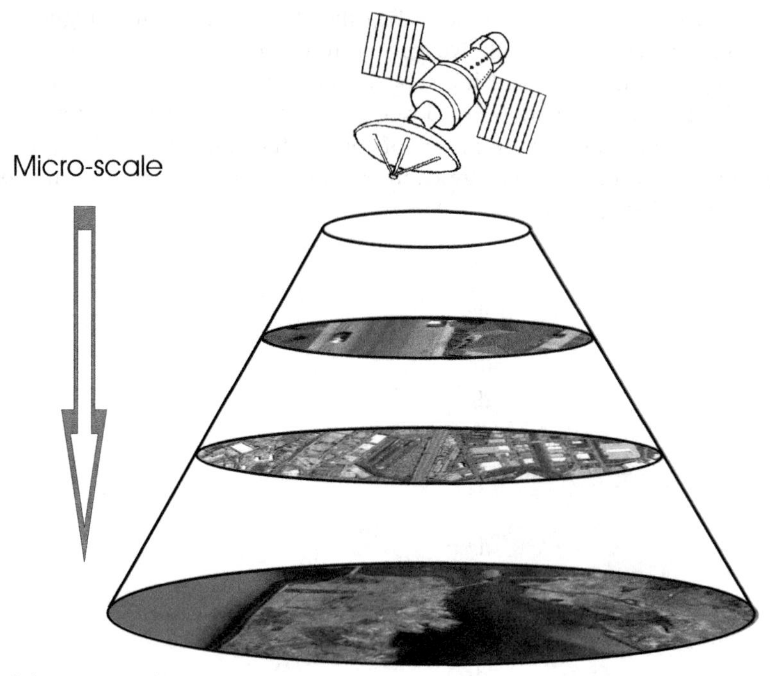

Fig. 2.4 The cone of resolution model

conceptualized as a regional scale level of detail where spatial resolution improves to a sharper representation of the surface. At this level, representative platforms may include the Landsat, IRS, SPOT, and Aster. The micro-scale introduces the finest definition of pattern and suggests a landscape perspective that would capture details present within a watershed or city. In the hierarchy of satellite remote sensing systems, micro-scale explains spatial resolutions at or below 5 m which are typically found on commercial platforms such as GeoEye, IKONOS, and QuickBird (Fig. 2.5).

Although scale is inherently an imprecise and elastic concept, its role is important and critical to the study of the environmental system (Gibson et al. 2000). First, scale defines the size at which the environmental structures exist and over what extent the environmental processes operate. This interpretation of scale attempts to present the "true" expression of environmental phenomena and recognizes that environmental processes are often scale-dependent, or at the very least, defined in part by a relative scale. However, there can be exceptions to this idea, particularly in the examples where patterns seen at one level of detail may also be observed at another and the possibility that environmental processes often operate at multiple scales. Its inexact nature, therefore, directs us to consider scale in a more pragmatic sense: analytical scale.

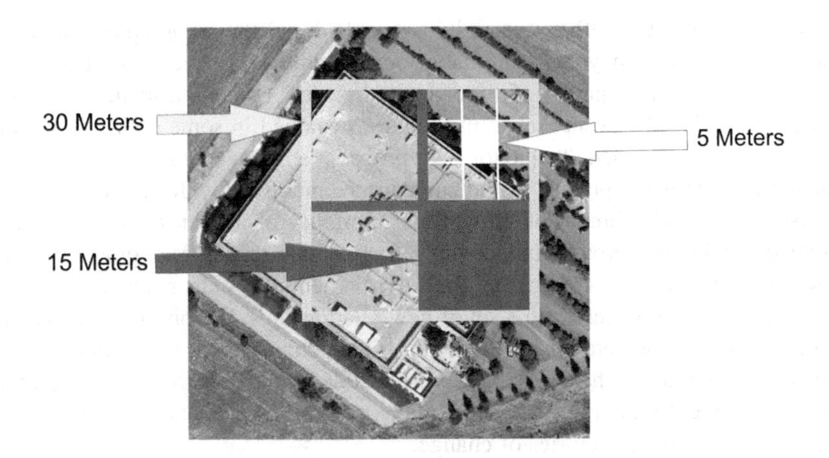

Fig. 2.5 Comparative spatial scale/sensor resolution relationships

Analytical scale refers to the size of the unit at which the problem under investigation is examined. This simple definition is useful in environmental remote sensing since it implies measurement and how measurements are aggregated for data analysis. In this context, analytical scale explains the scale of understanding; and with specific reference to satellite remote sensing, this scale is used to represent the surface whether as a raster (pixel) or polygon (object). Therefore, to observe and study the environmental system accurately, the scale of analysis must conform to the actual scale of the phenomenon, whether expressed over time or across space (Hudson 1992).

Identifying the correct scale can be problematic. Scale insensitivity introduces cross-level confusion, particularly when data at one scale is used to make inferences about phenomena at another or the narrower; an example of the "ecological fallacy" where aggregated data is used to make inferences about disaggregated patterns. In reality, the challenges imposed by the scale often require us to use the data at the "available" scale, which constrains an analysis to the units that are present in the data. While there may be no reasonable alternative, representing the environmental system (however defined) at the "available scale" may contribute to the loss of definition, particularly when the phenomena of interest do not conform well to the units imposed by the data. The result introduces an unavoidable level of error into our analysis, which limits the degree of confidence that can be ascribed to a solution.

Reconciling the issues of scale enables patterns to emerge that illustrate the important associations that bind the elements of the environmental system together. Identifying spatial pattern, therefore, not only supports an understanding of the system under study, but also provides clues that relate observed characteristics to underlying process, which highlight the dynamic nature of the environment (Dale 2002). Process, however, is difficult to capture. Taken broadly, the term suggests a sequence of events that actively shape and reshape the behaviors exhibited by the system of interest, whether it is a city, watershed, or other geographic entity. The sequence implied may be continuous across time and space, or discrete and observable in finite quanta of time. In both cases, there is also the underlying assumption that the events set in motion are sustained in some definable manner. Through the sustained "behavior" of process, a recognizable consequence is achieved, whether explained as the land surface modified by the geologic cycle of erosion, the establishment of growth and mortality of a plant community of the spread, or human–urban land cover over the landscape. The product of this sustained behavior is change, as the environmental system responds to and is transformed by a process into a set of new relationships. We observe the evidence of process as the generation of a new form in an environment where that form did not exist previously. The mechanisms responsible lurk behind the observed patterns and define the driving physical, economic, or social forces that propel the environment and fuel the trajectories of change.

Sensing environmental change shares a duality of purpose. At one level is the need to resolve a pattern; documenting the spatial expression of process through the contrasting patterns it reveals. At the more complicated level is the desire to infer the process from those patterns; deducing the driving forces that are actively at work within the environmental system. In both instances, the influence of time cannot be ignored, nor removed from the question. Temporal influences, whether explicit or implicit, remain a constant, although best viewed as a relative rather than an absolute quantity (Getis and Boots 1978). In some respect, sensing time in the environmental system is similar to watching an animation. Each frame in the sequence, like each image captured by our sensor, is a complete depiction of the scene at a specific instance in time. Set into motion, the individual scenes blend to characterize the change. The rate of motion between each frame describes, in a limited way, the pace at which a change takes place, and each individual frame influences how animated the action (process) appears. By examining one frame in the sequence, our interest is to describe the direction of motion and anticipate where in the subsequent frames action will take us (Getis and Boots 1978). Complications of course arise; particularly when processes are gradual or when new "actors" are introduced or leave the scene. Perhaps more frustrating to the goals of environmental remote sensing are those situations where a long interval of identical frames are encountered and no action (change) can be observed; begging the question: is the absence of change, change? While this analogy is simple, conceptualizing the idea of environmental change as an animation underscores the fact that, when sensing environmental process more often than not, our understanding is frequently limited by the available frames. In some cases, what we have may be sufficient to adequately capture the "action," although more typically we are left with an isolated or interrupted sequence that requires us to provide the missing context. Explaining environmental process, like viewing an animation, depends on (1) the subject matter, which in our case are the operative processes that direct environmental behavior (action) and (2) our ability to assemble each frame together into chain of events that complete the story.

The unfolding stories of environmental process that currently direct our concern are those new actors that introduce changes in the sequence that redirect the plot. Familiar examples of these active events include:

- *Deforestation* – the process of destroying or removing forest ecosystems through logging operations or burning
- *Desertification* – the degradation of land in arid and dry sub-humid areas
- *Environmental degradation* – uncharacteristic loss of habitat, biodiversity, or depletion of natural recourses contributing to ecological collapse
- *Erosion* – the process of removing sediment, soil, rock, and other material in the natural environment
- *Extinction* – the death of the last existing member of a species where there are no surviving individuals able to reproduce and create new generations

These environmental processes are complex, reflecting the influence of many casual factors that act on environmental systems. The causal mechanisms that contribute to the plot changes that confuse our animation have been neatly summarized by Goudie and Viles (2003) according to a set of:

- *Predisposing factors* – describing features of the natural or human environment that make a system vulnerable to stress (change)
- *Inciting factors* – defining stresses that trigger the change in a system
- *Contributing factors* – explaining the range of additional stresses that render a system's response more noticeable and acute.

Taken together, these factors conspire to direct the environmental system to a new state (frame) where we observe a transformation or a shift as human activities interact with a series of interlocking environmental responses.

The transformations characterizing environmental change can be subtle and slow to emerge, or dramatic and quick to materialize. In either case, they reflect the consequence of disturbances that alter material and energy flows within the environmental system. Here, the concept of a disturbance becomes a convenient way to connect environmental stress to actions that will display both temporal and spatial dimensions. In an environmental context, a disturbance describes an event causing change in the ecosystem that includes environmental fluctuations or destructive events. Along this implied continuum of events, disturbance may emanate from purely endogenous (internal) processes to those that are purely exogenous (external) (White and Picket 1985). Overall, landscapes may be disturbed by a range of actors from the physical consequence of strong winds, fire, flood, landslide, and lightening; the biological consequence of pests and pathogens; and the impacts of human and animal activities. In some cases, disturbances act at random within the landscape, while other events spread from a beginning point through the system over time. As a sensible quality, disturbances operate in a heterogeneous manner, since some features with the landscape are more susceptible to an event than are others. It is important to recognize that disturbance is an integral part of all environmental systems, and landscapes are defined in part by a common disturbance regime (pattern) (Gordon and Forman 1983). A disturbance regime represents the sum of types, frequencies, and intensities of disturbance through time in the landscape. When we observe the environmental system, the disturbance causes a given characteristic of an ecosystem (such as diversity, biomass, and nutrient levels) to exceed or fall below its common range of variation.

Landscapes subject to human-modifying actions are the changes by new disturbances introduced by economic and social forces. Human impact, expressed as a disturbance, however is discontinuous and unevenly distributed over the surface. Consequently, human disturbance regimes differ between landscapes and are superimposed on contrasting natural disturbance regimes. As a result, the landscapes produced by human modifications display a wide range of variability, often with sharp and distinct boundaries (Gordon and Forman 1983). The types of modified landscapes produced by human disturbances begin at the lower end of the "gradient" with natural vegetation such as grassland, rainforest, or desert produced by a natural disturbance regime void of significant human effects. Moving upward along this range are the areas recently exploited by human populations which are often characterized by scattered clearings in the natural land cover. Continuing along this gradient are the patterns that reveal managed landscapes where the majority of the surface appears to be composed on natural cover, but is controlled for human activities such as timber harvesting or livestock grazing. Control implies active management that introduces significant differences in species, energy, and nutrient cycles when compared to the natural vegetation. Cropland follows next in the sequence where planted vegetation dominates and may be intermixed with remaining sections of managed vegetation. Following next in this description of modified landscapes are the human settlement patterns characteristic of ex-urban and suburban development where managed vegetation has been reduced and the surface appears as a heterogeneous mixture of agriculture and urban forms. The final frame in this continuum describes urbanized areas where human use dominates. In this pattern, only small remnants of managed or cropland cover types remain visible. As this gradient of human impact is observed, key descriptors of pattern emerge as boundaries and edges shape and fragment the land surface into increasing levels of heterogeneity.

2.4 Patches and Progressions

From the altitude of a sensing platform, the Earth's surface appears as a mosaic of shapes and textures of varying configurations. These configurations assume arrangements that take on meaning in both an environmental and cultural context. This is the landscape and from an environmental perspective, it defines heterogeneous land areas composed of clustered, interacting ecosystems repeated in similar form across a discernable geographic extent. Delineating these surface arrangements is of fundamental interest in remote sensing, but a process that must be guided by an understanding of the mechanisms that contribute to their formation. Acting within the boundaries of these land surface arrangements are the geomorphic processes, colonization patterns, and local disturbances which, working in concert, produce distinctive, measurable units that display (Gordon and Forman 1983):

- *Structure* – spatial relationships among the landscape elements of energy, materials, and species relative to the size, shape, number, and type of these configurations
- *Function* – interactions among spatial elements in terms of energy, materials, and species flows among the elements
- *Changes* – alterations in structure and function over time

As a physical entity, this landscape reveals three universal characteristics: (1) patches, (2) corridors, and (3) matrix. In the language of landscape ecology these terms take on specific meaning. The term patch is defined as a relatively homogeneous area that differs from its surroundings. Patches serve as the basic unit of the landscape that change and fluctuate, a process called *patch dynamics*. When observed on remotely sensed imagery, patches have a definite shape and spatial configuration and can be described compositionally by internal variables such as number of trees, number of tree species, height of trees, or other similar measurements. Matrix defines the "background ecological system" of a landscape with a high degree of connectivity. Connectivity is the measure of how connected or spatially continuous a corridor, network, or matrix is. For example, a forested landscape (matrix) with fewer gaps in forest cover (open patches) will have higher connectivity. Within this explanation, corridors have important functions as strips of a particular type of landscape differing from adjacent land on both sides. When view in their entirety, a network emerges that defines an interconnected system of corridors forming a mosaic which explains the pattern of patches, corridors, and matrix that form the landscape. These building blocks of the landscape provide simple descriptors to express local influences that identify how landscapes are configured. These descriptors also account for the biodiversity patterns and natural processes that we observe (Dramstad et al. 1996). Thus, while the landscapes foundation reflects its background ecologic pattern, the local "neighborhood" forms as a configuration of patches, corridors, and background cover types revealing the matrix produced by natural processes as well as human activities that alter the mosaic. Alterations include the obvious and well-documented changes such as habitat fragmentation, and also include land transformations such as:

- Perforations
- Dissections
- Shrinkage
- Attrition
- Coalescence, each carrying significant ecological and human implications (Dramstad et al. 1996).

2.5 Sensing the Human Dimension

Although the visible alterations evidences in the environmental system induced by human activities have been well documented for over two decades (Mannion 1997; Roberts 1994; Stern et al. 1992), the causes promoting these alterations are more complex and less obvious. Therefore, while a satellite image may reveal patterns indicative of specific transformations in the landscape, the image alone carries little information concerning the how-and-why behind what is seen. The challenge in environmental remote sensing is to connect the patterns detected on the image to the decisions made that now characterize either the direct and purposeful alteration of the landscape or the unintended consequence of human decisions that have generated new, conditions that were not anticipated. This discontinuity rests at the core of environmental decision making and underscores the web of human behaviors and motivations that introduce themselves whenever choices are made, which affect the present or future state of the environmental system (Lein 1997; Chechile 1991). Decision making, however, does not take place in a vacuum. Rather the choices made describe a process driven by interconnected society needs and desires. The driving forces that direct human–environmental decision making fall into five broad categories:

1. *Population demand* – Each of us make demands on the environmental system for food, clothing shelter, and other services in support of our life styles. Greater numbers or increasing concentrations of people expand our ecological footprints and elevate demand for resources needed to sustain our activities.
2. *Economic growth* – The innate desire to improve our quality of life, provide for our needs and realize great opportunities, focus attention on the accumulation of wealth and capital formation to enhance our material standard of living. Expanding economic activity introduces environmental stressors, since the patterns of consumption contribute to both an expanding human footprint on increased consumption of natural resources and an elevated production of wastes and other energy and material residuals generated by these consumptive activities.
3. *Technological discovery* – Discovery impacts the environmental system through the innovations that enable wider exploitation of the resource base and through the types and characteristics of the waste residuals produced.
4. *Political institutions* – Taking the form of policy instruments that direct market influences and encourage social progress, governments and our increasing global political economy generate environmental outcomes by promoting (directly or indirectly) actions that damage environmental functioning, facilitating wider use of environmental resources and ignoring the environmental consequences.
5. *Cultural perceptions* – Individually and collectively, we are the product of values, beliefs, and attitudes that reflect our cultural teachings and experience. Through the lens of culture a world-view takes form, and our relationship to the environment becomes crystallized by the choices we make and the behaviors we follow.

Fig. 2.6 The spatial expression of urbanization

While none of these forces are sufficient alone to produce changes in the environmental systems, acting in combination they generate definable spatial events adequate in their scale and impact to alter properties of the landscape (Tillman and Lehman 2001; Vitousk 1992).

Incorporating the consequences of human decisions into the analysis of landscape heterogeneity begins with an appreciation of the spatially explicit actors that evidence the human dimension of our environmental system. Although the pathways followed by human activities are complex, they ultimately explain five distinctive decision-driven mosaics:

1. *Urbanization* – The decision to urbanize summarizes a human predilection with origins dating back over 15,000 years. As a spatial phenomenon, urbanization explains the transformation of land cover to a form and composition distinctly anthropogenic in nature, characterized by fragmented landscape dominated by asphalt, concrete, brick, and other manufactured materials. Morphologically, urban cover is typified by a terrain composed of angular forms assuming a planimetric arrangement that extends to a third dimension. As a pattern, urbanization is a physical element displaying a texture and extent wherein the concentration of structures, facilities, and people conspire to express economic and cultural influences that modify or replace "natural" form (Fig. 2.6).

Fig. 2.7 The spatial expression of agricultural intensification

2. *Agricultural intensification* – As a land cover pattern, agriculture is a mosaic of biological and physical patches within a matrix differentiated by settlement, cultivated land, and background cover that is defined as rural by virtue of its density and intensity. Geometrically, agricultural intensification is typified by a parallel structure and regularity of shape that conforms to land clearing practices and boundaries defined by land ownership. Intensification results in a progressive removal of existing landscape features with agricultural form (Fig. 2.7).
3. *Rangeland alteration* – Surface configurations of this variety explain land areas on which the climax or potential plant cover is composed of natural grasses, grass-like plants, and shrubs suitable for animal grazing and browsing. Rangeland areas are subject to limited management practices which may include deferred grazing, burning, or rotational grazing with little or no use of chemicals or fertilizers. Frequently subject to overstocking and fragmentation, semi-natural and natural rangelands are often adversely impacted by land degradation, loss of biodiversity, altered species connectivity, and intensification that retards recovery (Fig. 2.8).
4. *Deforestation* – Referring to the general process of forest clearing, deforestation characterizes a pattern of logging that expands progressively from an edge, a central cut strip, or patch. Although predicated on the presence of

Fig. 2.8 The spatial expression of rangeland alteration

large forested areas with low population density, effects emanate from policy decisions to open forested regions through instruments such as settlement programs, development projects, plantations, or other extractive industries. This suggests that deforestation may not be the exclusive consequence of timber harvesting. A related land pattern describes the re-establishment of forested areas, which may be planned (reforestation) or unplanned (afforestation). Afforestation is common to areas where soil degradation has occurred following farm abandonment or over cutting. Reforestation explains the large-scale planting of trees in a highly regular and systematic pattern of field-size units. Frequently, tree rows alternate with row crops during the early stages of these programs (Fig. 2.9).

5. *Corridors* – Visually identified as openings in an area that display highly linear patterns, a corridor develops either as the product of a human decision to construct features such as roads, power lines, rail lines, or irrigation canals or a lineation created by geologic and geomorphic factors. Typically, modification spreads and proceeds outward from the corridor on opposite sides penetrating through "natural" cover. Human constructions that create corridor features often include branching as a more complex linear network takes shape as a function of its design (Fig. 2.10).

Fig. 2.9 The spatial expression of deforestation

Fig. 2.10 The spatial expression of landscape corridors

2.6 Acknowledging Uncertainty

It would be convenient if the environmental system behaved in an unambiguous, consistent, and perfectly predictable manner. Unfortunately this is not the case. Rather, behaviors and processes descriptive of Earth's environments, while not random, are characterized by a complexity colored from a palette of deterministic, probabilistic, and stochastic relationships that vary across space and over time. As processes, disturbances, and natural perturbations in the environment evolve a landscape, even though the initial status or condition is known, there are many possible "realities" to consider, pathways to follow, and multiple potential future states of nature (Stewart 2000). This simple observation shapes our knowledge of the environmental system and invites a careful assessment of the uncertainties inherent to (1) our conceptualization of the environment, (2) the limitations of our knowledge, and (3) our inability to adequately resolve environmental process (Brown 2004). This observation also sustains our motivation to collect data and analyze information pertaining to the environmental system.

Uncertainty pervades all our attempts to ascertain absolutes with respect to the disposition of human–environmental interaction. As a concept uncertainty carries several connotations with important implications to the goals of environmental remote sensing (Regan et al. 2002). First, is the issue of epistemic uncertainty; an uncertainty associated with our present knowledge of the state of the environmental system. This form of uncertainty describes a "changeableness" that emerges due to limitations imposed by measurement devices, insufficient data, extrapolations, and interpolations as well as spatio-temporal variability. A second branch of uncertainty focuses the concern on the problem of linguistic ambiguity that describes the inexactness and vagueness introduced by language. This source of confusion is a product of our vocabulary and the presence of under-specific, ambiguity, and context-dependent terminologies. Both forms of uncertainty are problematic and develop from different sources. Furthermore, since they originate from difference sources, uncertainties are likely to compound. Therefore, identifying the main sources of uncertainty and exploring methodologies to control or minimize its impact are critical to an improved understanding of the environment. Several key sources of uncertainty with relevance to the environmental problem can be noted (Sutter et al. 1987; Regan et al. 2002), and each manifest in different ways:

- *Measurement uncertainty* – defines the limitations imposed by the observations techniques employed to measure environmental variables.
- *Natural variability* – explains behaviors in natural systems that are difficult to predict.
- *Inherent randomness* – identifies the limits of our understanding of process and the patterns that define environmental relationships.
- *Subjectivity* – Influence of judgment and its role in data interpretation can introduce bias, flawed reasoning, and misleading conclusions.
- *Linguistic imprecision* – language branded by concepts that are vague and inexact where the lack of specificity, clarity of meaning, and confusion in

definition contributes to generalities and misinterpretation and weakens communication.

Managing uncertainty, uncertain information, and recognizing its impact is central to the methods used to study the environmental systems. At this point, it is essential to realize that (Marjolein et al. 2002):

- Not all uncertainties can be adequately addressed with existing methods and tools.
- Uncertainty is usually treated as a marginal issue, an additional physical variable, or as a mathematical artifact.
- Little indication is provided relative to the magnitude or sources of uncertainty, and measures of uncertainty can be difficult to understand.

Dealing effectively with uncertainty in the context of environmental remote sensing moves beyond the technical proficiencies of image processing methods and requires the integration and synthesis of new conceptual knowledge together with a willingness to think with incertitude (Brewer and Gross 2002). Given that environmental processes are subject to forces above internal feedbacks, chance anomalies and deviations are as much a part of the environmental system as those aspects we understand (Faucheux and Froger 1995; Reckhow 1994).

Connecting our discussion of uncertainty back to the question of environmental sensing gives definition to the trends that accent our need for a better understanding of system behavior and change. Here, six major foci dominate and help frame environmental remote sensing investigations:

1. The impact of land use transitions
2. The rate of expansion of land use systems
3. The scale-dependent nature of changes in land
4. The reversibility of changes to the land
5. The locality of land change impacts
6. The overlapping, impact reinforcing, and mitigating nature of changes in land use

2.7 The Role of Measurement

To measure objects at a distance encapsulates the science of remote sensing. Our ability to understand the complexities of the environment remotely and manage the realities imposed by uncertainty is only as good as the measurement permits. We are reminded that measurement is nothing more than the use of numbers to describe data according to a set of rules. As such, measurement facilitates objective communication of objects and their attributes that can be readily manipulated conceptually. The key to useful measurement involves assigning numbers to object, events, and individuals that aptly characterize them in a precise and meaningful way. In remote sensing, where our measurements are made at a distance, the objects we sense do not readily lend themselves to numerical treatment. Rather, measurement builds from the isomorphic properties of our tool and the surfaces we seek to understand. From this relationship, insights are extended from one phenomenon (electromagnetic radiation) to the other (landscape) that creates an empirical situation, which can be expressed numerically.

Measurement is further informed by the distinction between the recorded observations and that which is analyzed (Amedeo and Golledge 1975). On the image, recorded observations represent a subset of the larger universe of observations that could potentially be made about the landscape. These recorded observations are derived from selected qualities that can be attributed to the objects we are interested in. Often this data is collected directly by measurement, but in the case of remote sensing, it must be translated into numerical terms before it can be realized as data. This phase of measurement is highly interpretative and directs attention at two problematic issues that are often overlooked:

1. The extent to which the numbers in the relationship are unique.
2. If the translation from the landscape (empirical situation) to its numerical definition retains the identity (uniqueness), order, and internal consistency of the original situation.

Observations of Earth's environments are further defined by the types of measurements made, all of which carry important implications for a remote sensing investigation that not only speaks to the overarching concern for data quality, but also to the larger question as to what the data actually reveal. In terms of types, we can explain a measurement as fundamental (primary) or derived. Primary measures explain measurements that record an existing property of an object. These define the distinguishing attributes of the object that separate it from other features in the scene. Derived measures are those, while expressed in numerical form, are defined on the basis of relationships between properties, such as a ratio or index. The rules used to produce a measure, therefore, affect its meaning, suggesting that the act of measurement is nontrivial since it establishes the basis of our understanding. In a digital world where files often appear as a "black box" read into software, this point is often lost, even though it impacts the simple things such as the stability of our measurements, their transferability from one situation to another, and calls to question concerns about uniqueness, comparability, representativeness, and utility. Each one is of significance; however, when combined they constrain what is observable, and ultimately, what becomes knowable. When considering measurement, it is also useful to make a clear distinction between facts and data. Such a distinction helps frame the problem and offers a more considered view of the remotely sense images. For practical purposes, a fact may be defined as a statement about some fundamental quality or quantity that is true regardless of where and when it was made. Data, by contrast, are not facts and are valid only for the time, place, and condition under which the observation was made (Jordan and Miller 1996). As we know from experience, a pixel captured for a given geographic location with a digital value of 31 on June 19 is not likely to enjoy that same value on August 8. To determine its "true" value, we would collect measures of that pixel over different times to predict its disposition; but even in this example, we realize that this number is an estimate and not a fact.

2.8 The Logic of Maybe

Despite the technological sophistication of remote sensing science, we can only observe limited aspects of and conditions with the environmental system. The measurements obtained through this technology are samples of dynamic processes influenced by the language of our perceptions as witnessed through the lens of contemporary logic. In the environmental sciences, that lens has been sharply focused on the concept of probability and the theory that reasoning with uncertainty can be accomplished by a set of tenants (rules) that impose order on measurement that illuminates the presence of chance in the situations we observe. Through this lens the realization of an event (*E*) is defined by the proportion of times that event occurs relative to the total number of observations. A condition observed in the landscape, however, generally describes an outcome (*A*) of an event (*E*) having occurred. This connection between outcome and event can be expressed as the conditional probability where,

$$P\left(\frac{A}{E}\right) = \frac{P(A - E)}{P(B)}.$$

The outcome, while never an absolute, becomes understandable by both its probability and the uncertainty (*U*), which we can express in simple terms as $U = (1 - P)$. Alternatively, we can apply mathematical expectation to "predict" the likelihood of event (*E*) from the set of variables (*X*) that we think explains it presence. According to this logic, the relationship takes the form:

$$E = a_1X_1 + a_2X_2 + \cdots + a_nX_n + \varepsilon.$$

In both examples, we are contending with estimations that attempt to manage uncertainty, place it into a more definable boundary, and resolve the problem by using a two-valued logic system in which our answers can be satisfied as either "true" or "false." This form of estimation and statistical representation has guided our study of the environment for well over a 100 years. This are produced problem-solving schemas based on conceptualizations that reflect the way we think things are and encouraged acceptance of the premise that our observations of the environmental system made remotely appear as fact. Consequently, the models we develop are fundamentally probabilistic in nature and should invite alternative conceptualizations based on an "ontology" of flux. This mindset moves us past the rules of probability and encourages a perspective where there are no facts, where time is a moving point, and reality is based on assumption. Through this lens, variability is an inescapable quality of what we observe; and through the application of approximate reasoning and critical thinking, problem-solving directs us to examine the "whole" of the problem as well as its components. The methods we device to assist us attempt to assess the effects of change on the whole, as one or more variables cascade their influence through the system. From this alternate frame of reference, sensing the environment is not simply the problems and their solutions, but also the processes involved. Environmental sensing, therefore, is not the application of a technology but the fundamental skills of:

- Problem identification
- Process reasoning
- Questioning basic premises, conclusions, and data
- Adaptive problem-solving
- Explanation of the problem, the solution, and the procedures involved

When these skills are married to the remote sensing technology, they create a method and style of questioning that may culminate in a single solution or as an intermediate step in a larger investigation. These skills and the methods they define are examined in the chapters to follow.

2.9 Summary

Remote sensing is often explained with an emphasis on the technical details that underlie this technology. How and where the methods of remote sensing connect to the study of the environment tend to be abstracted from these general principles. In this chapter, the question of how to study the environment remotely was undertaken. The goal of this chapter was to introduce the environmental system and its process–response relationships to identify the topics and targets germane to remote sensing data collection. The patterns, processes, and scale of environmental behaviors and the relations that define human–environmental interaction must be resolvable within the context of remote sensing technology. What are we looking for and how do we look become central questions in the effective use of satellite-based remote sensing when applied in the study of Earth's environments. By placing the environmental problem before the technology we can better appreciate the how the features of degradation, modification, and human alternation can be understood remotely and how the measurements obtained through our sensor systems can be employed to inform us of changes in the status of the environmental system and to improve our efforts to model environmental process in a proactive manner. In the chapter to follow, we will engage these intellectual activities and undertake a review of the sensor systems called upon to provide these measurements, and guide and support environmental solutions.

References

Amedeo, D. and Golledge, R. (1975) An Introduction to Scientific Reasoning in Geography, Wiley and Sons, New York.

Bissonete, J. and Storch, I. (2004) Landscape Ecology and Resource Management: Linking Theory With Practice, Island Press, Washington, DC, 480p.

Brewer, C. and Gross, L. (2002) Training Ecologists to Think with Uncertainty in Mind, Ecology, 84, 1412–1414.

Brown, J. (2004) Knowledge Uncertainty and Physical Geography: Towards the development of methodologies for questioning belief, Transactions of the Institute of British Geographers, 29, 367–381.

Chechile, R. (1991) Introduction to Environmental Decision Making, in Chechile, R. and Carlisle, S. (eds) Environmental Decision Making: A Multidisciplinary Perspective, Van Nostrand Reinhold, New York, 336p.

Dale, M. (2002) Spatial Pattern Analysis in Plant Ecology, Cambridge University Press, Cambridge, UK, 340p.

Dramstad, W., Olson, J., and Forman, R. (1996) Landscape Ecology Principles in Landscape Architecture and Land Use Planning, Island Press, Washington, DC, 80p.

Dury, G. (1981) An Introduction to Environmental Systems, Heinemann, Exeter, New Hampshire, 366p.

Ellis, E. and Ramankutty, N. (2008) Putting People in the Map: Anthropogenic biomes of the world, Frontiers in Ecology and the Environment, 6, 439–447.

Faucheux, S. and Froger, G. (1995) Decision-making Under Environmental Uncertainty, Ecological Economics, 15, 29–42.

Getis, A. and Boots, B. (1978) Models of Spatial Processes: An Approach to the Study of Point, Line and Area Patterns, Cambridge University Press, Cambridge, UK, 216p.

Gibson, C. Ostrom, E., and Ahn, T. (2000) The Concept of Scale and the Human Dimension of Global Change: A survey, Ecological Economics, 32, 217–239.

Gordon, M. and Forman, R. (1983) Landscape Modification and Changing Ecological Characteristics, in Mooney, H. and Gordon, M. (eds) Disturbance and Ecosystems: Components of Response, Springer, New York, 12–28.

Goudie, A. and Viles, H. (2003) The Earth Transformed: An Introduction to Human Impacts on the Environment, Wiley-Blackwell, Oxford, UK, 288p.

Hidore, J. (1996) Global Environmental Change: Its Nature and Impact, Prentice-Hall, Upper Saddle River, NJ, 263p.

Hudson, J. (1992) Scale in Space and Time, in Abler, R., Marcus, M., and Olson, J. (eds) Geography's Inner Worlds, Rutgers University Press, New Brunswick, NJ, 438p.

Huggett, R. (1995) Geoecology: An Evolutional Approach, Routledge, London, 344p.

Jordan, C. and Miller, C. (1996) Scientific Uncertainty as a Constraint to Environmental Problem Solving, in Lemons, J. (ed) Scientific Uncertainty and Environmental Problem Solving, Blackwell Science, Malden, MA, 91–117.

Lein, J. (1997) Environmental Decision Making: An Information Technology Approach, Blackwell Science, Malden, MA, 224p.

MacKenzie, F. (2010) Our Changing Planet: An Introduction to Earth System Science and Global Environmental Change, Prentice-Hall, Upper Saddle River, NJ, 576p.

Mannion, A. (1997) Global Environmental Change: A Natural and Cultural Environmental History, Prentice-Hall, Upper Saddle River, NJ, 400p.

Olson, D., Dinerstein, D., Wikramanayake, E., Burgess, N., Powell, G., Underwood, E., D'amico, J., Itoua,I., Strand, H., Morrison, J., Loucks, C., Allnutt, T., Ricketts, T., Kura, Y., Lamoreux, J., Wettengel, W., Hedao, P. , and Kassem, K. (2001) Terrestrial Ecoregions of the World: A New Map of Life on Earth, BioScience, 53, 933–938.

Reckhow, K. (1994) Importance of Scientific Uncertainty in Decision Making, Environmental Management, 18, 161–166.

Regan, H., Colyvan, M. and Burgman, M. (2002) A Taxonomy and Treatment of Uncertainty for Ecology and Conservation Biology, Ecological Applications, 12, 18–628.

Roberts, N. (1994) The Changing Global Environment, Wiley-Blackwell, New York, 544p.

Stern, P., Young, O., and Druckman, D. (1992) Global Environmental Change: Understanding the Human Dimension, National Academy Press, Washington, DC, 320p.

Stewart, T. (2000) Uncertainty, Judgment and Errors in Prediction, in Sarewitz, D., Pielke, R.A., and Byerly, R. (eds) Prediction: Science, Decision Making and the Future of Natures, Island Press, Washington, DC, 41–57.

Sutter, G., Barnthouse, L., and O'Neill, R. (1987) Treatment of Risk in Environmental Impact Assessment, Environmental Management, 11, 295–303.

Tillman, D. and Lehman, C. (2001) Human-caused Environmental Change: Impacts on plant diversity and evolution, PNAS, 98, 5433–5440.

Turner, M. and Gardner, R. (1994) Quantitative Methods in Landscape Ecology: The Analysis and Interpretation of Landscape Heterogeneity, Springer, New York, 556p.

van Asselt, M.B.A., and Rotmans, J. (2002) Uncertainty in Integrated Assessment Modeling, Climate Change, 54, 75–105.

Vitousek, P. (1992) Global Environmental Change: An Introduction, Annual Review of Ecology and Systematics, 23, 1–14.

White, P. and Picket, S. (1985) Natural Disturbance and Patch Dynamics: An Introduction, in Picket, S. and White, P. (eds), The Ecology of Natural Disturbance and Patch Dynamics, Academic Press, New York, 3–13.

Chapter 3
Sensors and Systems

Our capacity to understand the environmental system using remote sensing technologies is strongly dependent on the type and quality of information we obtain. As a data-driven process, understanding Earth's environments and unraveling the patterns of interaction between people and the planet have been greatly facilitated by an increasing array of sensor systems each with contrasting spatial, temporal and radiometric resolutions. When we compare the situation today to the recent history of satellite-based remote sensing, the expansion of sensor platforms, while offering an impressive variety of data products, has also necessitated a more considered view of the technology surrounding the remote data collection. As new sensing systems are deployed, greater attention must be given to their capabilities and how their designed features support specific information needs. This is a sharp contrast to space-based remote sensing of 40 years ago, when analysis and mapping was limited in scope to the "one-size fits all" realities of the Landsat multispectral scanner (MSS). In this chapter we will explore the evolution and advances in remote sensing satellite data acquisition. Our purpose here is threefold: (1) highlight the satellite data collection technology, (2) connect the capability of existing systems relative to the goals of environmental sensing, and (3) review these data products and examine their usefulness as data that feed the objectives of environmental analysis.

3.1 Data Acquisition Fundamentals

The use of remote sensing to study Earth's environments has progressed steadily over time (Qu et al. 2006; Thoely 2000). This evolution reflects both advancements in sensor technology and the desire to develop new data collection capabilities to address a growing list of environmental concerns. Eight distinct phases in the evolution of remote sensing can be noted (Melesse et al. 2007):

- *Phase 1: Airborne remote sensing* – This phase developed during the first and second world wars with the primary applications focused on surveying, reconnaissance, strategic land use mapping and military surveillance.

J.K. Lein, *Environmental Sensing: Analytical Techniques for Earth Observation*,
DOI 10.1007/978-1-4614-0143-8_3,

- *Phase 2: Early spaceborne systems* – explains a phase dominated by the launch of "proof of concept" satellites beginning with Russia's Sputnik and Explorer 1, introduced by the United States. These early examples contributed to the first meteorological satellite TIROS-1 that established the feasibility for continuous earth observation.
- *Phase 3: Era of spy satellites* – During the peak of the "cold war," satellites were introduced as a method of military surveillance hand on hard copy data formats.
- *Phase 4: Advanced meteorological satellites* – This phase introduced the methods of remote data collection in digital formats that could be analyzed using computer processing techniques and demonstrated the usefulness of global data coverage for environmental applications.
- *Phase 5: Landsat* – The launch of Landsat-1 in 1972, referred to at the time as ERTS (Earth Resource Technology Satellite) introduced the multisensor/multispectral scanning technology and the systematic collection of land surface data at an effective spatial scale for civilian earth surface observation. The success of Landsat-1 was followed by compatible multispectral platforms (Landsat 2 and 3) and ushered in advances in both spatial and spectral resolution systems deployed on board the Landsat Thematic Mapper series beginning in 1982.
- *Phase 6: Earth observing system (EOS)* – Beginning with the launch of the TERRA satellite in 1999, the EOS phase introduced several innovations to satellite remote sensing. Chief among these were frequent repeat coverage cycles, wider resolution capabilities and higher level processing to address a multiplicity of environmental applications.
- *Phase 7: New millennium* – This phase in the evolution of remote sensing is marked by the introduction of highly advanced test concept systems. These satellite sensors represent "next generation" systems such as EO-1, which carried the first spaceborne hyperspectral sensor and the Advanced Land Imager (ALI) into earth orbit; a less costly and superior replacement of Landsat TM technology.
- *Phase 8: Commercialization* – At this juncture, private industry ventures introduce innovative sensor designs and operational configurations that usher in very high spatial resolution satellites with pixels scales below 5 m^2. Systems of this variety include IKONOS, QuickBird and more recently GeoEye, together with revolutionary methods of data collection such as those introduced by the RapidEye constellation. Commercialization has also witnessed the introduction of microsatellite and nanosatellite technologies, encouraging the miniaturization of remote sensing platforms and the design of application-specific low-cost platforms.

Remote sensing systems are designed to record electromagnetic radiation emitted or reflected from an object at a distance (Maini and Agrawal 2007). The specific portions of the electromagnetic spectrum where these sensors operate is dictated by two important factors: (1) the availability of radiant energy to record and (2) the nature of surface objects whose physical, chemical and biological properties influence the manner by which radiant energy is received, absorbed, reflected and emitted. Those portions of the electromagnetic spectrum that facilitate reliable

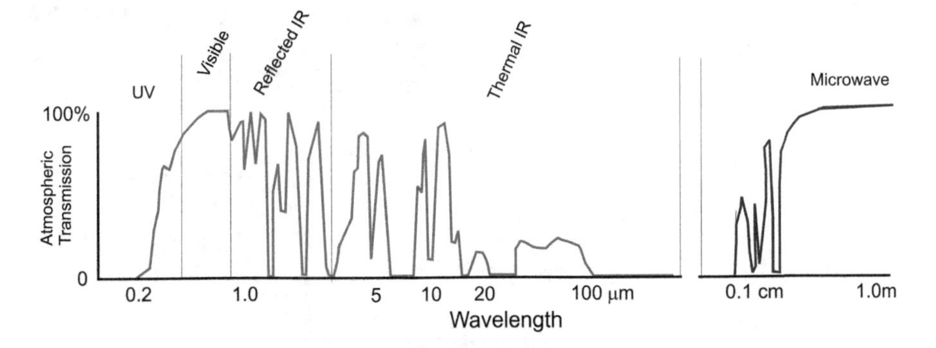

Fig. 3.1 Locations and patterns of atmospheric windows

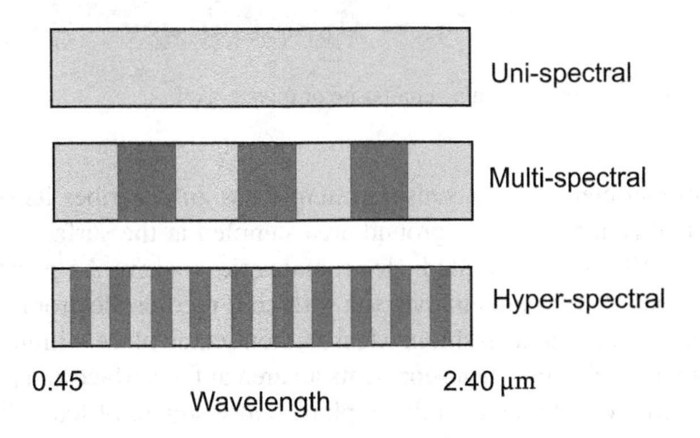

Fig. 3.2 The configuration of spectral bands for remote sensing application

measurement of radiant energy define "windows" that permit electromagnetic radiation to transmit from its source to a surface and then to the sensor (Fig. 3.1). The wavelengths of energy characterizing these windows, while explaining broad intervals described by unique frequencies and amplitudes, can be infinitely subdivided into smaller channels. Here, the number and the specific intervals designated to form a channel allow distinctive patterns and intensities of energy flux to become measureable. A sensor system can, therefore, be designed to measure electromagnetic radiations in a single broad interval of wavelengths, or in multiple narrow intervals within one or more openings in the window (Fig. 3.2). This design feature of a sensor system explains its spectral resolution; an attribute of the sensor that defines the total number of wavelength intervals, their width, and the specific portion of the electromagnetic spectrum that these intervals represent. To the environmental analyst, spectral resolution is a critical determinant of a sensor's applicability to a given problem, since the object, surface or condition the problem describes may not be detectable by the sensor. Therefore, knowing where to look, or more appropriately, where along the electromagnetic spectrum the feature of interest can be measured, is a nontrivial question. If this question can be answered, the selection of the appropriate sensor system and data product will be greatly simplified.

Fig. 3.3 The spatial patterns defining composite or mixed pixel

An additional element of a sensor system's design describes its spatial resolution. This quality measures the ground area sampled at the surface used to record the sensor's reflected of emitted electromagnetic energy. As a satellite sensor progresses along its orbital path over the surface it records electromagnetic energy in a systematic and regular fashion. Measurement takes place within the geometry of a pixel (picture element) that represents an area at the surface at a predetermined level of geographic detail. Generally explained as a square of fixed dimension, the spatial unit that a measurement of electromagnetic radiation explains may range from as coarse as 1 km × 1 km to as fine as 0.4 m × 0.4 m in area. In practical terms, a sensor's level of spatial resolution determines how well an object or a feature of the surface can be recognized by its distinctive geometric form. With reference to measurement, an object that fills the resolution "square" is observable as a discrete value. Should the object of interest fail to completely occupy the square, its measurement becomes a composite or mixture of whatever the other objects fall within that boundary (Fig. 3.3). This phenomenon is commonly referred to as the mixed pixel and represents a source of confusion and error that frustrates image classification.

Spatial resolution is analogous to the concept of scale and allows environmental investigations to be conducted at varying levels of spatial discreteness from the macroscale (resolution greater than 1 km), mesoscale (resolutions between 500 and 300 m) and microscale (resolutions less than 20 m). Because sensor systems tend to be categorized in terms of their spatial resolution, this value influences the selection of a sensor and the data product. Ideally, a level of geographic detail can be determined that will insure that the spectral measurements recorded by sensor will correspond closely to the object or feature of interest. In practical applications, however, an exact one-to-one correspondence can be illusive, particularly where the land surface is composed of complex entities. Under these circumstances it is

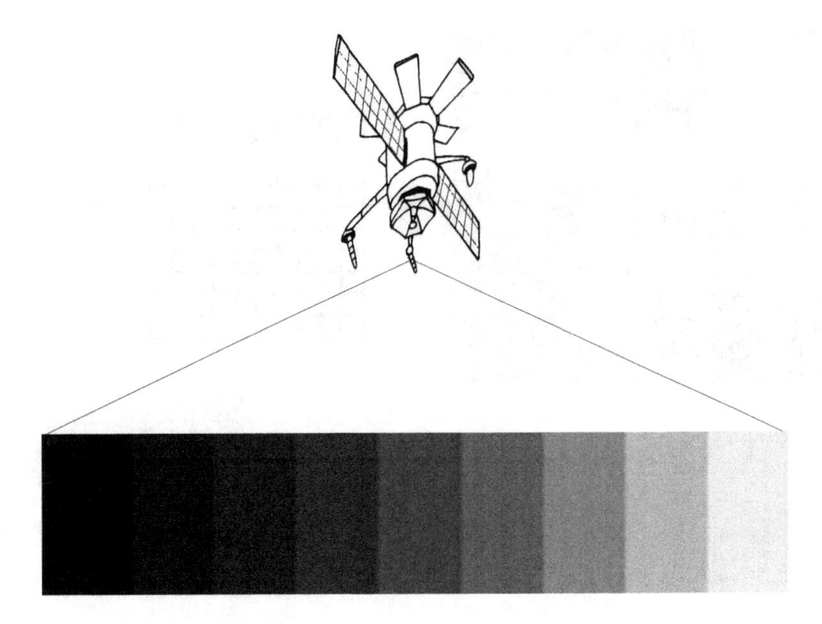

Fig. 3.4 The gray-scale concept

important to remember that a sensor provides a sample of the surface and as with any sampling schema there will always be error. We will explore the methods to compensate for error in later chapters. For now the goal is to select a resolution that matches the minimum mapping unit required for objects to emerge from the background clutter of the scene.

Related to both spatial and spectral resolution is the radiometric sensitivity characteristic of the sensor. This attribute of the sensor explains its measurement capabilities and provides a means to express the level of numerical detail used to record the intensity of the electromagnetic radiation the sensor captures. Typically, radiometric resolution can be conceptualized as a gray-scale continuum that quantifies the range of measurements available with a given wavelength interval (Fig. 3.4). As illustrated in Fig. 3.4, gray-scale quantification places radiometric resolution as a quantity that describes "brightness" as a function of wavelength per angular unit. This value, defined in terms of wavelength, spatial area, and intensity, is given as a real number commonly referred to as either the brightness value or a digital number, which is recorded for each pixel across all wavelengths and written to a file that becomes the digital image (Fig. 3.5). The range of brightness values that can be recorded by the sensor describes the sensors dynamic range. The greater the number of digits that can be used to translate this range into a gray scale, the greater is the capability of the sensor to capture (measure) subtle differences in the radiance characteristics of objects or features at the surface. In an environmental context this could be the difference between our ability to detect polluted water from clear water, or subtle contrasts in vegetation that may be indicative of stress due to drought or insect damage.

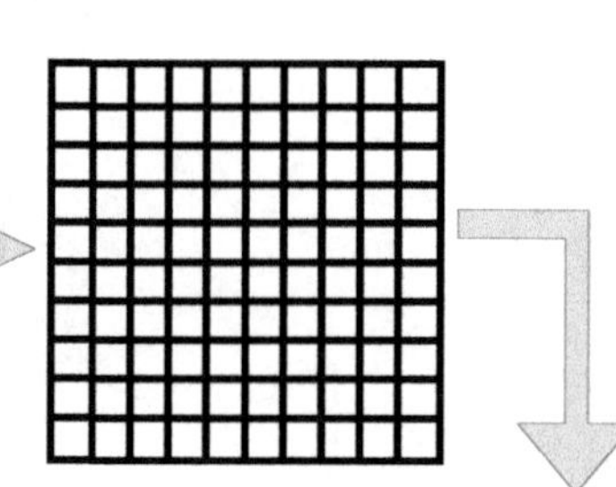

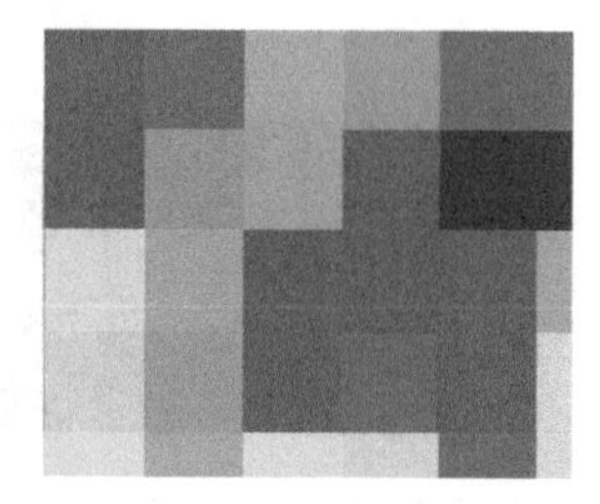

Fig. 3.5 Characteristics of a digital image

When taken together the spatial, spectral and radiometric resolution of a sensor determines how well an object or feature in the landscape can be imaged as a unique phenomenon. Features or conditions at the surface that fall below the thresholds of resolution cannot be sampled and therefore are not really present in the image. For example, objects that fall below the spatial resolution of the sensor can only be recorded as a composite or mixture of the features surrounding it. Similarly, should the spectral resolution of the sensor identify a wavelength in an interval that is too broad, a degree of specificity may be missing that fails to record the spectral properties of an object in sufficient detail to enable succinct differentiation. Finally, the radiometric characteristics of the sensor may not provide enough contrast to allow the feature at the surface to emerge as a measureable object. In each of these cases, measurement is confused and the digital number recorded to the image file may not be representative of meaningful data.

An equally critical aspect of the data acquisition question for successful environmental analysis is the representation of time. A single image acquired by a satellite sensor represents a "snapshot" of the surface at a fixed point of time. Often for environmental applications our interests are more process oriented and the temporal dimension is an important element of a study. Every sensor system placed into Earth's orbit has a revisit or repeat cycle that is a direct function of its orbital characteristics. The sensor's repeat cycle introduces a temporal resolution that can be used to formulate a time-sequenced set of measurements for a study. Exploiting this temporal dimension facilitates the representation of time-continuous processes and enables "progression over time" assessment of land surface behaviors and supports the data demands of environmental monitoring applications. The present constellation of satellites dedicated to earth observation describes revisit cycles as frequent as 3 days to upwards of 24 days. This suggests that under the most optimal conditions a location on the surface could be imaged between at least 15 times per year at a minimum and as often as 122 times per year. Satellites, such as the Landsat series, with operational histories of nearly 40 years, suggest a temporal archive of nearly 1,000 scenes for any location at the surface. However, obtaining a useable time sequence is easily perturbed by cloud cover and other environmental contaminants that reduce the number of useable scenes and impact data quality. Notwithstanding these limitations, time-stepped images can be acquired for nearly any environmental process over the history of satellite remote sensing that allows us to exploit the temporal dimension. This fact alone should encourage the continuity of satellite missions in order to preserve this essential attribute of satellite-based remote sensing (Hernanades 2005).

3.2 Footprints and Formats

While sensor resolutions in its various definitions remain perhaps the most important characteristic that distinguishes among satellite platforms, the application potential of this data source is further refined by two additional factors: (1) the sensors orbital parameters and (2) the formats employed to organize digital measurements of electromagnetic energy into computer processable files. Satellite orbits vary considerably with the individual facets of their design depending largely on purpose. Specific to the goals of environmental remote sensing focus can be directed toward the class of satellite systems referred to as earth observational satellites. This class of sensors describes systems maintaining geocentric orbits with altitudes that fall into one of the four categories:

1. *Low earth orbits* – Geocentric orbits with altitudes ranging from 160 to 200 km (100–12,400 miles). A satellite in low earth orbit typically completes one revolution every 90 min.
2. *Medium earth orbit* – Geocentric orbits with altitudes between 2,000 and 35,000 km.
3. *Geosynchronous orbits* – Geocentric orbits with a fixed altitude of 35,786 km, with an orbital period timed to follow the rotation of the Earth.
4. *High earth orbits* – Geocentric orbits with altitudes exceeding 35,786 km.

Satellite orbits may be further defined by their inclination. Earth observational satellites assume a near-polar, sun-synchronous path which combines altitude and inclination in a manner that sets the satellite's ascending or descending path over a given point on the surface as the same local mean solar time. This particular orbital design creates a surface illumination angle that will be the same each time the satellite makes an overpass with a "fixed" local time. By placing the sensor into this configuration a consistent illumination of the surface is provided which holds

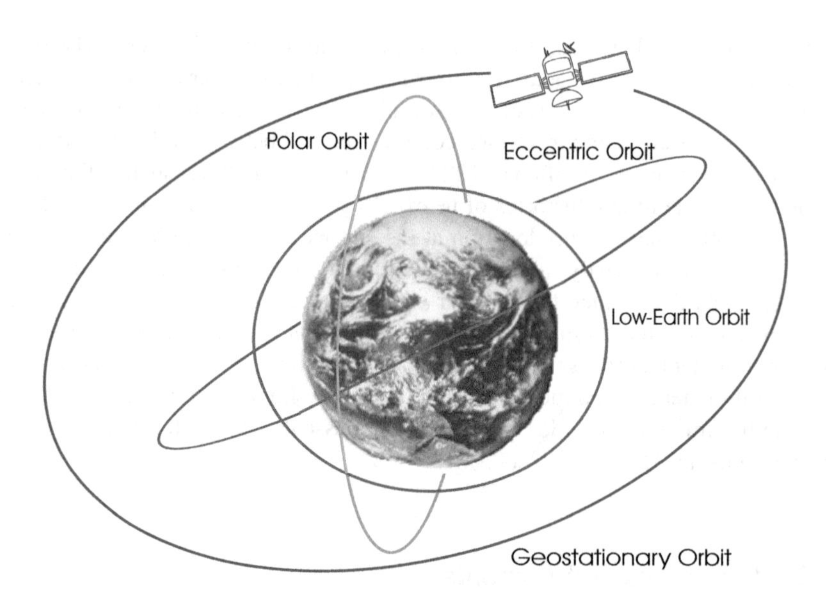

Fig. 3.6 General characteristics of satellite orbital geometries

shadows nearly constant on every satellite pass and produces some level of control as brightness is measured from the surface. The general characteristic of an earth observational orbit is illustrated in Fig. 3.6.

As a satellite progresses along its orbital track over the Earth, its altitude inclination and sensor architecture form a scan swath in which data are collected. The lateral dimension of this swath width explains the side-to side-ground distance the sensor record in during the satellites orbital pass. From the data users perspective swath width is the lateral distance that expresses the physical dimension of the satellites "footprint." In simple terms we can think of this as an area on the ground represented in one satellite image (Fig. 3.7). Since file formatting conventions used to store remotely sensed data are organized according to a grid or raster representation of geographic space, the swath width serves as a convenient device for establishing the areal coverage of the sensor. This physical extent of the data has important implications for environmental remote sensing since this footprint can determine the number of individual scenes that will be required to produce a complete coverage for a given study location, which can greatly influence other considerations when developing a remote sensing application. Sensor footprints vary in geographic extent from dimensions as little as 13 × 13 km to nearly as large as 2,400 × 6,400 km. In many respects taking these dimensions into consideration in the plan of a remote sensing investigation helps to match both the scale of geographic coverage and the elements of resolution with the functional scale required for the analysis (Fig. 3.8). If both are not thought through carefully, an analysis may not yield satisfactory results and errors related to over generalization, misidentification, and inaccuracy will dominate.

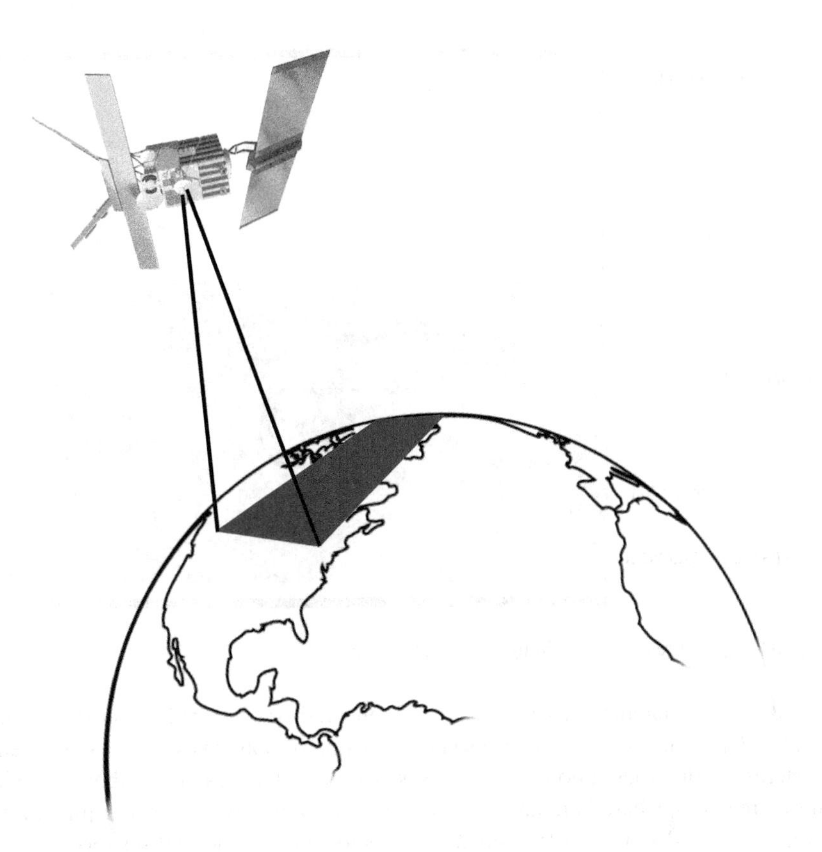

Fig. 3.7 Satellite imaging "footprints" or swaths

3.3 Sensor Characteristics

Satellite-based sensors have become the primary platform for carrying out environmental remote sensing activities (Ulbricht and Heckendorf 1998; Gulinck et al. 2000) due in large part to their capacity for

- Continuous data acquisition
- Frequent and regular revisit cycles
- Broad areal coverage
- Good spectral resolution
- Support of semiautomatic computerized processing and analysis

Based on the source of electromagnetic radiation used to activate the sensor electronics, remote sensing systems may be separated into two broad classes: (1) passive sensors and (2) active sensors. Passive sensors refer to systems designed to detect solar radiation reflected or emitted by objects at the surface. Active sensors describe systems designed to illuminate a surface using active artificial sources of radiation mounted on the platform. Sending a pulse of radiation toward the surface,

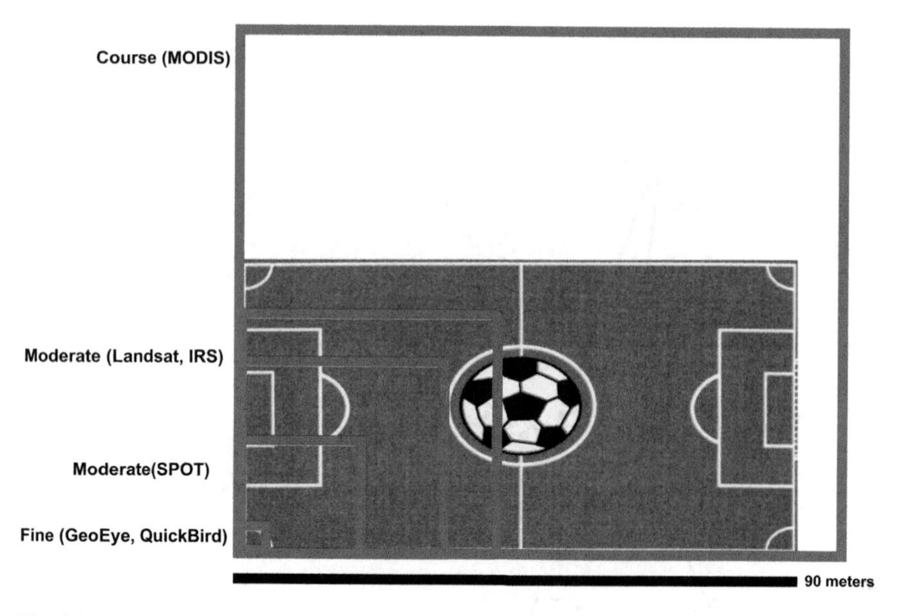

Fig. 3.8 Spatial resolution defining pixel spatial scale

active sensors capture this pulse as it is reflected or scattered by surface objects (Table 3.1). This table provides a selective listing of both passive and active sensor systems. Within these two general classes, sensor systems can be further subdivided into three groups based on the spectral regions they employ for data acquisition: (1) optical/infrared sensors, (2) microwave sensors, and (3) thermal sensors.

1. *Optical/infrared sensors* – This class of sensor uses the visible (0.3–0.7 μm), near infrared (0.72–1.30 μm) and shortwave infrared (1.3–3.0 μm) portions of the electromagnetic spectrum. Sensors of this design can be arranged according to their spatial resolution to better connect their functional configuration with the types of environmental applications they support. We can begin this overview working from low-resolution systems that capture spatial detail and comparatively coarse levels of resolution and continue with medium resolutions systems and culminate our discussion with those sensors that display high spatial resolutions.

   (a) *Low-resolution systems* – For the purposes of this review, low-resolution systems are defined as sensors that record reflected electromagnetic radiation as synoptic scales above 500 m to greater than 1,000 km. Typically sensors of this class capture data over large geographic areas which make them useful for examining macro-scale disturbances and environmental processes that exhibit wide areal extents. Satellite sensors of this variety include:

      - *The advanced very high resolution radiometer (AVHRR)* carried on board the National Oceanic and Atmospheric Administration (NOAA) series of polar-orbiting operational environmental satellites (POES). The AVHRR

Table 3.1 List of selected earth observational satellites and sensors

Satellites	Space agency	Equator crossing time + altitude	Launch date	Instruments	Status, applications and other information
ERS-2	ESA	10:30 (D) + 785 km	21/04/95	Altimeter SAR, SAR-wave ATSR-2 Scatterometer GOME	No onboard recorder since 06/2003. Data acquisition is ensured over ESA agreed acquisition stations; ATSR-2 anomaly since 02/2008; planned operations until mid-2011
OrbView-2 (former SeaStar)	NASA/ GeoEye	12:00 (D) + 705 km	01/08/97	SeaWIFS	Eight-channel imager for ocean color, vegetation and aerosols
Landsat-7	NASA	10:05 (D) + 705 km	15/04/99	ETM+	Well-calibrated, multispectral, moderate resolution, substantially cloud-free, sunlit digital images of continental and coastal areas and selected coral reefs
QuikSCAT (Quick Scatterometer)	NASA	06:00 (A) + 803 km	19/06/99	SeaWinds	Sea surface wind speed and direction data for global climate research and operational weather forecasting and storm warning; *stopped operating on 23/11/09 after antenna failure*
Terra	NASA	10:30 (D) + 705 km	18/12/99	CERES MISR MODIS MOPITT ASTER	Measurement of Earth's climate system, atmosphere, land, oceans and interactions with solar radiation
ACRIMSAT	NASA	10:50 (D) + 720 km	20/12/99	ACRIM 3	Active Cavity Radiometer Irradiance Monitor satellite measures total solar irradiance

(continued)

Table 3.1 (continued)

Satellites	Space agency	Equator crossing time + altitude	Launch date	Instruments	Status, applications and other information
Envisat	ESA	10:00 (D) + 800 km	1/03/2002	ASAR, RA-2 AATSR MERIS GOMOS MIPAS MWR SCHIAMACHY	MIPAS is operated at 80% of its duty cycle. GOMOS performs regularly with reduced azimuth range since 29/08/2005; RA-2: loss of secondary frequency (in S-band) in 01/2008, compensated with on-ground ionospheric corrections, operations extended up to 2013
Aqua	NASA	13:30 (A) + 705 km	04/05/02	AMSR-E AIRS HSB AMSU-A CERES MODIS	Collects data on Earth's water cycle, precise atmospheric, land and oceanic measurements, and interaction with solar radiation. AMSR-E provided by JAXA. HSB provided by INPE (no longer functional)
Resourcesat-1	ISRO	10:30 (D) + 817 km	10/2003	AWIFS	Land monitoring 4-channel camera
Aura	NASA	13:45 (A) + 705 km	15/07/04	HIRDLS MLS OMI TES	Comprehensive measurements of atmospheric chemistry and trace gases; HIRDLS, high-resolution dynamic limb sounder (infrared); MLS, microwave limb sounder; OMI, ozone monitoring instrument; TES, tropospheric emission spectrometer
Cartosat-1	ISRO	10:30 (D) + 618 km	05/2005	Carto-dem	High-resolution stereoimagery with two panchromatic cameras
Monitor-E	Roscosmos	10:30 + 550 km	27/08/05	LOS	Surface mapping, support to disaster management and monitoring the effects of pollution with two optical cameras; LOS, land observing satellite

DAICHI (ALOS)	JAXA	10:30 (D) + 691 km	24/01/06	PRISM, AVNIR-2, PALSAR	Mapping, precise land coverage observation, disaster monitoring, and resource surveying; ALOS, advanced land observing satellite
CALIPSO	NASA/ CNES	13:30 (A) + 705 km	28/04/06	CALIOP WFC IIR	Cloud-aerosol lidar and infrared pathfinder satellite observations for climate predictions
CloudSat	NASA/CSA	13:30 (A) + 705 km	28/04/06	CPR	Global cloud properties (applications: air quality, aviation safety, disaster management, energy and water management); CPR, cloud profiling radar
Compass-2	Roscosmos	450–500 km (79° incl.)	26/05/06	Radio frequency analyzer	Microsatellite for monitoring anomaly phenomena in the Earth's ionosphere
Resurs-DK	Roscosmos	360–690 km (70.4°incl.)	15/06/06	LOS	LOS, land observing satellite
HY-1B	CNSA	10:30 (D) + 798 km	04/2007	OCTS	Ocean monitoring; Four band CCD. camera; OCTS, ocean color and temperature scanner

Table 3.2 Characteristics of the advanced very high-resolution radiometer (AVHRR)

Swath width	2,399 km	
Spatial resolution	1.1 km	
Spectral resolution	Wavelength (μm, NOAA-6, 8, 10, 12)	Wavelength (μm, NOAA-7,9,11)
Channel 1	0.58–0.68	0.58–0.68
Channel 2	0.725–1.10	0.725–1.10
Channel 3A		1.58–1.64 (NOAA 15, 16)
Channel 3B	3.55–3.93	3.55–3.93
Channel 4	10.50–11.50	10.3–11.3
Channel 5	Channel 4 repeated	11.5–12.5
Orbit type	Sun-synchronous circular, PM orbit	
Revisit cycle	Daily	

sensor is a broad-band, four- or five-channel scanning radiometer designed to record reflected electromagnetic radiation in the visible and near-infrared portions of the spectrum and emitted electromagnetic radiation in the thermal infrared regions between 3.55 and 12.50 μm. The general specifications of the AVHRR system are given in Table 3.2. The first AVHRR sensor was a four-channel radiometer deployed on the TIROS-N meteorological satellite in 1978. This initial system was improved to a five-channel instrument (AVHRRR/2) placed into orbit on NOAAA-7 in 1981. The latest AVHRR sensor (AVHRR/3) is a six-channel instrument launched into orbit on NOAA-15 in 1998. The main objective of the AVHRR program is to provide radiance data for the study of clouds, land-water boundaries, snow and ice extent , ice and snow melt inception, day and night cloud distribution, temperatures of radiating surfaces and sea surface temperature estimates. Over its history, a wide range of environmental applications have utilized the AVHRR system including studies focused on agricultural crop assessment, ecosystem disturbance, drought evaluation, fire detection, land cover analysis, large area mapping and the evaluation of regional-scale and continental-scale snow cover analyses.

Data from the AVHRR sensor are acquired with a wide-field scanning system that permits global-scale coverage on a daily basis with a 1.1 km spatial resolution. The sensor also provides a data stream at 4 km ground resolution that is achieved by sampling and averaging the full 1.1 km data on board the satellite. Several data products derived from the AVHRR platform are of particular value to environmental remote sensing investigations. These include the AVHRR normalized difference vegetation index (NDVI) composite and the global land cover characterization (GLCC) data set. The NDVI composite are greenness maps produced by calculating the NDVI using the red (0.6–0.7 μm) and near infrared (0.7–1.1 μm) bands of the sensor. Weekly and bi-weekly composites are

Table 3.3 Characteristics of the Defense Meteorological Satellite Program operational linescan system (DMSP–OLS)

Swath width	3,000 km
Spatial resolution	2.2 km
Spectral resolution (μm)	
Visible	0.40–1.10
Infrared	10.0–13.4
Photomultiplier tube	0.47–0.95
Orbit type	Sun-synchronous near-polar orbit
Revisit cycle	Daily

produced from multiple AVHRR observations that have been composited together to create nearly cloud-free images depicting maximum vegetation health and vigor at scales suitable for regional investigations. The GLCC product is a series of global land cover classification data sets based on the unsupervised classification of 1.1 km AVHRR 10 day NDVI composites (http://edc2.usgs.gov/glcc/background.php).

- *The Defense Meteorological Satellite Program Operational Linescan System (DMSP-OLS)* – This sensor is a polar orbiting satellite supported by the U.S. Air Force. The DMSP-OLS is a sensor designed for cloud imaging with two spectral bands; a visible near-infrared band operating within the 0.47–0.95 μm spectral range and a thermal band at 10.0–13.4 μm. The system has a 2.7-km spatial resolution and an image swath width of 3,000 km. The satellite system was officially acknowledged and declassified in 1972, and imagery was made available to the scientific community shortly thereafter. Since this declassification, the DSMP instrument has been repeatedly upgraded. The OLS sensor is the most recent of these upgrades, orbiting the Earth 14 times a day with a night-time overpass between 20:30 and 21.30 h (10:30–11:30 PM). As a function of its design and orbit the OLS sensor can detect visible light sources as low as 10–9 W/cm^2/steradian which is approximately 4 orders of magnitude greater than the NOAA-AVHRR system. This feature of the OLS has been shown to provide unique capabilities for detecting urban footprints, gas flaring, biomass burning and for tracking vessels in the oceans (Elvidge et al. 2001) (Table 3.3).
- *OrbView-2/SeaWIFS* – This sensor is housed onboard the GeoEye Corporation's OrbView-2 satellite. The sensor is a follow-on experiment to the coastal zone color scanner and began its operations on September 18, 1997. The sensor covers eight spectral bands in the 0.40–0.88 μm spectral range with a 1.1-km spatial resolution in the local area coverage (LAC) mode and 4.5 in the global area coverage (GAC) mode. Sea-viewing wide field-of-view sensor (SeaWIFS) was designed specifically to monitor ocean characteristics, such as chlorophyll-a concentrations and water clarity. Because the sensor has the capability to view every square kilometer of a cloud-free ocean every 48 h, the data captured by the satellite can provide measurements of the abundance of ocean biota,

Table 3.4 Characteristics of the SeaWIFS sensor

Swath width	2,800 km
Spatial resolution	1 km
Spectral resolution (μm)	
Visible	0.40–1.10
Infrared	10.0–13.4
Photomultiplier tube	0.47–0.95
Orbit type	Sun-synchronous polar circular orbit
Revisit cycle	16 days

oceanic primary productivity, and global biogeochemistry. To date, the SeaWIFS sensor has been used to examine the magnitude and variability of chlorophyll, primary productivity of marine phytoplankton and to assist with the detection and timing of spring algae blooms. The general specifications of SeaWIFS are given in Table 3.4.

(b) *Medium resolution systems* – Medium resolution systems are defined as the sensors that record reflected electromagnetic radiation at synoptic scales between 100 and 500 m. Typically, sensors of this class capture data over mesoscales which makes them useful for examining localized patterns of disturbance and regional level environmental processes. Satellite sensors of this variety include:

- *Moderate Resolution Imaging Spectrometer (MODIS)* – An instrument placed into orbit on board the TERR and AQUA satellites, the MODIS sensor has been designed to observe the entire Earth in a 1–2-day revisit cycle with a 2,330-km swath width. The sensor captures surface radiance in 36 spectral bands covering the interval between 0.40 and 14.38 μm (Table 3.5). MODIS also images the surface at three distinct spatial resolutions with bands 1 and 2 recording surface radiation at 250 × 250 m, bands 3–7 at 500 × 500 m and bands 8–36 at a low-resolution scale of 1,000 × 1,000 m. The contrasting spectral and spatial ranges enable MODIS data to be processed at several different levels and facilitate the creation of 44 different standard data products. A selection of those germane to environmental remote sensing include:
  - *MODIS Level 1A radiance counts (MOD 01)* – A level 1A data set contains counts for the 36 MODIS channels that are used primarily for geolocation, calibration and processing.
  - *MODIS Level 1B calibrated geolocated radiance (MOD 02)* – This data set contains calibrated and geolocated at-aperture radiances for the 36 bands generated from MODIS Level 1A sensor counts. Visible, short-wave infrared (SWIR) and near-infrared measurements are made during daytime only whereas the thermal radiances (TIR) are measured continuously by the sensor.

Table 3.5 Characteristics of the MODIS sensor

Swath width	2,330 km		
Spatial resolution	250 m (bands 1–2) at nadir	500 m (bands 3–7) at nadir	1,000 m (bands 8–36) at nadir
Spectral resolution (μm)			
Spectral range	0.4–14.4		
Reflected bands	0.405–0.965		
Emitted bands	3.66–14.385		
Equatorial crossing time	10:30 AM		
Orbit type	Sun-synchronous polar circular orbit		
Revisit cycle	1–2 days		

- *MODIS normalized water-leaving radiance (MOD 18 and MOD 37 (MODIS aerosol optical depth)* – These Level 2 and Level 3 products contain ocean water-leaving radiances for bands 8–14. Normalized water-leaving radiance is used to estimate chlorophyll-a concentrations and the ocean primary productivity.
- *MODIS aerosol product (MOD 04)* – This monitors the ambient aerosol optical thickness over the oceans and portions of the continents. The MODIS aerosol product is used mainly to study aerosol climatology, and sources and sinks of specific aerosol types.

• *European Space Agency's Envisat (environmental satellite)* – This instrument was launched on March 1, 2002 and carries an array of nine sensor elements with a 35-day repeat cycle (Table 3.6). The medium resolution imaging spectrometer (MERIS) is one of the main instruments on this satellite. MERIS is composed of five detector elements each equipped with a push-broom spectrometer that provides data in 15 spectral bands. This system has an intrinsic spatial resolution of 300 m and was designed primarily to observe the color of the ocean and derive estimates of chlorophyll and suspended sediments. The sensor is also useful for terrestrial applications using vegetation transforms such as the NDVI.

(c) *High-resolution systems* – Satellite systems included in this category define spatial resolutions between 15 and 80 m. Many of these sensors share similar spectral and radiometric characteristics as well as repeat cycles falling between 16- and 20-day intervals. Satellites of this class include

• *Landsat TM and ETM+* – The Landsat program was the first earth resource satellite system with the first generation Landsat satellite launched on January 23, 1972. Over the course of the Landsat program succeeding satellites evolved toward higher spatial and spectral resolutions. Presently,

Table 3.6 Characteristics of the Envisat MERIS sensor

Swath width	1,150 km	
Spatial resolution	Ocean: 1,040 × 1,200 m	Land: 260 × 300 m
Spectral resolution	15 spectral bands	
Spectral range (μm)	0.39–1.40	
Equatorial crossing time	10:00 AM	
Orbit type	Sun-synchronous polar circular orbit	
Revisit cycle	3 days	

Table 3.7 Characteristics of the Landsat TM sensor

Swath width	185 km	
Spatial resolution	Multispectral	Thermal
	30 m	120 m
Spectral resolution (μm)		
Band 1	0.45–0.52	
Band 2	0.52–0.60	
Band 3	0.63–0.69	
Band 4	0.76–0.90	
Band 5	1.55–1.75	
Band 6	10.40–12.50	
Band 7	2.08–2.35	
Equatorial crossing time	9:45 AM	
Orbit type	Sun-synchronous near-polar orbit	
Revisit cycle	16 days	

Landsat series of satellites have acquired high-resolution multispectral data on a systematic and repetitive basis with an archive boasting 40 years worth of earth observational data (Table 3.7).

The orbital characteristics of Landsat satellites 1–3 are provided in Table 3.8. These satellites describe instruments deployed into sun-synchronous near-polar orbits with two on-board sensors: a return beam vidicom (RBV) and a 4-channel MSS. The RBV system suffered crippling malfunctions and only a limited number of images were acquired. The MSS however was more robust, recording reflected electromagnetic energy in four spectral bands covering the interval between 0.5 and 1.1 μm with a 185-km swath width. Each detector on the MSS sampled a 57 × 79 m ground area which was resampled to a nominal resolution of 79 × 79 m.

Table 3.8 Characteristics of the Landsat satellite series

(a) Landsat MSS sensor (Landsat 1–3)			
Band number (L1–L3)	Band number (L4–L5)	μm	Resolution
4	~2 (0.52–0.60 μm)	0.5–0.6	68 m × 83 m
5	~3 (0.63–0.69 μm)	0.6–0.7	68 m × 83 m
6	~4 (0.76–0.90 μm)	0.7–0.8	68 m × 83 m
7	~4	0.8–1.1	68 m × 83 m
8	~6 (2.08–2.35 μm)	10.41–12.6	68 m × 83 m
(b) Landsat TM sensor			
Swath width	185 km		
Spatial resolution	Multispectral	Thermal	
	30 m	120 m	
Spectral resolution (μm)	Wavelength (μm)		
Band 1	0.45–0.52		
Band 2	0.52–0.60		
Band 3	0.63–0.69		
Band 4	0.76–0.90		
Band 5	1.55–1.75		
Band 6	10.40–12.50		
Band 7	2.08–2.35		
Equatorial crossing time	9:45 AM		
Orbit type	Sun-synchronous near-polar orbit		
Revisit cycle	16 days		

In 1982 the MSS system was replaced by a sensor referred to as the thematic mapper (TM). The TM sensor was placed into orbit on Landsat 4 and 5. Both of these satellites retained similar orbital characteristics of the earlier Landsat platforms; however, in order to match the swath width of the Landsat 1–3 the altitude of the TM system was lowered, giving the TM series a 16-day repeat cycle. The TM series has improved spectral, radiometric, and spatial resolution, as well as the inclusion of a thermal band. On April 15, 1999 a third generation of the Landsat series was launched into orbit. This new satellite carried the enhanced thematic mapper (ETM+), a system with two noteworthy improvements over the Landsat TM. First, the ETM+ added a panchromatic band (Band 8) with 15 m spatial resolution. Secondly, the ETM+ included two thermal infrared channels (Band 6a and 6b) with 60 m resolution. In addition to these sensor enhancements, Landsat 7 ETM+ was placed into an orbit that allows the sensor to precede the TERRA satellite by 30 min along a common ground track. This orbital configuration, referred to as a "sensor train," is an important advancement in earth observation science, since it allows numerous sensor platforms to record surface phenomenon in a near-synchronous fashion. With sensors following along in close temporal spacing opportunities exist for data fusion between Landsat ETM+ and the ASTER, MODIS and HYPERION instruments.

Table 3.9 Characteristics of the SPOT sensor

Swath width	60 km	
Spatial resolution	Multispectral mode	Panchromatic mode (P)
	20 m	10 m
Spectral resolution (μm)		
Band 1	0.50–0.59	
Band 2	0.61–0.68	
Band 3	0.79–0.89	
PAN	0.51–0.73	
Equatorial crossing time	10:30 AM	
Orbit type	Sun-synchronous near-polar orbit	
Revisit cycle	26 days	

The Landsat ETM+ on board the Landsat 7 satellite experiences a scan-line corrector failure on May 31, 2003. Although the system is still operational, the failure significantly reduced the utility of the data. Data are still collected using this sensor with missing data optionally filled in using other prefailure Landsat data selected by the user. An estimated 22% of any ETM+ scene is lost due to the scan line corrector failure. The gap-filled product option is an attempt to compensate for this loss.

- *Satellite Pour L'Observation de la Terra (SPOT) Satellite* – The SPOT series of earth observational satellites consist of four high-resolution platforms (SPOT 1–4) and a recent very high-resolution satellite (SPOT 5). Initiated by the French space agency CNES (Centre National d'etudes Spatiales), the first SPOT satellite was placed into orbit in 1986. This system introduced several enhancements to sensor design including the addition of two identical high-resolution visible sensors capable of delivering stereoscopic imagery, and panchromatic and multispectral operational modes with 10 m (panchromatic) and 20 m (multispectral) spatial resolutions (Table 3.9).

SPOT data is recorded along a 60 km swath width with a 26-day repeat cycle. The SPOT 4 satellite, launched in 1998, features several improvements over its predecessors including the addition of a mid-infrared band at 1.58–1.75 μm to its high-resolution visible and infrared sensor and a vegetation monitoring instrument called the VMI. Perhaps the most significant advancements to the SPOT series are those found on SPOT 5, a platform launched on May 4, 2002. The SPOT 5 satellite has two high-resolution instruments that provide spatial resolutions of 2.5–5 m in panchromatic mode and 10 m in multispectral mode. The vegetation instrument on board SPOT 5 (Vegetation 2) also offers continuity for environmental applications initiated by SPOT 4 functionality. The specifications of the SPOT 5 sensor are listed in Table 3.10.

Table 3.10 Characteristics of the SPOT 5 sensor

Swath width		60 km	
Spatial resolution	Multispectral	SWIR	PAN
	20	10	5
Spectral resolution (μm)			
		0.50–0.59	
		0.61–0.68	
		0.79–0.89	
		1.58–1.750	
Equatorial crossing time		10:30 AM	
Orbit type		Sun-synchronous near-polar orbit	
Revisit cycle		2–3 days	

- *TERRA – ASTER* – Referred to as the flagship of the Earth Observing System (EOS) series of satellites, TERRA represents a new generation of sensors that carry into orbit a complement of five synergistic instruments including
  - CERES – Clouds and Earth's radiant energy system
  - MISR – Multiangle imaging spectroradiometer
  - MODIS – Moderate resolution imaging spectrometer
  - Mopitt – Measurement of pollution in the troposphere
  - ASTER – Advanced spaceborne thermal emission and reflection radiometer

  Launched on December 18, 1999 into a sun-synchronous orbit with an equatorial crossing time of 10:30 AM, the TERRA system began collecting data on February 24, 2000.

  The ASTER sensor is significant in its design; covering a wide spectral region with 14 bands from the visible to the thermal infrared with both high spectral and radiometric resolutions. In addition, ASTER has a backward looking near-infrared band that provided stereoscopic coverage. Overall ASTER consists of three different subsystems (Table 3.11). The visible and near-infrared (VNIR) is a three band instrument with 15 m spatial resolution plus the addition of a backward looking detector. The shortwave infrared (SWIR) instrument images across 16 bands at 30 m resolution. Lastly, thermal infrared (TIR) data are collected over five bands with a spatial resolution of 90 m. The main features of these subsystems can be outlined briefly as follows:

  - The VNIR instrument is comprised of two sensor assemblies and produces the highest data rates using the three ASTER imaging subsystems.

Table 3.11 Characteristics of the ASTER sensor

Swath width	60 km		
Spatial resolution	VNIR	SWIR	TIR
	15 m	30 m	90 m
Spectral resolution (μm)			
VNIR			
Band 1	0.52–0.60		
Band 2	0.63–0.69		
Band 3N	0.78–0.86		
Band 3B	0.78–0.86		
SWIR			
Band 4	1.60–1.70		
Band 5	2.14–2.18		
Band 6	2.18–2.22		
Band 7	2.23–2.85		
Band 8	2.29–2.36		
Band 9	2.36–2.43		
TIR			
Band 10	8.12–8.47		
Band 11	8.47–8.82		
Band 12	8.92–9.27		
Band 13	10.25–10.95		
Band 14	10.95–11.65		
Equatorial crossing time	10:45 AM		
Orbit type	Sun-synchronous near-polar orbit		
Revisit cycle	16 days		

- The SWIR is a single fixed aspheric refracting sensor that operates six SWIR channels and a pointable mirror to allow coverage of any surface location over the 16-day repeat cycle.
- The TIR operates within 8.125 and 11.5 portion of the electromagnetic spectrum using a pointing mirror system.

Overall, 15 data products can be derived from the ASTER system including a Level 1B product with radiometric and geometric coefficients applied, a Level 2 radiance at sensor converted to temperature product and a Level 3 digital elevation model (DEM) produced by correlation of nadir and aft-looking Band 3 data.

- *IRS – Indian Remote Sensing Program* – India launched its first civilian satellite in March 1988. This date marked the beginning of a remote sensing program that presently collect data from eight satellite systems (Table 3.12) The IRS program is the largest constellation of remote sensing currently in operation. Among these systems are: the IRS-1B satellite, launched in 1991 with four spectral bands and 72 × 36 m spatial resolution, a 148 km swatch width and 22 day repeat cycle, the IRS-1C system

Table 3.12 Satellites and sensors of the Indian remote sensing system

Resolution (m)	Sensor	Satellite
360	OCM	IRS-P4
180	WIFS	IRS 1C, IRS 1D, IRS P3
72.5	LISS-I	IRS 1A, IRS 1B
56	AWIFS	IRS P6
36.25	LISS-II	IRS 1A, IRS 1B
24	LISS-III	IRS 1C, IRS 1D
5	PAN, LISS-IV	IRS 1C, IRS 1D, IRS P6
2.5	PAN	IRS P5
0.8	PAN	CARTOSAT-2

Table 3.13 Characteristics of the IRS-1B sensor

Swath width	148 km
Spatial resolution	72.5 m
Spectral resolution (μm)	
Band 1	0.45–0.52
Band 2	0.52–0.59
Band 3	0.62–0.68
Band 4	0.77–0.86
Equatorial crossing time	9:40 AM
Orbit type	Sun-synchronous near-polar orbit
Revisit cycle	22 days

with three sensor instruments (PAN, LISS-III and WiFS) and a 24 day repeat cycle and IRS-P6, a three sensor platform with spatial resolutions ranging from 5.8 m in panchromatic mode, 23.5 m in LSS-3, and 56 m in AWIFS capturing surface reflectance across a 740 km swath width (Table 3.13). The LISS (Linear Imaging and Self-Scanning Sensor) is a high-resolution multispectral instrument capable of operating in either multispectral mode with a spectral sensitivity between 0.53 and 0.86 μm or a mono-chromatic mode nominally set to the 0.62–0.68 μm spectral region. The AWiFS sensor operates in four spectral bands covering the range 0.52–1.70 μm. Data products generated by this system fall into five broad categories:

- *Scene-based standard products*
- *Scene-based georeferenced products*
- *Map-based products*
- *Floating geocoded products and*
- *Ortho-rectified geocoded products*

- *EO-1* – The National Aeronautics and Space Administration (NASA) EO-1 satellite was launched on November 21, 2000 as part of a 1-year technology demonstration mission (Digenis 2005). Strong advocacy from

Table 3.14 Characteristics of the ALI sensor

Swath width	37 km	
Spatial resolution	Multispectral	PAN
	30 m	10 m
Spectral resolution (μm)		
Band 1[a]	0.433–0.453	
Band 1	0.45–0.515	
Band 2	0.52–0.60	
Band 3	0.63–0.69	
Band 4	0.77–0.80	
Band 4[a]	0.84–0.89	
Band 5[a]	1.20–1.30	
Band 5	1.55–1.75	
Band 7	2.08–2.35	
PAN	0.48–0.69	
Equatorial crossing time	10:03 AM	
Orbit type	Sun-synchronous near-polar orbit	
Revisit cycle	16 days	

[a] Denotes analogous Landsat band

the remote sensing community to continue data acquisition from EO-1 led to an extension of the mission. EO-1 is presently chartered to collect and distribute data from it two main instruments: the advanced land imager (ALI) and the hyperspectral sensor (HYPERION). The ALI sensor, identified as the first earth observing instrument to be flown under NASA's New Millennium program, employs sophisticated optics and highly integrated multispectral and panchromatic spectrometers designed to produce Landsat-type imagery. The ALI has nine multispectral bands with 30 m spatial resolution and a 10-m panchromatic band (Table 3.14). When compared to the Landsat TM and ETM+ systems, ALI has three additional bands covering the intervals 0.433–0.453 μm, 0.845–0.890 μm, and 1.20–1.30 μm. However, unlike Landsat, the ALI sensor does not contain a thermal band. A standard ALI scene is 37 km wide and 42 km long with an option to increase scene length to 185 km.

Perhaps the most significant advancement introduced by the EO-1 system is the HYPERION imager. HYPERION is a push-broom hyperspectral sensor consisting of 220 10 nm bands that cover the electromagnetic spectrum from 430 to 2,400 nm. As a hyperspectral instrument, HYPERION represents a new direction for earth observation by offering resolution of surface characteristics across an expanded number of spectral bands (Pignati et al. 2009). Since hyperspectral data is acquired in a continuous fashion, HYPERION supports more complex forms of environmental analysis traditionally unavailable using the standard multispectral scanning

Table 3.15 Characteristics of the HYPERION sensor

Swath width	7.5 km
Spatial resolution	30 m
Spectral resolution (μm)	
220 spectral bands, 10 nm increments	0.40–2.5
Equatorial crossing time	10:03 AM
Orbit type	Sun-synchronous near-polar orbit
Revisit cycle	16 days

technology. The general specifications of the HYPERION instrument are given in Table 3.15. As noted in Table 3.15, the sensor can image a 7.5 × 100 km land area per scene, providing detailed spectral mapping across 220 channels maintain high radiometric accuracy. The ALI and HYPERION instruments onboard the EO-1 platform have a descending equatorial crossing time of approximately 10:00 AM with the EO-1 satellite flying in sensor-train formation with Landsat 7 and TERRA.

(d) *Very high-resolution systems* – This family of sensor systems are characterized by spatial resolutions below 5 m. Sensors that fall within this class are unique in that the majority of these satellites represent commercial ventures into the realm of spaceborne remote sensing data acquisition. Because the commercial systems are supported by "for-profit" private industries, they are less constrained by government policies and directed more toward satisfying market demands for spatial data products; products that were traditionally supplied by the aerial survey industry. The era of the planned commercial satellite began with the successful launch of the IKONOS-2 satellite in 1989. Since that date there is an expanding list of commercial entries into the marketplace (Williamson and Baker 2004). However, despite government policies that have facilitated market opportunities, the commercial remote sensing industry remains a high-risk venture in search of long-term stability. These limitations notwithstanding, the "hyperspatial" resolution offered by these platforms, often on the order of sub-meter scale, provide data in the near visible and near-infrared portions of the spectrum that encourage the use of these systems for cartographic mapping and specialized environmental applications. This section provides a selective review of those very high-resolution systems with demonstrated environmental utility.

- *IKONOS-2* – The IKONOS system was launched by Lockheed-Martin, Inc. for Space Imaging Corporation in September, 1999. The satellite contains both a 1-m 11-bit panchromatic sensor and a four-band 11-bit multispectral sensor with 4 m spatial resolution. The general characteristics of the IKONOS satellite are given in Table 3.16. In general this data combination facilitates applications that require detailed

Table 3.16 Characteristics of the IKONOS sensor

Swath width	11.3 km	
Spatial resolution	Multispectral	PAN
	4.0 m	1.0 m
Spectral resolution (μm)		
Band 1	0.44–0.51	
Band 2	0.50–0.59	
Band 3	0.63–0.69	
Band 4	0.75–0.85	
PAN	0.45–0.90	
Equatorial crossing time	10:30 AM	
Orbit type	Sun-synchronous polar-circular orbit	
Revisit cycle	3 days	

Table 3.17 Characteristics of the QuickBird sensor

Swath width	16.5 km	
Spatial resolution	Multispectral	PAN
	2.44 m	0.61 m
Spectral resolution (μm)		
Band 1	0.44–0.52	
Band 2	0.52–0.60	
Band 3	0.63–0.69	
Band 4	0.76–0.90	
PAN	0.45–0.90	
Equatorial crossing time	10:30 AM	
Orbit type	Sun-synchronous polar-circular orbit	
Revisit cycle	3 days	

and highly accurate data. However, unlike civilian satellite programs, the cost of IKONOS data, as with other commercial systems, must be carefully factored into the application when developing an investigation. To date, IKONOS data has been applied to problems ranging from precision agriculture to various forms of urban analysis and planning (Van Delm and Gulinck 2011).

- *QuickBird-2* – The QuickBird satellite provides the largest swath width of any present commercial satellites. The satellite collects both multispectral and panchromatic imageries concurrently and a pan-sharpened composite product is available at sub-meter resolution. The main features of the QuickBird system are listed in Table 3.17. The spectral wavelength of the multispectral and panchromatic bands covers the interval from 0.45 to 0.90 μm. The spectral resolution of the multispectral band corresponds to the first four Landsat ETM+ channel and the panchromatic band is also similar to that found of the ETM+ instrument. When compared to the IKONOS sensor, QuickBird-2 has improved spatial resolution (0.61 m in panchromatic mode and 2.4 m in multispectral) with a 16.5 km swath width.

Table 3.18 Characteristics of the GeoEye-1 sensor

Swath width	15.2 km	
Spatial resolution	Multispectral	PAN
	1.65 m	0.41 m
Spectral resolution (μm)		
Band 1	0.45–0.51	
Band 2	0.51–0.58	
Band 3	0.65–0.69	
Band 4	0.78–0.92	
PAN	0.45–0.80	
Equatorial crossing time	10:30 AM	
Orbit type	Sun-synchronous polar-circular orbit	
Revisit cycle	3 days	

QuickBird imagery is available at three different processing levels: (1) raw data that allows users to process imagery for individual needs, (2) standard imagery with radiometric and geometric corrections applied, and (3) orthorectified with terrain correction applied. As a source of environmental data, QuickBird imagery has been applied to problems ranging from change detection to thematic mapping, environmental assessment and hazard analysis (Laba et al. 2008).

- *OrbView-3* – This satellite was launched in June 2003 for Orbimage, Inc. The satellite has 1 m spatial resolution in panchromatic mode and 4 m in multispectral with 8 km swath width. One-meter viewing enables the accurate discrimination of fine-scale objects while the multispectral sensor supports a range of environmental characterization applications. Unfortunately the satellite experienced a malfunction of its main sensor system on March 4, 2007 and is no longer operational.
- *GeoEye* – Described as the next-generation platform, GeoEye-1 was launched in September 2008. This sensor is capable of acquiring image data at 0.41 m resolution in panchromatic mode and 1.65 m resolution in multispectral mode (Table 3.18). The sensor is able to record ground areas from side-to-side and front-to-back, and its revisit period is approximately 3 days or less. A second satellite in this series, GeoEye-2, is scheduled to launch prior to 2015 and will boast a spatial resolution of 25 cm.

2. *Microwave sensors* – Microwave remote sensing is a dynamic field driven in large by advances in sensor design and improvements in data processing techniques (Tsang 1985; Woodhouse 2004). Microwave systems operate in the 1 cm to 1 m wavelength band. For environmental remote sensing investigations, microwave systems are unique in that microwave radiation can penetrate through clouds, haze, and dust; rendering microwave sensors weather independent. In addition, microwave systems can operate during both daylight and

nighttime conditions and are not constrained by solar illumination conditions. Perhaps their most attractive feature is that microwave sensors can provide information products that cannot be generated using visible or infrared sensors. However, the need for sophisticated data analysis and poorer sensor resolution due to the use of longer wavelengths are some of the disadvantages.

Microwave remote sensing platforms can be either passive or active in design. Passive sensors typically detect the microwave radiation emitted from an object. Here, the object is characterized on the basis of the microwave energy received by the sensor. This energy is related to the temperature, moisture content and physical features of the object in question. Active microwave sensors provide their own source of microwave radiation to illuminate the surface. An active sensor, therefore, records the brightness of the energy returned to it and the object is detected by the intensity of the microwave energy scattered back to the receiver. The intensity of this backscatter depends on several properties of the surface such as

(a) Slope
(b) Roughness
(c) Dielectric constant of the material composition of the surface (a function of moisture content)
(d) Geometric orientation of the object in relation to the active microwave pulse
(e) Land cover contrasts corresponding to the juxtaposition of soil, vegetation and human-made materials within the sensors field of view

Other factors that influence backscatter include frequency, polarization and the angle of incidence of the incident microwave beam.

Although not as numerous in number as optical/infrared sensors, there are several noteworthy microwave sensors with environmental applicability: Radarsat-SAR, Envisat-ASAR, and ERS-SAR.

(a) *RADARDAT-SAR* – Radarsat is an earth observational satellite developed by the Canadian Space Agency. Designed to support environmental monitoring applications such as sea-ice monitoring, flood mapping, oil spill detection, and mineral prospecting, Radarssat-1 and Radarsat-2 complete 14 earth orbits per day with a 24 day revisit interval. The heart of the Radarsat system is its advanced synthetic aperture radar (SAR) sensor. The SAR sensor is an active microwave instrument that sends pulsed signals to the surface and processes the received reflected return energy. The SAR therefore provides its own microwave illumination that enables day or nighttime data collection. In addition, the SAR, using a single-frequency C-band radar, can steer its beam in swath from 35 to 500 km wide at spatial resolution ranging from 10 to 100 m. The operational specifications of Radarsat-1 and Radarsat-2 are provided in Table 3.19.
(b) *Envisat-ASAR* – Envisat was placed into a sun-synchronous orbit on March 1, 2002. Maintained by the European Space Agency, Envisat deploys an advanced synthetic aperture radar system (ASAR) operating in the

Table 3.19 Characteristics of the RADARSAT-1 sensor

Swath width	100 km (standard imaging mode)
Spatial resolution	30 m
Spectral resolution	C-band,5.3 GHz
Equatorial crossing time	6:00 AM
Orbit type	Sun-synchronous, near-polar circular orbit
Revisit cycle	6 days

Table 3.20 Characteristics of the Envisat-ASAR sensor

Swath width	58–110 km
Spatial resolution	30–150 m
Spectral resolution	C-band,4–8 GHz
Equatorial crossing time	10:30 AM
Orbit type	Sun-synchronous polar-circular orbit
Revisit cycle	35 days

Table 3.21 Characteristics of the ERS-SAR sensor

Swath width	100 km
Spatial resolution	26 m
Spectral resolution	C-band,5.3 GHz
Equatorial crossing time	10:00 AM
Orbit type	Sun-synchronous polar-near circular orbit
Revisit cycle	3 days

C-band in conjunction with other instruments on the Envisat platform. The ASAR sensor can operate as a conventional SAR or as a scanning SAR. When functioning in scan mode, the SAR is a strip mapping sensor that images in one of seven predetermined swaths (Table 3.20).

(c) *ERS* – The ERS system is the predecessor to the Envisat-ASAR sensor. There are two predominate satellites of this series, ERS-1, which was launched in 1991 and ERS-2, which was launched in 1995. The ERS program was designed to provide global measurements of sea wind, waves, and ocean ice using active and passive systems. The main instruments onboard the ERS include the active microwave instrument (AMI), a C-band SAR, the ATSR, a passive long-track scanning radiometer and GOME – a passive spectrometer deployed on ERS-2 to monitor atmospheric ozone (Table 3.21).

3. *Thermal infrared systems* – Thermal remote sensing satellites utilize the mid-infrared (3–5 μm) and the long-infrared (8–14 μm) portions of the electromagnetic spectrum. At these wavelengths, imagery is derived from the thermal radiation emitted by an object as a function of its temperature. Through the

use of thermal sensors, materials on the surface can be distinguished based on differences in temperature and emissivity (the relative power of a surface to emit heat by radiation: the ratio of the radiant energy emitted by a surface to that emitted by a blackbody at the same temperature). Since the incoming solar radiation is absorbed and reemitted from various surface arrangements, surfaces will differ in their ability to absorb and reemit this energy. When viewed as an image, this pattern of emitted energy resembles a black and white photograph in which bright areas represent relatively warm objects and darker areas relatively cooler ones. Moving from relative terms to precise measurement of actual thermal conditions can be problematic and requires careful calibration and correction of measured radiances to consistent physical properties. Ease of calibration can be complicated by factors such as cloud cover, the geometry of the vegetative canopy, and antecedent precipitation. Despite these challenges, the potential for thermal infrared (TIR) remote sensing as a source of environmental information is significant (Quattrochi and Luvall 1999; Weng 2009).

There are several thermal sensors deployed on earth observational satellites including

(a) Landsat 3 MSS (237 m spatial resolution)
(b) Landsat TM (Band 6 with 120 m spatial resolution)
(c) Landsat ETM+ (Band 6 with 60 m spatial resolution)
(d) AVHRR – three thermal bands at 1 km spatial resolution
(e) EOS Aqua

- AIRS – 240 bands covering 3.7–15 at 13.5 km spatial resolution
- MODIS – 16 thermal bands at 1 km spatial resolution

(f) EOS TERRA

- MODIS – 16 thermal bands at 1 km spatial resolution
- ASTER – 5 thermal bands at 90 m spatial resolution

Thermal data obtained from these sensors have found useful applications in agricultural water stress analysis, urban heat island studies, mineral mapping, ocean surface analysis, environmental contamination investigations, and weather forecasting.

3.4 Summary

Efforts to characterize environmental systems are strongly dependent on the type and quality of information we obtain using remote sensing technologies. Understanding Earth's environments and unraveling the patterns of interaction between people and the planet have been greatly facilitated by an increasing array of sensor systems each with contrasting spatial, temporal, and radiometric resolutions. In this chapter, a selective review of satellite-based sensors were focused on the capacity of these systems to provide: (1) continuous data acquisition, (2) frequent and regular revisit cycles, (3) broad areal coverage, and (4) good spectral resolution. Given the expanding array of sensor platforms anticipated in the future, earth observation capabilities will expand the types of data collected and the formats available to derive information products through remote sensing image processing techniques. Knowing which sensor system to employ for a given application problem can be more challenging. In this chapter, a set of basic specifications were suggested that can help direct selection process and simplify image processing activities.

References

Digenis, C. (2005) The EO-1 Mission and the Advanced Land Images, Lincoln Laboratory Journal, 15, 161–164.

Elvidge, C., Imhoff, M., Baugh, K., Hobson, V., Nelson, I. , Safran, J., Dietz, J. and Tuttle, B. (2001) Night-time lights of the world: 1994–1995 ISPRS Journal of Photogrammetry and Remote Sensing, 56, 81–99

Gulinck, H., Dufourmont, H., Copin, P. and Hermy, M. (2000) Landscape research, landuse policy and Earth Observation, International Journal of Remote Sensing, 21, 2541–2554.

Hernanades, D. (2005) Possible Scenarios for Future Missions in Earth Observation, Acta Astronautica, 56, 9–15.

Laba, M., Downs, R., Smith, S., Welsh S., Neider, C., White, S., Richmond, M., Philpot, W. and Baveye, P. (2008) Mapping invasive wetland plants in the Hudson River National Estuarine Research Reserve using quickbird satellite imagery, Remote Sensing of Environment, 112, Issue 1, 286–300

Maini, A. and Agrawal, V. (2007) Satellite Technology Principles and Applications, Wiley, New York, 340p.

Melesse, A., Weng, Q., Thenkabail, P. and Senay, G. (2007) Remote Sensing Sensors and Applications in Environmental Resources Mapping and Modelling, *Sensors,* 7, 3209–3241

Pignatti, S., Cavalli, R., Cuomo, V., Fusilli, L., Pascucci, S., Poscolieri, M. and Santini, F. (2009) Evaluation of Hyperion capability for land covers mapping in a fragmented ecosystem: Pollino National Park (Italy) case study. Remote Sensing of Environment, 113, 622–634

Thoely, N. (2000) Digital Processing of Earth Observation Images, Survey in Geophysics, 21, 209–222.

Tsang, L. (1985) Theory of Microwave Remote Sensing, Wiley, New York, 613p.

Qu, J. Gao, W., Kafatos, M. Murphy, R. and Solomonson, V. (Eds) (2006) Earth Science Satellite Remote Sensing Vol 1 and 2, Springer 335p.

Quattrochi, D. and Luvall, J. (1999) Thermal infrared remote sensing for analysis of landscape ecological processes: methods and applications Landscape Ecology, 14, 6577–6598.

Ulbricht, K and Heckendorf, W. (1998) satellite images for recognition of landscape and landuse changes, ISPRS Journal of Photogrammetry and Remote Sensing, 3, 235–243.

Van Delm, A. and Gulinck, H. (2011) Classification and quantification of green in the expanding urban and semi-urban complex: Application of detailed field data and IKONOS-imagery *Ecological Indicators*, 11, 52–60

Weng, Q. (2009) Thermal infrared remote sensing for urban climate and environmental studies: Methods, applications, and trends, ISPRS Journal of Photogrammetry and Remote Sensing, 64, 335–344

Williamson, R. and Baker, J. (2004) Current US remote sensing policies: opportunities and challenges, Space Policy, 20, 109–116.

Woodhouse, I. (2004) Introduction to Microwave Remote Sensing, CRC Press, Boca Raton, FL., 208p.

Chapter 4
Fundamentals of Image Processing

Previously, we defined remote sensing as the art and science of obtaining measurements of objects without direct physical contact. This familiar definition when extended to the topic of the environment can be reworded slightly to focus measurement on the elements and features that characterize the environmental system. Measurement in this context implies the collection of data, numerical symbols that explain how a surface emits or reflects electromagnetic radiation in a characteristic pattern as recorded by a sensor. Data recorded in this manner are nothing more than an array of values that only becomes useful when transformed into some more meaningful information. When applied to activities that describe remote sensing, information forms specialized type of knowledge that is communicated or received concerning a particular circumstance that is of interest to us. Generating information is, therefore, directed and purposeful, following a systematic method that is transparent and repeatable. In this chapter, we will examine the fundamental principles and methods that guide the transformation of remotely sensed data into information. Through this review, we can explore how these methods connect to the analysis and assessment of Earth's environmental system and gain a clearer understanding of the value of remote sensing technology when applied to this task.

4.1 Information: Defined

In the context of information technology, the concept of information has been explained as data that has been processed into a meaningful form to the recipient and is of real or perceived value in current or prospective actions and decisions (March and Smith 1995; Davis and Olson 1985). Dissecting this definition a little, we can gain a little more insight on the nature of information and establish the thought and reasoning that encourages its production. First, the word derives from the Latin *informare*, which means to give form, shape, or character to something. Today, the word signifies knowledge and aspects of cognition such as meaning,

J.K. Lein, *Environmental Sensing: Analytical Techniques for Earth Observation*,
DOI 10.1007/978-1-4614-0143-8_4, © Springer Science+Business Media, LLC 2012

instruction, communication, and representation. When data are translated to information, the ordered sequence of symbols communicates and assists us in categorizing our environment in a practical way (Ackoff 1989; Brodie and Brodie 2009; Zins 2007). In a purely functional sense, one purpose of information is to control action and guide the selection of what action to take (Scarrott 1989). From the perspective of the remote sensing analyst, translating data to information requires achieving a match between the information wants and needs of a "client" and the knowledge hidden in the data. This suggests that information is sensitive and purposeful, allowing its "user" to think about how reality is being represented and how thought is being communicated. Information is also a device to help understand, reduce, or remove uncertainty, particularly as it relates to vexing issues such as:

1. The lack of clarity of information
2. Uncertainty about the cause and effect relationships
3. The situation factors that influence decision making
4. The outcome of a decision
5. The ability to assign "probabilities" to an outcome or an event.

When considered in the abstract, information embodies three distinguishing characteristics that can be used to guide information extraction:

- *Process* – information as process concentrates of the transfer and flow of knowledge, such that when someone is informed what they come to know is changed.
- *Knowledge* – as knowledge, information explains understanding selectively communicated to focus on a particular fact, subject, or event.
- *Thing* – information ultimately assumes a representation of knowledge and ideas as a physical object such as a map, document, or file.

Although highly situational, information is further refined by its relevance; a point that is easily overlooked but essential in order to realize the full potential of remotely sensed data. Relevance summarizes to several key ingredients and for an analysis to be informative one or more of these properties should be in evidence:

- Topical
- Timely
- Current
- Useful
- Novel
- Precise
- Complete
- Accurate
- Authoritative
- Reliable.

Extracting information is not simply the action of entering command via a keyboard into a software environment, but a thought process with clear goals directing our actions. The end result is this thing called information, which we can now appreciate as a concept that can assume a variety of meanings. Perhaps the most useful explanation is the role of information in influencing how decisions are made. Here, while the task of extraction can be described as the application of various image processing algorithms on a data set, the end product (information) serves to reference five more fundamental requirements (Dehons et al. 1988; Lein 1997):

1. Information as a commodity that has value
2. Information as communication that transfers understanding and meaning
3. Information as facts devoid of context
4. Information organized according to established rules
5. Information as knowledge that facilitates the intellectual capability, to extrapolate beyond facts and data, to draw conclusions.

When an algorithm executes and produces a "result," how well it agrees with the requirements stated above will determine how well this product "informs." The question then becomes one of how best to achieve this "result" to yield information that is meaningful and communicates critical environmental understanding.

4.2 Explanation and Classification

When remotely sensed data are received, they are in essence context-less measurements written to a file, placed onto a medium of storage, and disseminated. Transforming these data into information describes the process of identifying patterns in the measurements that represent larger themes. Pattern, as used here, explains the statistical characterization of the numerical data. In measurement, a statistical regularity emerges that is not attributed to chance or randomness, but the properties of the surface that communicate. In a data set, measurements form associations and relationships to each other and the objects they attribute. Similarities in these numerical arrangements connect to the object of interest in specific ways, provided that these arrangements can be sufficiently understood. Understanding necessitates enforcing logic on the data that enables "pattern" to emerge by imposing order on measurement in such a manner that information suddenly materializes out of "noise." In remote sensing that logic is called classification.

Classification is a familiar topic in science. We can define classification as a unifying operation by which objects are recognized, differentiated, and understood according to rules or a logic that impose order on measurement. The objective of this process is to produce an arrangement or structure on measurements that assembles objects into categories to illuminate important relationships among them; simplifying measurement into a new dimension that effectively communicates meaning. This fundamental activity reduces the complexities inherent to measurement and yields a simplified generalization of data with understanding as its primary goal. As a step in the process of information extraction, classification is an indispensible device that:

- Assists in the identification of pattern
- Organizes information

- Supports prediction and inference
- Offers metrics to examine how objects change over time
- Enables objects to be symbolized and labeled.

Recognizing that remote sensing is essentially an exercise in resolving spatial data into pattern, classification becomes a means of separating "signal" from "noise" and the mechanism by which information is extracted from the image.

As a mechanism for structuring the data into something more meaningful, how a classification is achieved requires careful consideration. While ultimately dependent on purpose, organizing observations according to a set of common characteristics can be based on numerous criteria. In the example of remote sensing, those criteria are the measured reflectance or emittance values recorded by the sensor as observed for the pixels that constitute the image. The method or logic selected to impose order on these data can either be:

- *Exogenous* – with categories fixed according to threshold values related to the data
- *Idiographic* – with categories chosen with respect to specific aspects of the data, such as natural breaks
- *Serial* – with categories defined by direct mathematical relationships such as equal intervals, standard deviations, or percentiles, or
- *Arbitrary* – based on predefined nominal categories or categories established by a classification system in which categories form on the basis of decision rules or observation, such as the USGS/Anderson system of land cover classification.

Ideally the goal of any system of classification is to account for as much of the variation in the data as possible that effectively communicates its theme, whether that is expressed in terms of land cover, climate, soil type, or some other environmental attribute. To realize appropriate fit with the observed data, the categories of any classification system should be clearly defined, mutually exclusive and collectively exhaustive; however, no classification system is error-free or perfectly transferable to every situation encountered.

With a focus on the concept of pattern and the structuring of numerical data, the classification problem in remote sensing centers on statistical pattern recognition and categorization methods that translate numerical relationships into thematic information (Friedman and Kandel 1999; Duda et al. 2001). Statistical pattern recognition describes a family of techniques in which individual observations are organized into groups based on quantitative measures and relationships. Recognizing that patterns in this context are simply regularities in data, the classification problem in remote sensing reduces to the task of deriving "functions" that assign objects (pixels) to the classification labels, which we assume are correct. Here, it is convenient to think of a "function" as nothing more than a set of rules (logic) that forms a statistical decision region in the measurement boundaries established by our data (Fig. 4.1). These rules guide the process of allocating pixels to those regions in measurement space that are identified with a specific category of interest. The rules, therefore,

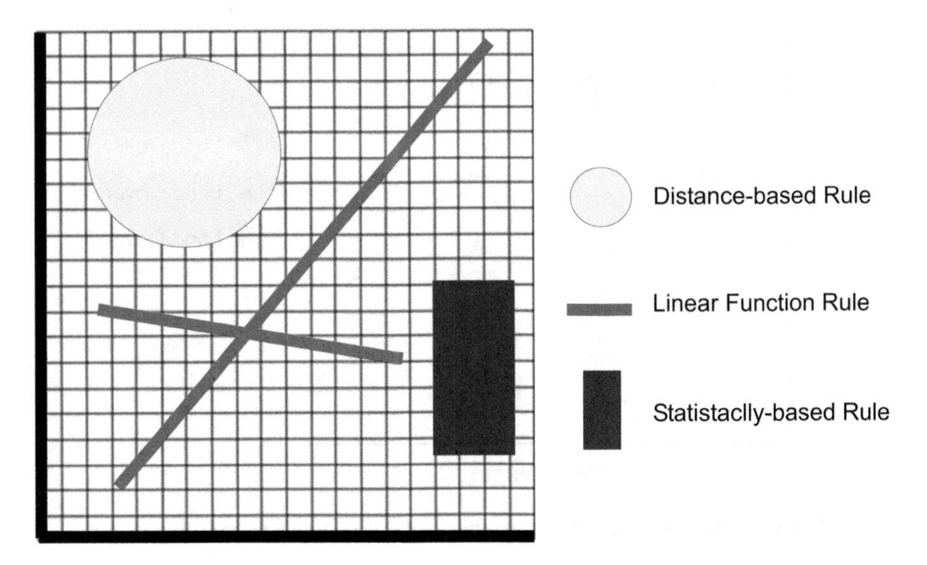

Fig. 4.1 Decision rule schemas in relation to measurement space

implement a specific strategy that is based on statistical principles; concepts such as dispersion, central tendency, or probability. Using these rules, the spectral measurements recorded by the sensor are translated into information (e.g., the named categories of a land cover classification or the moisture stress levels of means to define drought conditions). Success in achieving a translation relies largely on our skills at recognizing pattern.

4.3 Recognizing Pattern

We can define a pattern in a remotely sensed image as a characteristic arrangement of electromagnetic radiation as recorded by a sensor within one or across many spectral bands. Objects at the surface, owing to their physical, bio-chemical, and material composition interact with electromagnetic radiation and produce a measureable response. This response, commonly referred to as a spectral signature, suggests that under ideal conditions, objects possessing similar material compositions should, in theory, display similar response patterns relative to the wavelength(s) of electromagnetic radiation they receive (Fig. 4.2). By identifying the parameters of electromagnetic radiation that define a pattern, the type, nature, or disposition of the object can become known. The relationship between the categorization of an object based on the labels with assigned meaning and the form and spectral composition of its pattern, however is complex and fraught with confusion. When specified only by wavelength, the pixels that compose an image form observable groupings that explain a common similarity in their brightness value in multispectral space. These groupings are considered natural patterns in the data, since they distinguish separations that result uniquely from how the scene was measured by the sensor (Fig. 4.3). As a sole product

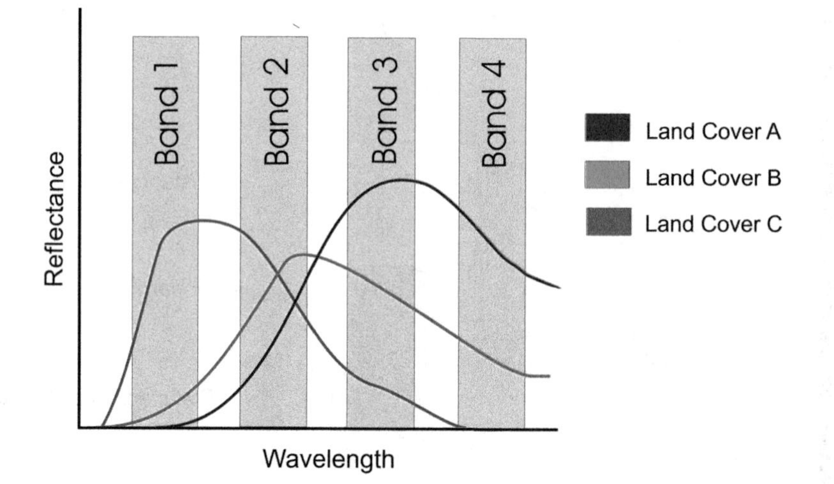

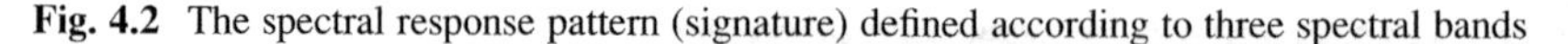

Fig. 4.2 The spectral response pattern (signature) defined according to three spectral bands

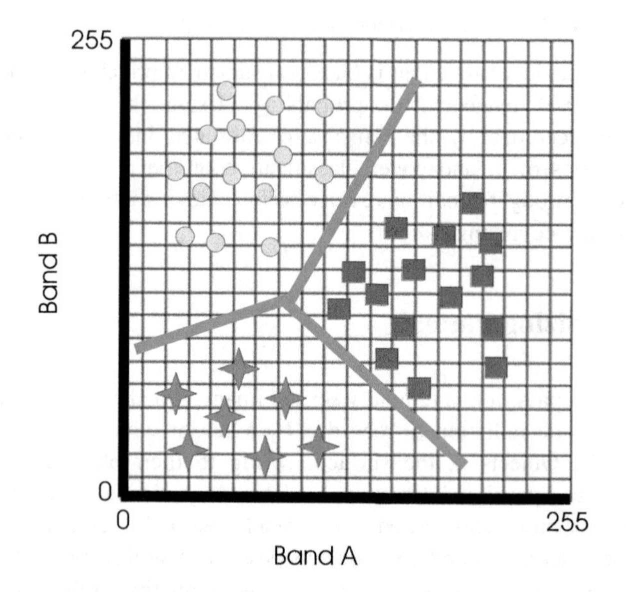

Fig. 4.3 Spectral patterning expressed in relation to spectral bands

of measurement, these groupings explain classes that are exclusively spectral in nature. In other words, they are the consequence of how electromagnetic radiation interacted with the objects that comprise the surface. Confusion enters into their categorization when we recognize that these natural patterns may or may not correspond to meaningful thematic information. They fail to translate, if the natural patterns do not convey information in a practical sense.

Meaning is ascribed to a pattern only by connecting the numerical arrangements obtained via the sensor to a categorization of features within the scene, which is not a number, but rather a construct that communicate information in a semantic context. Constructs form the foundations of the informational categories that drive information extraction. Information classes are those categories of interest that the analyst is actually trying to identify in the imagery, such as different kinds of crops, different forest types or tree species, different geologic units or rock types, and so on. The connection between spectral pattern and informational class, however, is rarely an unambiguous, one-to-one fit. Rather, due to contrasts in material composition, biochemical, or physical attributes, a single informational class my exhibit a wide range of variation in its spectral value. This source of confusion introduces uncertainty into the pattern recognition problem, since a single information class may correspond to several spectral groupings formed out of slight, but significant variations in a feature's spatial arrangement, status, or environmental setting. To illustrate the potential uncertainty and its impact on information extraction, consider the example of an agricultural field. To the eye, a typical agricultural field is easily recognizable by its shape and arrangement, but spectrally, identifying an agricultural field depends on factors such as crop type, crop cycle, phenology, and cultural setting that will render a single definitive spectral pattern impossible to acquire. The challenge for both classification and pattern recognition is to rationally and systematically organize disparate spectral patterns into consistent and realistic informational classes as accurately and reasonably as possible.

4.4 Classification Schemas

There are two fundamental approaches in statistical pattern recognition to connect spectral groupings to informational categories. Selecting between the two depends largely on purpose as directed by one's prior knowledge of the study area, presence of predefined informational classes, complexity of the problem, and issues related to time and operational costs (Lu and Weng 2007; Richards and Jia 2006). The first method is referred to as unsupervised image classification. This technique relies on the application of statistical clustering algorithms to identify "natural" patterns in the multispectral data set. As a methodology, unsupervised classification requires very few inputs into the classification process. Instead, this method relies on the algorithms to select classes based exclusively on the spectral properties of the pixels expressed over the relevant spectral bands. The general method of unsupervised classification is illustrated in flowchart form in Fig. 4.4. Data clustering algorithms implement one of several grouping strategies (Duda and Canty 2002). The most common include:

1. Hierarchical clustering
2. Partitioning clustering or
3. Spectral clustering.

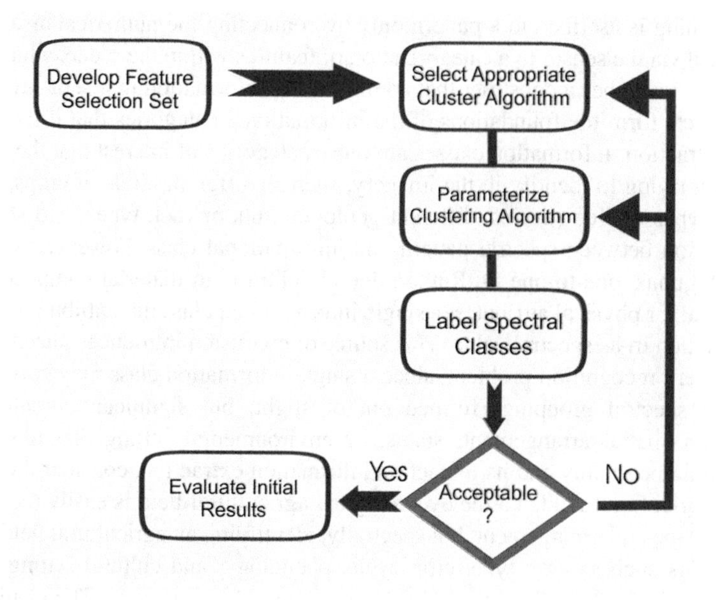

Fig. 4.4 The general method of unsupervised image classification

Hierarchical clustering is a strategy used to group pixels to form a hierarchy based on a common similarity metric. In the majority of examples that metric is distance, expressed according to measures, such as Euclidean distance or Mahalanobis distance, which is used to form a decision rule that organizes measurement space and guides the allocation of pixels into group structures. Alternatively, clustering methods may employ similarity metrics, such as correlation (Table 4.1). The clustering algorithm progressively merges individual pixels into larger agglomerations based on the logic that pixels closest to each other numerically represent the same feature. Clustering continues merging clusters together based on the same idea of "closeness" until all groups form into a single cluster (Fig. 4.5). Partitioning algorithms attempt to decompose the data set into a series of disjoint groups all at once based also on criteria established as a function of distance. The spectral clustering approach takes a somewhat different strategy and develops clusters in a manner similar to principle component analysis. This strategy uses the spectrum of the similarity matrix of the data set to perform a "dimensionality reduction" which in essence greats a smaller number of groups based on the how well individual pixels correlate (Lee 2002).

For the majority of remote sensing applications, unsupervised classification tends to employ partitioning or hierarchical methodologies. However, as a means of exploratory data analysis, classification, and discovery, any algorithm may reveal hidden associations in the imagery and provide useful structure to guide further study. Exploration is often useful particularly where the results may

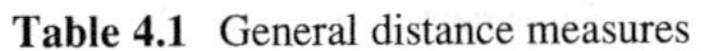

Table 4.1 General distance measures

Metric	Definition
Euclidean distance	The distance between two points described by the length of the path connecting them. This "ordinary" distance is similar to what would be measured with a ruler, as solved using the Pythagorean formula
Chebyshev distance	The Chebyshev distance between two points is the maximum distance between the points in any single dimension. The distance between points $X = (X_1, X_2, \text{etc.})$ and $Y = (Y_1, Y_2, \text{etc.})$ is computed using the formula: $\text{Max}_i \lvert X_i - Y_i \rvert$ where X_i and Y_i are the values of the ith variable at points X and Y, respectively
Mahalanobis distance	Mahalanobis distance is based on correlations between variables by which different patterns can be identified and analyzed. It is a useful way of determining *similarity* of an unknown sample set to a known one. It differs from Euclidean distance in that it takes into account the correlations of the data set
Bhattacharyya distance	Measures the similarity of two discrete or continuous probability distributions and is related to the Bhattacharyya coefficient which is a measure of the amount of overlap between two statistical samples or populations
Manhattan distance	Also known as the city-block distance, this metric measures the rectilinear distance between two pixels. Computed by the length of edges between points that must be traversed to get from a to b within a grid

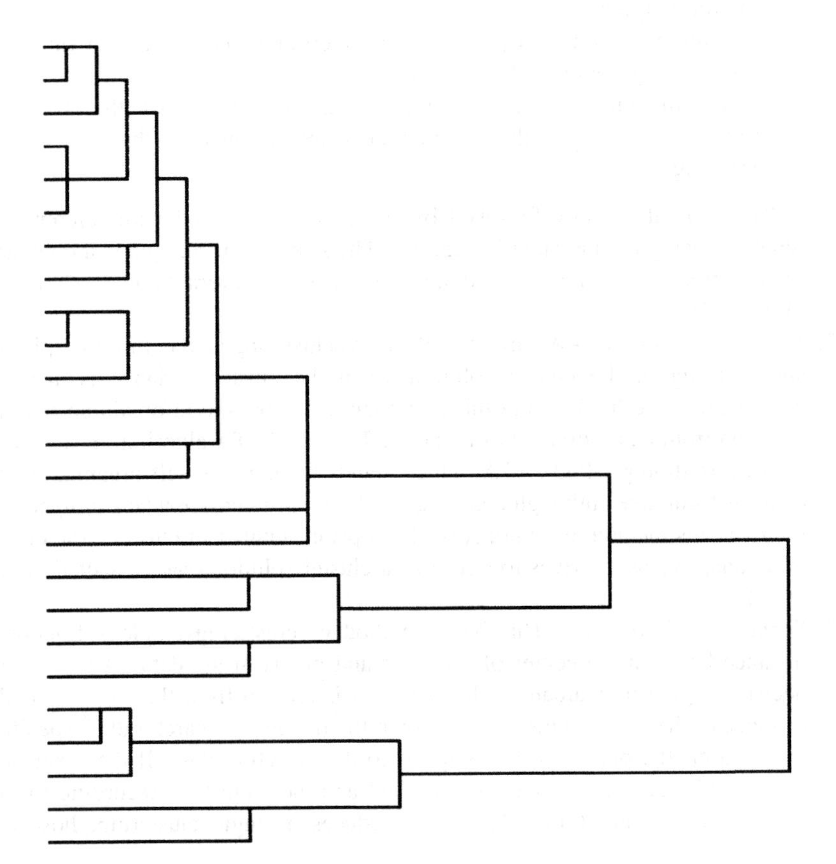

Fig. 4.5 Generalized pattern of hierarchical clustering

contribute to the definition of a more formal classification scheme or where there is evidence to suggest the presence of categorical anomalies that clusters may detect.

A selection of the algorithms most frequently employed in unsupervised image classification includes the following.

1. *K-means clustering* – K-means clustering is a partitioning method. The decision rule implemented by this algorithm allocated N data points in M-dimensional measurement space into K clusters. Each cluster is parameterized by a vector $\mathrm{m}^{(k)}$ called its mean. Clusters form by minimizing the sum of squares distances between a pixel and the corresponding cluster center. The algorithm calculates an arbitrary center of an initial cluster using a predetermined number of pixels to estimate a mean. Clustering proceeds to iteratively refine the initial solution until a "best fit" is achieved. Typically, the cluster solution is directed by establishing values for the:

   (a) Maximum number of clusters to be created
   (b) Convergence thresholds to stop the process once cluster means fail to change beyond a lower percentage limit
   (c) Maximum radius of a cluster defining the distance to the nearest cluster to control merging
   (d) Maximum iterations to perform that determines the number of times the algorithm cycles through the data
   (e) Minimum number of pixels in a cluster that specify whether a cluster remains valid or whether its members are recruited to the next nearest grouping

   The general strategy followed by the K-means method with reference to remote sensing is illustrated in Fig. 4.6. The K-means technique is a comparatively simple partitioning method; however, there are several known weaknesses (Table 4.2).
2. *ISODATA clustering* – A variant of K-means clustering is to permit the splitting and merging of the cluster solution when the cluster variance is above a prespecified threshold. Using this approach, it is possible to obtain an optimal partition from an arbitrary starting point. The ISODATA algorithm is an example of this strategy. ISODATA, which stands for iterative self-organizing data analysis technique, although computationally more intensive when compared to the K-means method, does not solve for a priori number of clusters, rather it is more adaptive as it strives to optimize a cluster solution (Memarsadeghi et al. 2007).
3. *Sequential clustering* – This basic method of clustering employs Euclidean distance between the center of a cluster and pixels in the data set to create a decision region to evaluate each pixel as it is read in from the data set. If the distance to the nearest cluster is less than the maximum search radius specified by the user, the pixel will be assigned to that spectral class. If the pixel falls beyond this radius, the pixel is classified as a new cluster. According to this strategy, each run of the algorithm produces a single clustering; however,

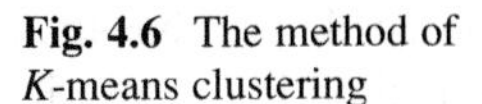

Fig. 4.6 The method of K-means clustering

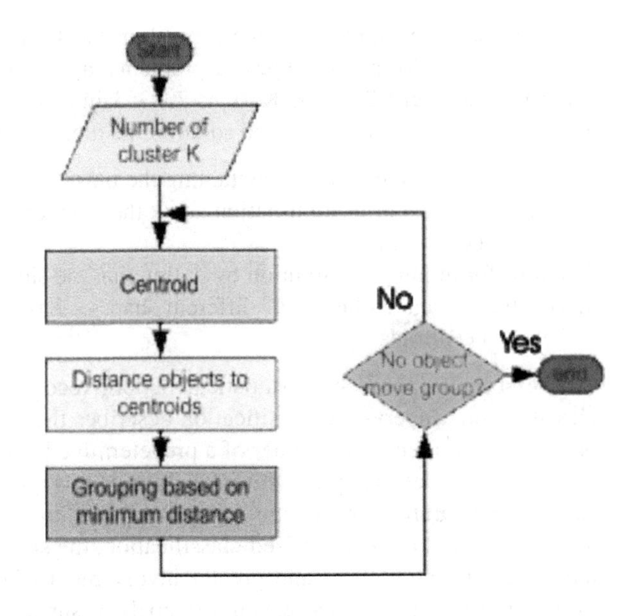

Table 4.2 Limitations of K-means clustering

Preprocessing (image analysis and normalization):
Which genes (variables) are used
Which samples are used
Which distance measure is used
Which algorithm is applied
How to decide the number of clusters K
K-means clustering has many weaknesses:
When the numbers of data are not so many, initial grouping will determine the cluster significantly
The number of clusters, K, must be determined before hand
We never know the real cluster, using the same data, if it is inputted in a different order may produce different cluster when the number of data is a few
Sensitive to initial condition. Different initial condition may produce different result of cluster. The algorithm may be trapped in the local optimum
We never know which attribute contributes more to the grouping process since we assume that each attribute has the same weight
Weakness of arithmetic mean is not robust to outliers. Very far data from the centroid may pull the centroid away from the real one

because pixels are read in sequential order, useful results will depend on the parameters selected to control its execution. Compared to other clustering solutions, sequential clustering is most sensitive to the value set for the maximum radius of a cluster.

A wide assortment of clustering algorithms has been introduced, each has their own advantages for certain types of problems, and no one method is necessarily better than another (Zhao and Karypis 2003; Jain et al. 1999; Gan et al. 2007). To achieve a successful and unbiased solution the method selected should:

1. Identify one unique pattern containing the most "natural" clusters
2. Not assume any prior information about the number, shape, or internal distribution of the clusters
3. Allow for refining the solution by optimizing the similarity measures
4. Be able to handle clusters of different shapes, densities, and unequal distances between centroids.

The second approach to statistical pattern recognition is called supervised classification. Supervised classification describes the example where the desired pattern is identified as a member of a predetermined class. The solution develops from the set of data samples whose labels and class types are known and spectral patterns (signatures) are assembled to lend definition to the known categories. When compared to unsupervised classification, the supervised approach requires more steps in the process and greater involvement of the analyst to generate a meaningful result. For this reason, the goals of supervised classification and its contribution to information extraction differs, since there is a greater reliance on prior knowledge throughout. When the problem of information extraction is considered, searching for a pattern or anomaly in the image where there is little guidance to fall back on, unsupervised classification, with its focus on "natural" classes inherent to the data, offers appropriate solution. In these situations, deciding on what the data reveals follows after the data has been processed. Once produced, the natural classes require naming. For example, in an environmental application, a class of pixels may emerge that corresponds to conditions, such as a field damaged by corn blight, a stand of trees impacted by Gypsy moth, or a water body contaminated by illegal dumping. Typically, these are not the expected condition of an existing land cover or land use category, and the clusters that form are outside the "norm." Conversely, the supervised method, by placing a premium on the analyst's prior knowledge, functions well when informational classes are known and the problem is well specified. In these instances, the analyst must locate representative examples of the known target(s) of interest and use those examples to produce statistical patterns that can be recognized elsewhere across the image by the classification algorithm.

With prior knowledge of the study area, an understanding of the phenomena under investigation and a system of classification with well-defined terminology, supervised image classification can be extremely useful for detecting similar instances of a given land surface condition or arrangement. The general procedures followed when conducting the supervised image classification are illustrated in Fig. 4.7. The process follows an eight step procedure that culminates with a map product that conveys thematic information of value to the user.

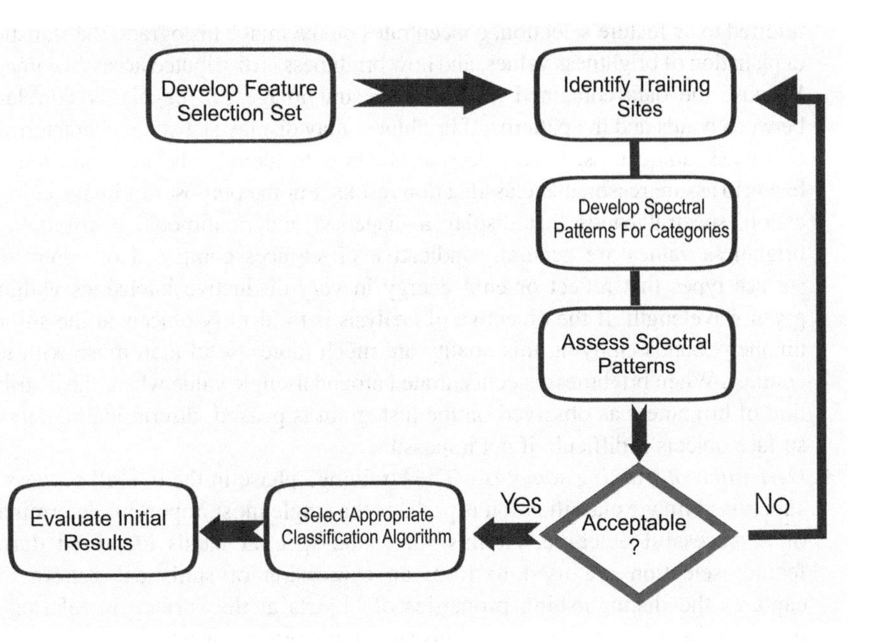

Fig. 4.7 The general method of supervised classification

1. *Determination of classification goals* – Supervised classification begins with a clear understanding of the purpose that drives the investigation as well as a carefully thought-out classification scheme that will be applied to the data. Traditionally, supervised classification has been used in land cover mapping where the classes that define specific land cover types are well defined. Examples of this include the United State Geological Survey (USGS) Anderson land use/land cover classification system and its derivatives (Anderson et al. 1976). When a well-designed system of classification is in use, the analyst must decide how well the labels for the land cover types defined by the classification scheme fit both the purpose of the study and the nature of the scene as evidenced by the data. When a preexisting classification system does not exist or when the problem is sufficiently unique, the analyst must develop and define problem-specific class labels and aptly characterize the nature of the scene. With a functional classification schema, the process of information extraction simplifies to the problem of visually identifying examples of each class of interest in the data and deriving spectral patterns that represent each class distinctly.
2. *Data selection and simplification* – Once the goals motivating the classification problem are understood focus shifts to the question of data and the selection of those spectral bands in the imagery that contribute meaningful "pattern" to the analysis. Although this step is also relevant to the unsupervised classification process, it plays a more vital role here by both reducing the number of bands required for computer processing, while at the same time facilitating the creation of representative spectral response patterns (signatures) for the categories of interest. Band selection, also

referred to as feature selection, concentrates on the image histogram, the statistical explanation of brightness values, and how brightness is distributed across the image. Because the data contained in a multispectral image can be highly correlated between bands and the patterns of brightness may display statistical characteristics that mask uniqueness, band selection attempts to identify the most informative bands to assure reasonable classification results. For the purposes of image classification, spectral bands that display a dispersed and multimodal distribution of brightness values are generally indicative of surfaces composed of contrasting surface types that reflect or emit energy in very distinctive intensities within a given wavelength. If the objective of analysis is to identify objects at the surface uniquely, bands enjoying this quality are much more useful than those with less contrast. When brightness is concentrated around a single value where the distribution of brightness as observed on the histogram is peaked, discriminating among surface objects is difficult, if not impossible.

3. *Derivation of training samples* – The "training" phase in the overall process of supervised image classification is perhaps the single most important determinant of a successful outcome. At this stage, the spectral bands identified during feature selection are used to form an expression of statistical pattern that captures the distinguishing properties of objects at the surface in relation to their classification label (name). This step in the process is where the element of supervision enters into information extraction. Using knowledge of the scene, the analyst systematically identifies areas that are assumed to be representative examples of the classes of interest that motivate the investigation. Training involves collecting data values of the pixels that fall within these examples. These pixels are then used to calculate descriptive statistics for each of these named samples that will form the expression of pattern. Generally the statistical parameters calculated for these regions in the image are the mean, standard deviation, and variance/covariance. Taken together, these statistics serve as the fundamental definition of an object's spectral signature. Therefore, to be useful, these spectral patterns should characterize the spectral behavior exhibited by the object as uniquely as the data permits. As a statistical summary, careful evaluation of these patterns is required in order to ascertain their "uniqueness." Because signature training is essentially an exercise in spatial sampling, how the sample pixels are acquired from the imagery influences the reliability of the spectral response patterns obtained. Recalling that the goal of supervised classification is to extend the sample of known signatures to the remaining unknown pixels that make up the image, training success depends on six interrelated considerations (Mather 2004; Swain and Davis 1978):

   (a) *Pixel quality per sample* – Although this value depends on the spatial prevalence of the informational classes involved, the more extensively an object/land type is distributed over the image, the more pixels are needed in order to obtain a representative sample. As a general rule of thumb, for every individual example on the object/land type selected, at least 25 pixels should included for the delineating polygon.

(b) *Size on the delineating polygon* – Acquiring a spatial sample involves inscribing a polygon on the image that envelopes the object of land type of interest through the process of digitizing. The polygon should be large enough to accurately delineate the geometric form of the object without overlapping onto other neighboring features. When digitizing, care must be taken to follow the physical shape of the object well within the bounding pixels.

(c) *Sample location* – Because a spectral response signature must extend their pattern over the expanse of the image, samples should be drawn from known locations throughout the study area and from all possible variations specific to the informational class. By sampling at locations across the scene, as the spectra properties of the object modify geographically, this natural variability can be integrated into the spectral signature. For example, the broad information class "forest" can be explained by several different tree species under contrasting environmental conditions. A single "signature" derived from one example of forest cover in the image is unlike to characterize all other instances of forest in the scene.

(d) *Number of samples taken* – The actual number of sample polygons needed to adequately represent a specific informational category varies with both the complexity of the geographic area involved and the nature of the land type as defined by the classification system. In examples of general uniformity at the surface and generality in terms of class definition, it does not require an extensive number of sample polygons, simply because radiometric contrast is likely to be negligible. Uniformity suggests that within sample differences in brightness values the possible range of brightness values can be assumed to be small. However, as complexity increases, both in terms of geometric arrangement and inherent variability, the number of samples will need to increase as well. Increasing the number of sample polygons has the effect of "averaging" and, therefore, maintains better sensitivity with respect to the physical properties of the object/type and the ambiguities that may be introduced by language in the interpretation of classes.

(e) *Internal Consistency and uniformity* – As with any exercise in sampling, concerns pertaining to the representativeness of the statistical characterizations dominate. Ideally, for the polygon samples taken from the image, the pixels that comprise the sample should display a uni-modal distribution for each spectral band being used with little deviation from their mean value. When examined for consistency, pure pixels should dominate the sample and the influence of mixed pixels would be limited to only those instances were object/land type complexity makes mixtures of brightness values unavoidable. Unfortunately, perfection in training is illusive. Consequently, error and uncertainty will propagate through the analysis and place a premium on quality control.

(f) *Quality assessment and evaluation* – To effectively place an unsampled pixel in the image to its appropriate information class, the sample-derived signatures should be dissimilar. Evaluating signature separability requires a metric, such as distance, to measure and quantify its divergence. Statistical

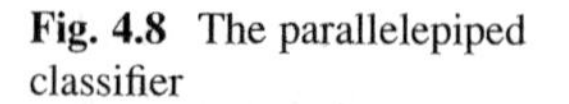

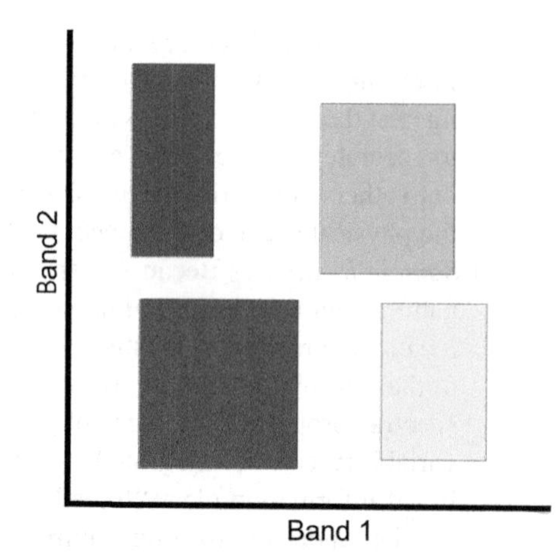

Fig. 4.8 The parallelepiped classifier

overlap will suggest confusion and indicate the need to improve the sample(s) by retraining.

Once a satisfactory set of spectral response patterns has been produced for the information classes in interest, the supervised classification process can proceed to the task of choosing and executing a classification algorithm.

4. *Classifier selection and implementation* – The classification algorithms that guide supervised image classification employ statistical decision rules based on the training signatures to produce a thematic map. Each rule is a strategy for partitioning multispectral measurement space into regions that correspond to the informational classes as defined by their statistical patterns. The algorithms perform a per-pixel "test" based on the vectors of brightness values that characterize the pixel across the bands selected for analysis. The test essentially determines which pixels most logically fit into one of the categories as expressed by its decision region. Assigning pixels as members of a class based on this singular evaluation of a brightness vector can be accomplished in one of four principal ways (Hudak and Brockett 2004; Nangendo et al. 2007; Tso and Mather 2009):

   (a) *Parallelepiped Classification* – The parallelepiped classifier defines a decision region by establishing minimum and maximum boundaries based on the standard deviation from the mean for each selected information class. The conceptual logic of the parallelepiped design is illustrated in Fig. 4.8, assuming the configuration of an n-dimensional box. The decision rule, using the region inscribed by the parallelepiped, implements a simple boolean Yes/No logic where a given pixel whose category is unknown either falls within the parallelepiped and becomes a member of that category, or does not. The success of this classifier is based in part on the thresholds used to set the dimensions of the parallelepiped. If the decision boundary is too narrow, the algorithm is overly selective and restrictive, and pixels that should be named to a given class are not assigned. Conversely, if the decision region is

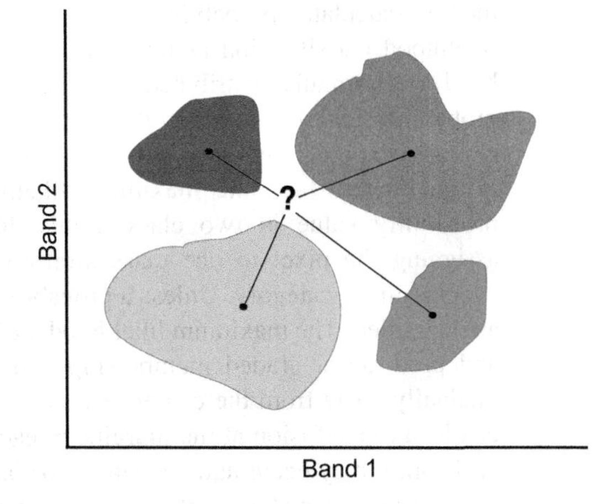

Fig. 4.9 The minimum distance classifier

too broad overgeneralization is common, and pixels that should not be assigned to a given class are erroneously included. Although computationally efficient, parallelepipeds often overlap, and a pixel may satisfy the decision criteria for more than one informational class. In such cases, that pixel is assigned to the first class it falls within and produces yet another source of error and confusion.

(b) *Minimum Distance to Means* – The minimum distance to means decision algorithm determines each pixel's "distance" from the class means and assigns them to the closest class. The algorithm evaluates the mean vectors for each informational class from the training data and calculates the distance to each mean vector for every pixel in the scene. All pixels are classified to the closest class unless the user specifies standard deviation or distance thresholds, in which case some pixels may be unclassified if they do not meet the selected criteria. In most cases Euclidean distance is used, although other distance metrics can be employed. The logic is comparatively simple; a pixel (X) is considered a member of class (B), if and only if its spectral distance to the mean of class (B) is the shortest when compared to all other classes. This relative selection criterion requires a distance threshold to establish an n-dimensional hypersphere in measurement space to form a decision boundary and to reconcile what brightness value constitutes the "shortest" distance (Fig. 4.9). The pixel whose label is unknown, falling within the hypersphere is assumed to be a member of the class. Once again thresholds are critical to producing a useful classification, and error and uncertainty can develop based on how these thresholds manipulate the decision region.

(c) *Maximum likelihood classification* – this algorithm exploits probability theory to determine whether a pixel is a member of a given informational class. Unlike previous statistical classifiers that rely on decision boundaries to partition measurement space in class regions, the maximum likelihood method calculates probability density functions for each class. Maximum likelihood classification assumes that the statistics for each class in each band are normally distributed. The decision rule requires computing the probability for every pixel in the scene as a member of each informational class. Each pixel is assigned to the class where its probability of membership is the highest (i.e., the maximum likelihood). If a pixel has an equal probability value in two classes, this dilemma is reconciled by either assigning the pixel to one class arbitrarily or assigning the pixel to an "unclassified" category. Unless a probability threshold is selected, all pixels are classified. The maximum likelihood method is computationally intensive and produces a graded membership pattern where probability decreases gradually away from the center of each class. This "distance decay" effect can lead to confusion at the margins of each category where areas of equal probability may occur and contribute to classification uncertainty and error.

(d) *Nearest-neighbor classification* – nearest-neighbor classification implements a decision strategy known as proximity search. The decision rule forms as an optimization problem solved by computing the distance from a pixel to its closest training class. Class assignment can be based on one of several logics:

- *Linear search* – which computes the distance from the training class centroid to every pixel.
- *Space partitioning* – a procedure that iteratively bisects the decision space into regions based on approaches such as the *k*-dimension tree algorithm.
- *Locality-sensitive grouping* – a method that organizes pixels into "buckets" based on distance criteria and probability constraints.
- *K-nearest neighbor* – assigning pixels to the "closest" class in measurement space based on a majority vote of its neighbors.

5. *Initial evaluation* – Each statistical classifier will produce a thematic map depicting the spatial pattern produced by the informational classes under investigation. Since each classifier implements a different decision strategy in producing a result, the spatial characterization of a given thematic representation will vary and no one method is necessarily better than another for a given task. With minimizing error as the goal, a cursory examination of the results obtained in classification is needed to assess the expected loss of information as a specific classification algorithm assigns pixels based on its initial parameterization (Yom-Tov 2004). This phase of the information extraction process is not directly concerned with the thematic accuracy of the map product, but rather the general performance of the classifier. Whether maximum likelihood, parallelepiped or another method was chosen, the algorithm may classify all the sample regions that were used to collect spectral signatures, but not perform well on the unknown areas. Ideally, the classifier should be able to generalize from the training samples without over-fitting (Yom-Tov 2004).

To begin this initial review it is important to establish simple criteria to guide the evaluation. Typically the following points should be considered:

(a) What is the extent of error produced by the algorithm.
(b) What is the computational cost and memory requirements involved in the selected method.
(c) How difficult is it for the user to implement the classifier.
(d) Can the user gain insight about the problem from an examination of the results.
(e) On visual inspection are their noticeable gaps in the results that indicate error.

In terms of performance, the parallelepiped algorithm is comparatively easy to understand, computationally simple, but tends to be the least useful except for specialized applications requiring a limited number of informational classes. The minimum distance approach is relatively easy to understand, but is more computationally intense. However when compared to the parallelepiped method, the minimum distance strategy tend to yield more useful results upon initial parameterization. The most complex classifier is the maximum likelihood algorithm. Although its decision rule is more challenging to understand and its computing requirements are very intensive, it produces the most useful results. Ultimately, the initial success of any of the methods reviews will depend on factors such as image quality, the quality of the training samples supplied to the classifier, and the appropriateness of the system of classification used to produce the informational categories.

6. *Post Classification refinement* – Following the satisfactory execution of the classifier, slight thematic and geometric incongruities will remain in the information product. These processing artifacts are due in part to the inherent spectral variability encountered by the algorithm when conducting a classification as well as external modifiers related to localized environmental irregularities induced by factors such as soil moisture contrasts, discontinuities in tree cover, and local shadowing. Visually, these artifacts may manifest as a speckled appearance or as broken patterns, gaps, and inconsistencies in the resulting thematic map. When such artifacts are encountered it may be desirable to "smooth" or generalize these patterns, removing or reducing their occurrence, and thereby enhancing the fidelity of the map product and its esthetic appeal.

   Post-classification operations are employed to generalize the thematic content, showing only the dominant and assumed to be correct depiction of the information present on the map data. Post classification includes several operations:

   (a) *Post-classification smoothing/filtering* – Filtering methods act to selectively emphasize or suppress information at different spatial scales over the image.
   (b) *Majority/minority analysis* – This undertakes a search within a "window" set by the user to change spurious pixels into either the majority of pixel

in a category within the window or to the class defining the minority of pixels.

(c) *Clumping* – It groups adjacent similarly classified areas together using basic morphological operations. Because thematic output often suffers from a lack of spatial coherency, clumping classes can effectively remove gaps in the classification. Clumping first performs a dilation operation and then an erode operation within a predefined window to close gaps and fill holes.

(d) *Sieving* – Isolated pixels in a classification can be a source of confusion. Sieving operations use a "blob-grouping" technique to remove isolated pixels. Blob-grouping is performed by examining pixels with a predefined window to determine if neighboring pixels for a consistent class that can incorporate an isolated case. If neighboring pixels satisfy that condition, the isolate pixel is replaced into this neighborhood.

(e) *Merger and combination* – Operating on the class values alone, merging and combination techniques allow the analyst to selectively re-code and manually reassign pixels in one class to another existing category. This can be useful when a land type or object is explained by two classes that can be merged into one single expression.

7. *Accuracy assessment* – Accuracy assessment determines the "applied" value of the information created during the image classification process and quantifies the degree of confidence users of the thematic product can ascribe to its contents. Accuracy assessment remains an active area of research in remote sensing and a detailed literature presenting the various methods available for establishing the "correctness" of both supervised and unsupervised classification has been introduced (Foody 2002; Liu et al. 2007; Congalton and Green 2009; Stehman 1997). In each of the method introduced into practice, the goal is to objectively quantify how well the thematic content of a classified pixel agrees with the actual identity of that pixel on the ground. While specifics in the technique may vary, establishing confidence in the product and understanding the presence of error in the resulting information simplifies to a comparison test of the assumed "true" value or observed class identity of a pixel as witnessed by a set of reference data against the method-produced result based on what the classifier predicts for a random sample of pixels. Before the actual methods for calculating accuracy (error) are reviewed, a brief examination of the key questions that direct the how an accuracy test is conducted helps frame this essential aspect of the image classification process.

   (a) *What is an error?* – In this phase of the image classification process, it is important for the analyst to think carefully and critically about the methods used to generate the information product. Here, consideration must be given to an assessment of all the possible things that may have gone wrong methodologically and what aspects of this process remained out of the analyst's control. Recalling limitations due to training, spectral and spatial resolution, sensor and environmental anomalies helps to clarify what error might look like in the thematic map and where in it may have originated in the process of classification. Common problems include:

- *Misclassification* – defining pixels that are not or cannot be correctly labeled. Errors of this type may be the product of the classification system in use and its lack of fit with the spatial or spectral resolution of the data. Misclassification may also result from the natural heterogeneity of the scene and the presence of an overabundance of mixed pixels.
- *Processing* – inaccuracies in the results may also be attributable to processing error either resulting from poor preprocessing and image rectification or due to limitations in the classification algorithm.
- *Contamination* – contamination describes the influence of the atmosphere or environment and their ability to introduce "noise" into the data. Noise can produce variability in radiometric response.

Overall, when considering error, Campbell (2006) reminds us that:

- Errors are not distributed over the image at random, but display a degree of systematic, order occurrence in space
- Errors tend not to be assigned at random to the various classes on the image but are associated preferentially to certain categories
- Erroneously assigned pixels are not spatially isolated but tend to group in areas of varied size and shape
- Error may display specific spatial relationships to land units such as at boundaries or within the interior of land unit polygons

(b) *What is the ground truth?* – The reference data used to anchor the accuracy test is central to the credibility of an assessment. Reference data is also the most challenging and problematic aspect of this step in the analysis. Reference data are those examples we assume contain the "on ground" conditions captured by the sensor. They are the actual values or categories we hope to have identified in the processing of the remotely sensed imagery. To be useful, reference data must define a "reality" that is free of any form of bias and is, therefore, representative of the theme. Any misinformation or misrepresentation of these data degrades the confidence in the quality of the assessment and threatens the validity of the information product. Selecting representative examples of "true" ground conditions can be a decision that invites controversy; however, it is also a decision that must content with the practical constraints imposed by time, cost factors, and resource availability. Common sources of reference data used to anchor an accuracy test include:

- *Higher spatial resolution imagery* – relying on visual interpretation, this source of reference data involves the acquisition of aerial photographs or satellite imagery of a finer resolution than that used in the analysis. These sources provide the corroborating evidence that the class as labeled on the method-produced image is a reasonable assignment. To be credible, the reference must be close enough in time to the date of the image, at a scale that permits visual interpretation, and displays the surface characteristics germane to the thematic content of the classified scene.

- *Site visitation* – collecting reference data by means of field visits has been made more precise by the use of GPS, but this method suffers from several drawbacks including scene instability, site inaccessibility, and logistical complexity. For some environmental applications, it may be necessary to synchronize the collection of field reference data with the overpass of the satellite sensor. This requirement can add additional complexity to data acquisition.
- *Reference maps* – using existing maps as a source of ground truth is a common practice, and often it may be the only viable approach to collecting reference data. When existing maps are used, they need to be at a comparable spatial scale in relation to the image, correct, and appropriate with respect to topic, design, and content. Furthermore, maps are generalizations that contain their own unique sources of error. Often the accuracy of a published map is unknown, and its compilation sources may be difficult to ascertain and trace.

(c) *What is a sample?* – The accuracy test is an exercise in comparing a sample of pixels on the image to their exact locations in reality. The sample schema used to acquire data to populate an assessment must assume that the data have been selected at random and without biased. As assortment of spatial sampling designed are available to use when collecting the reference data, each employs a different strategy with different statistical properties (Montello and Sutton 2006).

- *Simple random sampling* – is a sample design where pixels are selected (without replacement) from the image such that any pixel had an equal chance of being chosen. Typically, selection is based on a random number generator or similar device that draws row and column coordinates for the image that forms the sample locations.
- *Systematic samples* – is a sampling design where pixels are selected in a regular fashion which may begin with a randomized start (systematic random sampling).
- *Stratified random sample* – is a sampling method where the image is divided into sections based on predesignated criteria, and samples are taken at each section in this framework (stratum).

(d) *Doing the math* – Once a sampling strategy has been chosen, accuracy assessment proceeds by comparing the sample-selected pixels to the "real-world" compliment.

Fig. 4.10 The general form of an error matrix

Land Cover Type	A	B	C	D	E
A	11	4	12	0	6
B	8	33	0	5	20
C	0	0	41	7	13
D	9	3	0	17	0
E	4	19	8	0	26

As with any sampling methodology, careful consideration must be given to the actual size of the sample and its statistical validity. Typically sample size can be estimated by the formula:

$$N = \frac{Z^2 pq}{E^2},$$

where p is the expected percentage of accuracy, $q = 100 - p$, and E is the allowable error.

Based on this approximate sample size, N, error can be quantified by tabulating an error matrix or by calculating specific map accuracy statistics to produce an accuracy report (Congalton and Green 2009).

The error matrix is a common method of assessing classification accuracy. The matrix is a table of numbers defined by the rows and columns that represent the number of sample pixels (or polygon) assigned to a given category relative to the actual category as confirmed by the reference data (Fig. 4.10) Configured in this manner, the rows across the table express the classification as derived from the remote sensing analysis, while the columns explain those same classes as evidenced by the reference (ground truth) data. Once the sample is fully tabulated, the matrix facilitates the calculation of several useful statistics including:

- *Overall classification accuracy* – determined as the sum of the diagonal elements on the matrix divided by the total number of pixels in the sample.
- *User's accuracy* – expressing the probability that a given pixel will appear on the ground as it is classed by the image. This value also defines errors of commission, explaining the percentage of pixels that should have been assigned to a given category but actually belong to another class. This value is calculated as the percentage correct for a given row, divided by the total for that row.

- *Producer's accuracy* – describing the percentage of a given class that is correctly identified on the map. Here, omission errors, indicating pixels that should have been assigned to a class but were not, are calculated by the percentage correct for a given column divided by the total for that column.
- *The kappa (K_{hat}) coefficient* – a measure of agreement between the derived classification image and the reference data. This estimate of agreement is calculated by subtracting the estimated contribution of chance agreement to the agreement as observed based on the formula:

$$K_{hat} = \frac{(\text{observed} - \text{expected})}{(1 - \text{expected})}.$$

According to this relationship, observed agreement is based on the overall classification accuracy, and the expected accuracy is a function of the chi-square distribution. The statistic is computed as:

$$K_{hat} = \frac{\left(N\sum_{i=1}^{r} x_{ii} - \sum_{i=1}^{r} x_{i+}x_{+i}\right)}{\left(N^2 - \sum_{i=1}^{r} x_{i+}x_{+i}\right)},$$

where r is the number of rows in the matrix; x_{ii} is the number of observations in row i and column i; x_{i+} and x_{+i} are the marginal totals for row i and column i, respectively; and N is the total number of observations.

Interpretation of K_{hat} is not unlike a correlation coefficient where values moving beyond +0.75 represent strong agreement, values between +0.5 and 0.75 suggest moderate agreement, and values falling below +0.5 indicate suggestively declining agreement between the classification image and the ground reference information.

8. *Applying the information* – This final stage in the remote sensing process directs our attention to the actual use of the thematic product as a source of information. Here, consideration is given to how well the thematic product communicates to the intended user and how well the product delivers based on the goals and purpose of the study. The applied value of remote sensing rests in the capacity for this technology to support decision making by reducing the impact of uncertainty that surrounds any decision problem. The quality of the information contained in the thematic product extends beyond the treatment of accuracy and envelops those aspects of information that adds value to a decision. In this context, the valued added nature of remotely sensed information depends on how well it:

   (a) Documents important spatial patterns
   (b) Illuminates alternatives
   (c) Frames the problem

(d) Records and stores information
(e) Communicates uncertainty
(f) Presents the findings

Because the products of a remote sensing investigation often take the form of a map, the principles of effective graphic communication warrant due consideration. In the majority of environmental applications, the intended audience is likely not a technical expert in remote sensing; therefore, the "pretty picture from space" must inform, not simply impress. This often overlooked element of the communication question requires maintaining a level of clarity and simplicity that agrees with the user's level of technical sophistication. Above all, the analyst, while manipulating the remotely sensed data, should avoid two common ethical pitfalls:

(a) *Cooking the data* – retaining only those points that fit the theory and discarding others and
(b) *Trimming the data* – smoothing of irregularities to make the data look extremely accurate and precise.

All of the data contained in the processed satellite image are real data and good faith, effort should be made to explain outlying situations in an objective and unbiased manner. Ideally, the results of an environmental investigation conducted via remote sensing offers the user insight, not only into the nature and significance of what has been discovered, but also the level of efforts involved in achieving the results. Informing the audience efficiently means capitalizing on that medium of communication that takes the user the shortest amount of time to understand.

4.5 Summary

Image processing and analysis can be explained as the "act of examining images for the purpose of identifying objects and judging their significance." The image analyst studying the remotely sensed data attempts to, through logical process, detect, identify, classify, measure, and evaluate the significance of physical and cultural objects, their patterns, and spatial relationship as evidenced in the image. A wide compliment of image processing and analysis techniques have been developed to aid the interpretation of remotely sensed data and to extract as much information as possible from the images. The choice of specific techniques or algorithms to use depends on the goals of each individual project. In this chapter, a sample of the more common method based on the principles of statistical pattern recognition were examined. Recognizing pattern is essential to the extraction of information from an image, and from this perspective statistically based techniques guiding the supervised and unsupervised approaches to image classification were reviewed. How effective these methods are depends on the accuracy they can achieve. Issues surrounding error and accuracy were examined that culminated this procedural review with a focus on how the information obtained from a remote sensing exercise can be communicated to its intended audience.

References

Ackoff, R. (1989) From Data to Wisdom, Journal of Applied Systems Analysis, 16, 3–9.

Anderson, J., Hardy, E. and Roach, J. (1976) A Land Use Land Cover Classification System for Use With Remote Sensor Data, USGS Professional Paper 964, United States Geological Survey, US Government Printing Office, Washington, DC.

Brodie, I and Brodie, L. (2009) A knowledge-information-data concept model for engineering education, Australasian Journal of Engineering Education, 15, 137–114.

Campbell, J. (2006) Introduction to Remote Sensing. The Guilford Press, 626p.

Congalton, R and Green, K. (2009) Assessing the Accuracy of Remote Sensing Data: Principles and Practice, Taylor and Francis,

Davis, G.B. and Olson, M.H. (1985). Management information systems. Conceptual foundations. structure, and development. New York: McGraw-Hill.

Dehons, A., Horne, E.; and Cronenweth, S. (1988). Information science: An integrated view. Boston: G. K. Hall.

Duda, T. and Canty, M. (2002) Unsupervised classification of satellite imagery: choosing a good algorithm, International Journal of Remote Sensing, 23, 2193–2212.

Duda, R., Hart, P and Stork, D. (2001) Pattern Classification, Wiley, New York, 680p.

Foody, G. (2002) Status of Land Cover Classification Accuracy Assessment, Remote Sensing of Environment, 80, 185–201.

Friedman, M. and Kandel, A. (1999) Introduction to Pattern Recognition, Imperial College Press, River Edge, NJ., 329p.

Gan, G. Ma, C. and Wu, J. (2007) Data Clustering: Theory, Algorithms, and Applications (ASA-SIAM Series on Statistics and Applied Probability) SIAM, Society for Industrial and Applied Mathematics, 466p.

Hudak, A., Brockett, B. (2004) Mapping ire scars in a southern African savannah using Landsat imagery, International Journal of Remote Sensing, 25, 3231–3243.

Jain, A., Murty, M., Flynn, P. (1999) Data Clustering: A Review, ACM Computing Surveys, 31, 264–323.

Lee, T. (2002) Unsupervised Image Classification, Segmentation and Enhancement Using ICA Models, IEEE Transaction on Image Processing, 11, 270–279.

Lein, J. (1997) Environmental Decision Making: an information technology approach, Blackwell Science, Malden, MA. 213p.

Liu, C., Frazier, P. and Kumar, L. (2007) Comparative Assessment of the Measures of Thematic Accuracy, Remote Sensing of Environment, 107, 606–616.

Lu, D. and Weng, Q. (2007) A Survey of Image Classification Methods and Techniques for Improving Classification Performance, International Journal of Remote Sensing, 28, 823–870.

March, S. and Smith, G. (1995) "Design and Natural Science Research on Information Technology." Decision Support Systems, 15, 251–266.

Mather, P. (2004) Computer Processing of Remotely Sensed Images: an introduction, Wiley, New York, 292p.

Memarsadeghi, N., Netanyahu, N.S. and LeMoigne, J. (2007) A Fast Implementation of the ISODATA Clustering Algorithm, International Journal of Computational Geometry and Applications, 17, 71–103.

Montello, D. and Sutton, P. (2006) An Introduction to Scientific Research Methods in Geography, Sage Publications, 303p.

Nangendo, G., Skidmore, A., Oosten, H. (2007) Mappin East African topical forest and woodlands – a comparison of classifiers, ISPRS Journal of Photogrammetry and Remote Sensing 61, 393–404.
Richards, J. and Jia, X. (2006) Remote Sensing Digital Image Analysis: An Introduction, Springer, 437p.
Scarrott, G. (1989) The nature of information. The Computer Journal 32, 262–326.
Stehman, S. (1997) Selecting and Interpreting Measure of Thematic Classification Accuracy, Remote Sensing of Environment, 62, 77–89.
Swain, P and Davis, S. (1978) Remote Sensing: he Quantitative Approach, McGraw-Hill New York, 396p.
Tso, B. and Mather, P. (2009) Classification Methods for Remotely Sensed Data, CRC Press, Boca Raton, FL. 376p.
Yom-Tov, E. (2004) An Introduction to Pattern Classification in Bousquet, O (ed) Machine Learning 203, LNAI 3176, Springer-Verlag, Berlin, 1–20.
Zhao Y., Karypis G. (2003) Clustering in the life sciences. *In* M Brownstein, A Khodursky, eds, Functional Genomics: Methods and Protocols. Humana Press, Totowa, NJ.
Zins, C. (2007) Conceptual Approaches for Defining into Knowledge, Journal of the American Society of Information Science and Technology, 58, 479–493.

Chapter 5
Sensing Uncertainty

Image extraction methods, employing statistically based image classifiers, implement decision rules that partition spectral space into mutually exclusive categories. These methods perform well, particularly when measurements are precise and our conceptualizations of environmental process are unambiguous. In those specific situations, using Boolean logic to delineate crisp search spaces, support by traditional expressions of probability sufficiently capture items of interest reliably enough to permit their representation as thematic content. However, there are situations we encounter in the study of the environment where our ability to define the processes or parameters in exact terms is impossible and the environmental conditions we wish to categorize display a level of complexity that defies clear and certain classification. The conceptually simple problem of mapping degraded land or a deforested area are examples that illustrate the presence of "maybe" in our search for information and underscores the frustrations that follow from the inherent vagueness of our definitions and our adherence to ridged models that fail to conform to the lack of certainty that surrounds most environmental issues. In this chapter, we will explore approaches to image classification that departs from strict statistical methodologies. In our exploration, we will examine how these different models exploit the uncertainty and produce very different realizations of environmental conditions, presenting information in ways that communicate differently in the context of decision making.

5.1 Embracing Imprecision

Well-recognized human–environmental processes such as land degradation, deforestation, drought, urban sprawl, and other chronic events that produce observable patterns on the land surface are also conditions that suggest a continuum of states that transition from one circumstance to another with a range of possible examples in between (Bennett 2001). Similarly, the boundaries separating these transitions are not likely to be clear and simple to identify in every instance. These examples and numerous others serve as useful reminders

J.K. Lein, *Environmental Sensing: Analytical Techniques for Earth Observation*,
DOI 10.1007/978-1-4614-0143-8_5,

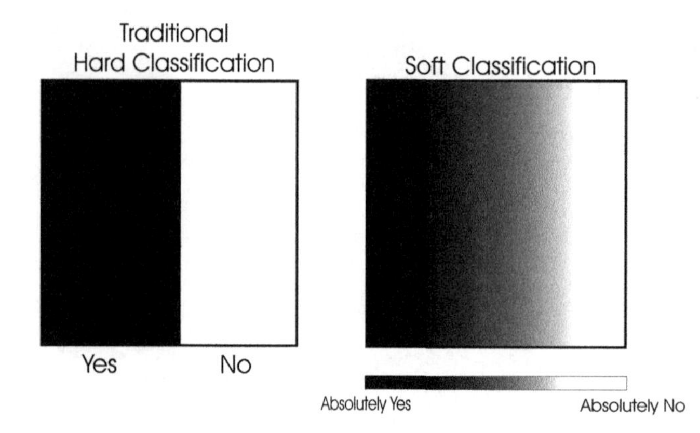

Fig. 5.1 Hard versus soft classification logic

where our explanations and definitions may imply dynamic environmental progressions, yet our classification algorithms can only recognize a pixel as either in a category or not in a category. Membership in an intermediate state or condition cannot be adequately recognized, and those locations that fall into the abyss of "maybe" are lost, even though those pixels can contribute useful information. The pixel with a spectral response pattern that is neither "forest" nor "soil" does not fit within the two-value logic of deforested or not deforested. Its status or label is unknown and subject to misidentification, and uncertainty is the result. Placing this pixel into the "unknown" category removes it from further consideration, and we lose a degree of explanatory power that could have improved our ability to portray the pattern of land in the transition from purely forest to perfectly barren.

In the study of human–environmental interaction, particularly those examples where human impact induces patterns of change that may be continuous over area, but highly variable with respect to intensity, capturing the continuum and representing uncertainty, not as an expression of error, but as a characteristic inherent to process, provides new forms of thematic information (Robinson 2003). Sensing the uncertainty exploits the logic of "maybe" to support the detection and mapping of modulations, diffusions, and spatiotemporal perturbations in a more realistic and intuitive manner (Eastman and Laney 2002; Nachtegael et al. 2007). This approach to satellite remote sensing is based on soft classification strategies that suggest decision rules built on the premise that a pixel can belong to more than one category; a sharp contrast to the statistical classification approach discussed in the previous chapter, where "hard" separations between categories provided the structure from which thematic maps were produced. The distinction between "hard" and "soft" methods of image classification is perhaps better understood using an illustrative example (Fig. 5.1). As suggested in the illustration, hard classifiers partition measurement space according to a logic that applies firm (crisp) boundaries to separate the informational classes. Parallelepipeds, distance-based search spaces, and delineations based on a maximum probability value ascribe to this logic and provide a framework to organize pixels into thematic information. Using these devises, the decision is simple; if the pixel as defined by its brightness value falls within the boundary, then that pixel has its appropriate label. The delineation, from a classification perspective, is clear and unambiguous with no exceptions. Soft classification explains a logic that preserves ambiguity in the data and employs it to allow a wider range of possibilities when categorizing the pixel. Possibility permits pixels to enjoy membership in more than one class or permits the "evidence" available as training signatures to demonstrate that a pixel belonging to more than one condition or class is a plausible solution. In Fig. 5.1, this alternative conceptualization is depicted as the gray-scale where the gradational shift in tone captures the "maybe" surrounding the classification problem and creates a decision boundary that is less ridged. Operating inside this "gray" area, the candidate pixel, whose label is unknown, can be identified by where it falls along this continuum. Thematic information adopts this logic, and the classification is characterized as a continuous surface where each label in the classification system defines a unique terrain depicting the spatial arrangement of membership over the study area. In a manner similar to the raster data model common to a geographic information system (GIS), each thematic class becomes a layer that can be subject to further analytical manipulation.

Although the idea of a gray-scale is a useful way to visualize this logic, the continuum describing "maybe" is quantified to a range of numbers that identifies the value of the pixel within its various class definitions. This value not only attributes the pixel, but also lends itself to digital representation in a computing environment. Hard classification methods assign a pixel a value symbolizing its class. For example, a water pixel may be given a label (value) of 1, and nonmembership in that category would be 0. Soft classification techniques might assign the water pixel a value of 0.78 for the class "water" and allocate the remaining 0.22 to some other class, such as sand or even an unknown category. By accepting this lack of perfection, soft classification methods enable this water pixel to emerge as information rather than relegating it to an unknown or misclassified category. Water, like many other land surface features are often not perfect, but influenced by extraneous factors. A pixel that it 0.78 water is more likely that quality than it would be some other characteristic. From the analyst's perspective, this acceptance of "maybe" translated to a potentially higher level of thematic correctness, whereas to the end user of this information, expanding the representation of a class membership from the yes/no model to a range of possibilities introduces the opportunity to adjust the decision boundaries defining classes, thereby incorporating the subtleties of a phenomenon into its spatial representation (Ibrahim et al. 2005; Foody 1999). This less deterministic approach to thematic representation overcomes the limitations imposed by the two-value logic, accommodates class mixtures, within-class spectral variability, and reduces the adverse consequences of uncertainty.

Soft classification strategies were introduced into remote sensing from the field of soft computing (Zadeh 1994; Gomez and Montero 2008). As suggested previously, the primary tenants of traditional hard classification are precision, certainty, and rigor. In soft classification, the guiding premise is that precision and certainty carry a "cost" and computation, reasoning, and decision making should embrace a wider tolerance for imprecision, uncertainty, and partial truth. The collection of soft computing techniques that have been introduced into remote sensing offer a means to achieve greater tractability, robustness, and solutions at a lower cost. In many respects, we can consider the role of soft computing/classification in environmental remote sensing as an example of knowledge discovery, involving the process whereby we sift through large amounts of data searching for patterns attempting to derive knowledge from that data. Considering a multispectral or hyperspectral image, often dominated by complex mixtures of surface types and classification systems imposing logics based on language concepts rooted in the vagueness of our definitions, methods that can assist us in making sense of the data, and transforming it to meaningful information, is critical to effective problem solving (Fayyad et al. 1996).

5.2 Soft Classification Strategies

As a process, soft classification begins by developing an understanding of the application domain and the goals directing the investigation. Data, in this context, are a set of facts (reflectance values in an image), and a pattern is an expression in language describing a model we wish to apply to the data. An overview of this process is suggested in Fig. 5.2. The discovered pattern should be valid with some level of certainty, but to be useful these patterns should also be novel and understandable to the end user. For a pattern to be considered knowledge, it must meet the criteria of being interesting in a domain-specific way. The geospatial variables acquired by means of satellite remote sensing suggest emerging patterns that require categorization, avoiding in the process, information loss, or the arbitrary imposition of artificial boundaries. Because the spatial patterns tend to exhibit spatial continuity and high autocorrelation with nearby features, soft classification alternatives become appropriate by

- Facilitating the identification of representative patterns
- Enhancing the selection of optimal spatial boundaries
- Summarizing pattern in a more intuitive manner

Three of the more common soft classification techniques applied to remotely sensed data are: (1) fuzzy classification, (2) evidence-based classification based on Dempster–Shafer theory, and (3) classification based on Bayesian probability. Each of these methods has its own unique facility for illuminating uncertainty and modeling the imprecision inherent to thematic extraction.

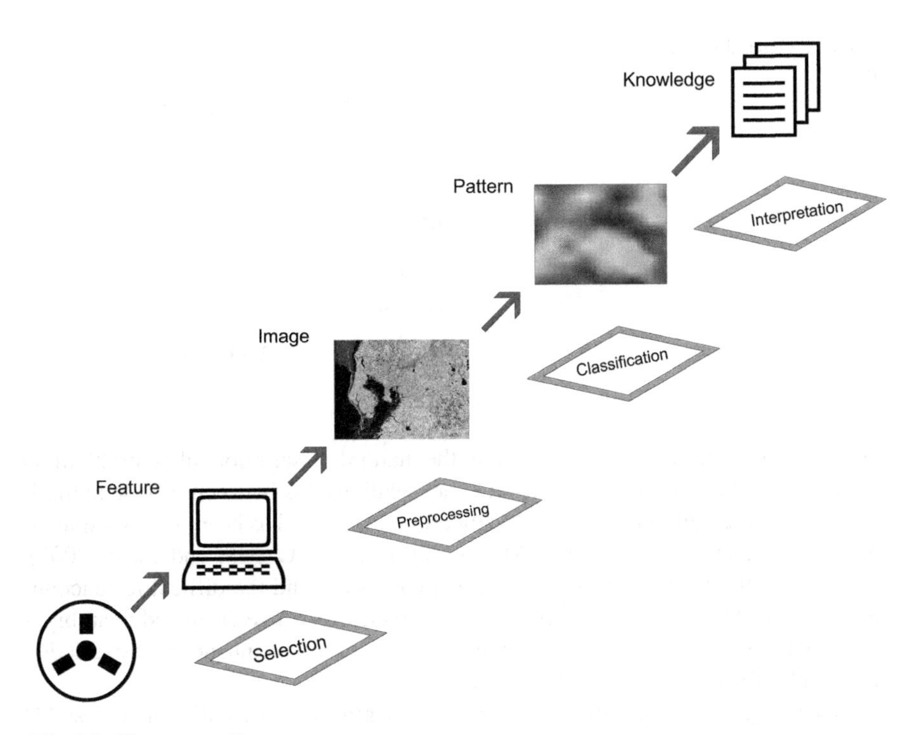

Fig. 5.2 The pattern discovery process

5.2.1 *Fuzzy Classification*

Fuzzy classification is based on the principles of fuzzy logic. Fuzzy logic represents a departure from traditional Boolean thinking and presents a useful means to model imprecise modes of reasoning and to discover solutions (approximate answers) to questions based on data that are inexact, incomplete, or unreliable (Zaheh 1988; Klir and Yuan 1995). The underlying theme of this logic system is the concept of fuzziness, which has been defined as a type of imprecision or vagueness that originates in language and characterizes events, phenomena, and features that cannot be precisely defined or measured (Leung 1988; Metternicht 2003; Foody 1992; Fisher 2010). According to classical two-value logic systems, the disposition of a pixel must conform to either one of two possible states with respect to a defining class. Fuzziness accepts the proposition that a pixel, owing to the imprecision inherent to its defining class, may be better explained by a degree of membership in that defining category. Therefore, as an element of a set of conditions (categories) that are fuzzy, belonging to one class is no longer restricted to a yes or no state, but instead expresses a compliment of partial membership that explains the possibilities as to where that pixel falls. The major advantage of this theory of

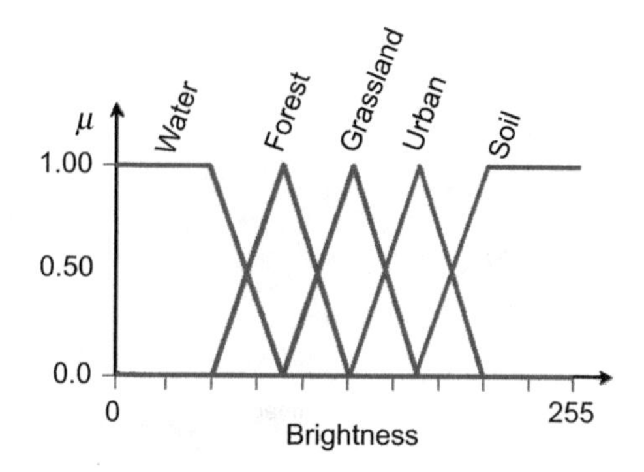

Fig. 5.3 General schema of a fuzzy set

possibilities is its ability to represent the natural description of a problem in linguistic terms rather that in terms of exact relationships between precise numerical values. As a method of image classification, fuzziness has been widely explored with useful results (Wang 1990; Melgani et al. 2000; Ozesmi and Bauer 2002). Given the heterogeneity of the environmental system and the difficulties encountered when attempting to discriminate between classes where clean and unambiguous spectral separations are not possible, the fuzzy solution enhances the extraction of useful information from our imagery.

Accepting the premise that a pixel can enjoy graded participation in a class (A) that is inherently fuzzy, a pixel (p) assumes the property:

$$A = \{[\mathrm{X}, \mu_{\mathrm{a}(p)}] : p = X\},$$

where $\mu_{\mathrm{a}(p)}$ explains the degree of or level of membership of pixel (p) in class (A). Defining category (A) as a fuzzy class requires deriving a membership function (μ_{a}) to assign each pixel in the image its degree of compatibility in that category. According to the principles of fuzzy logic, membership in a fuzzy class can range from 0.0 to 1.0 where 0.0 describes the condition of perfect nonmembership in class (A) and 1.0 defines the example of perfect membership in that class. The interval between 0.0 and 1.0 expresses the degree to which a pixel can be explained as a member of class (A). Symbolically a membership function can be expressed as

$$\mu_{\mathrm{a}(p)} = \begin{bmatrix} 1 & \text{if } x >= \text{ an upper threshold} \\ \left(\frac{x - \mathrm{low}}{\mathrm{range}}\right) & \text{if low} < x > \text{upper} \\ 0 & \text{if } x < = \text{a lower threshold} \end{bmatrix}.$$

A fuzzy set defined by $\mu(x)$ can be represented graphically in a variety of ways. Figure 5.3 illustrates one example of membership for five hypothetical land cover categories expressed across the recorded brightness levels of the sensor.

As suggested in Fig. 5.3, the practical application of fuzzy logic in image classification hinges on the derivation of membership functions that are used to extract thematic information from the scene. The methods by which these functions are obtained is not always made explicit which can frustrate the use of fuzzy classification and contribute to contradicting interpretations of what "fuzziness" communicates in a given study. Neglecting clarification about the origin of membership can make it difficult to verify that a procedure is performed as intended. Ideally, the fuzzy condition derived from the membership function expresses the degree of compatibility with its defining class or concept. These functions, to maximize the usefulness of fuzzy classification, should be comparatively easy to calculate and described by a relatively small number of meaningful parameters. The parameters in a remote sensing investigation are the spectral response patterns that describe the thematic information we wish to extract from the image. To transform spectral patterns into fuzzy representations of their class definitions, the raw image data has to be "fuzzified." There are several ways to convert the spectral patterns into a fuzzy set, and different approaches may be necessary depending on the numerical or linguistic variable under consideration. To illustrate this point consider the example where the goal is to represent the thematic class "urban." Fuzzification requires establishing the set of brightness values that satisfy the condition "urban." These are essentially the signatures for that land cover type, but also induced are the possible ranges of brightness that would be considered "urban" based on the analyst's judgment and experience. In mathematical terms fuzzification can be expressed as a mapping F from a variable X to a fuzzy set μ with a membership function $\mu(x)$ such that:

$$F : X \rightarrow \mu(x).$$

The conditions fuzzification must satisfy considerations related to

- *Continuity* – such that small changes in the value of the spectral response data do not yield large changes in the resulting membership grade
- *Uniqueness* – implying that each spectral response value has a unique membership grade
- *Range* – requiring that the values defining the spectral pattern for a given problem cover the entire [0,1] range of the membership function
- *Monotonic rising* – suggesting that a higher value of spectral response has a greater membership grade

To appreciate the challenges that surround reasoning with fuzzy concepts, we can return to our example of the linguistic variable "urban" in a land cover classification system. The problem of determining membership in the urban class begins by evaluating two important definitional questions:

1. What is the intensity of land development that identifies the condition "urban" land cover?
2. What is the spectral response value(s) that characterizes a level of intensity or use that agree with the condition "urban" land cover?

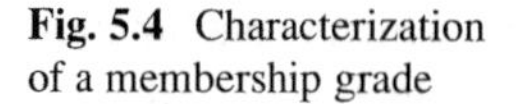

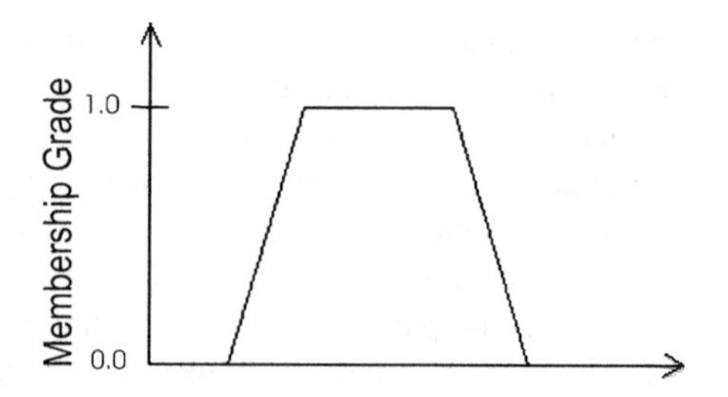

Fig. 5.4 Characterization of a membership grade

Obtaining answers to these fundamental questions can be accomplished in several ways. Initially, we could employ "expert" judgment, perhaps going so far as to survey a group of "experts" to solicit an opinion. From here, a set of values could be assembled from which the frequency of digital values being scored as "urban" would then be used to produce an empirical membership range of the form

$$\mu_{\mathrm{a}(x)} = \frac{\text{Number of positive responses to "}x\text{" is urban}}{n},$$

with x describing the range of digital numbers from 0 to 255 in the image and n the number of experts solicited in this example. Based on the frequency of responses, a membership function could be defined that would assign the range of digital values along the 0.0–1.0 continuum (Fig. 5.4). Using expert judgment to establish the thresholds for a fuzzy class offers a degree of flexibility that can model uncertainty by incorporating a priori knowledge into the classification process. Expert judgment also offers the opportunity to modify class membership based on contrasting assumptions about the data and the classification schema being used. Judgment-based methods may also utilize subjective assignment of membership grades for training sites used to define spectral signatures for a class. This type of fuzzy portioning can be employed to assign each pixel a weight proportional to its degree of membership (Wang 1990; Townsend 2000).

In general, using frequencies or employing direct estimation methods to manually craft membership functions produce tractable solutions to the image classification problem; however, they are not without limitations (Watanabe 1979; Turksen 1991; Kempton 1984). Although the appropriateness of adopting a fuzzy classification strategy is ultimately determined by the accuracy of the resulting thematic product, certain deficiencies can be noted. First, the manual generation of membership functions relies on the subjective interpretation of classification labels. When considering the concepts such as "eroded," "degraded," "stressed," "disturbed," connecting these concepts to absolute conditions in the environmental can be difficult. Second, placing a membership score or similar valuation based on a brightness value without a clear rationale risks confusion and introduces the potential for methodological inconsistencies. Finally, judgment-based approaches are entirely dependent on the experience of the experts called upon to deliver membership grades. Provided the expert has an understanding of the conditions

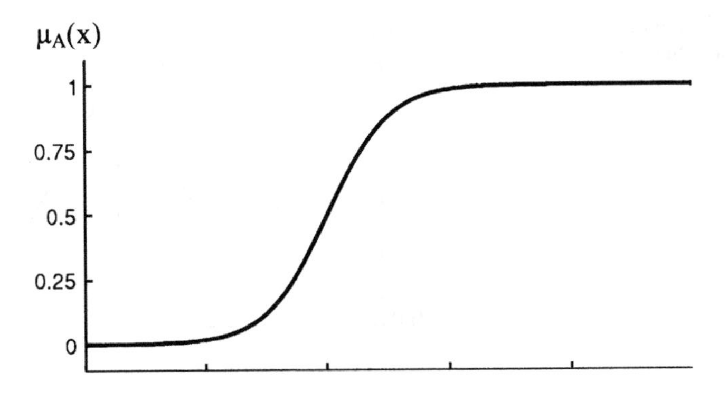

Fig. 5.5 Sigmoid membership function

that define the upper and lower limits for membership in a class, reasonable functions can be obtained. In the absence of experience, the resulting membership grades may not be meaningful.

Alternatively, membership functions can be developed from trigonomic relationships that approximate the concept of condition. The choice of function will depend on how the shape of the trigonomic relationship models the nature of the concept to be fuzzified. Functions can range from simple linear representations to approximations based on the Gaussian distribution. Linear representations define a straight line membership function that can define two states; increasing or decreasing. Sigmoid representations, also known as the *S*-curve, are often used when natural processes display a "history dependent" progression that and either accelerates or declines. This membership function is defined using three parameters: (1) its zero membership value, (2) its complete membership value (max), and (3) its crossover (inflection) point. This *S*-curve function is illustrated in Fig. 5.5. Other functions include

- *Triangular representations* – perhaps the most common representational form, this function is defined using three variables where

$$\mu_A(X,a,b,c) = \begin{bmatrix} 1 & \text{if } x \text{ is greater than or equal to } a \\ \left(\frac{x-a}{b-a}\right) & \text{if } a \le x \le b \\ \left(\frac{c-x}{c-b}\right) & \text{if } b \le x \le c \\ 0 & \text{if } x \text{ is less than or equal to } c \end{bmatrix}.$$

  The parameters *a*, *b*, and *c* represent the *x* coordinates of the three vertices of $\mu_A(X)$ in fuzzy set *A*. This representation is illustrated in Fig. 5.6.
- *Trapezoidal representation* – this function is defined using four points, two to depict points in the surface (*b*/*c*) and two for the fuzzy slope of the surface (*a*/*d*) (Fig. 5.7).
- *Shouldered representation* – that the form from both the triangular and trapezoidal shapes, shouldered sets present no upper or lower boundaries, and the shape

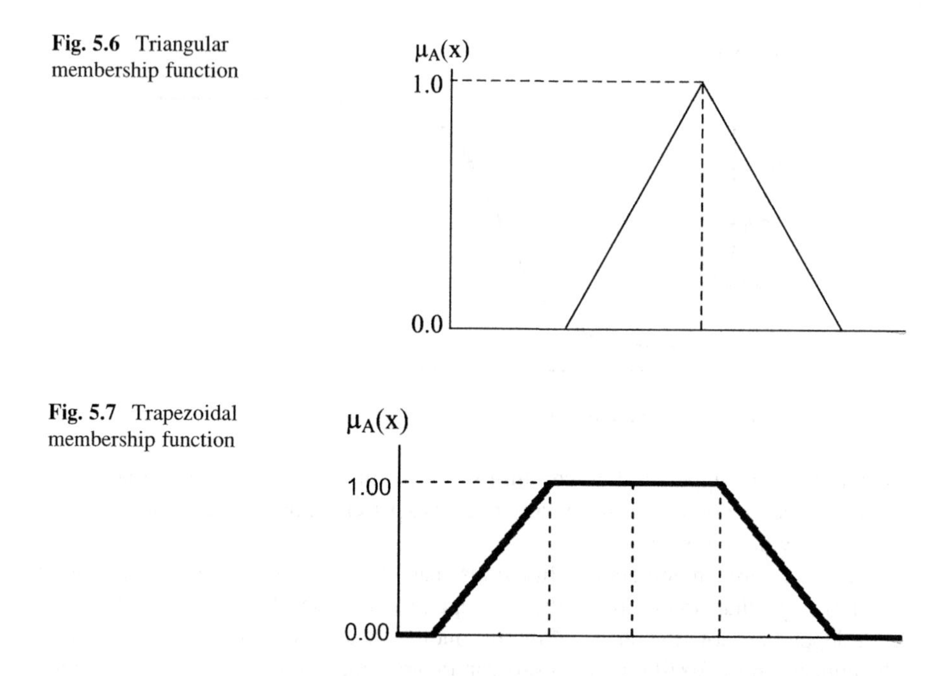

Fig. 5.6 Triangular membership function

Fig. 5.7 Trapezoidal membership function

generally assumes the expression of a truncated trapezoidal fuzzy set. The shoulder in this representation defines the edges of the condition or variable.

- *Gaussian representations* – defines sets whose membership grades approximate a normalized Gaussian function with 0 mean and a standard deviation of δ. Membership according to this function is defined by

$$\mu_A(X, c, s, m) = \exp\left[\left\{-\frac{1}{2}\left(\frac{x-c}{s}\right)\right\}^m\right],$$

where c equals the center of the distribution, s represents the width, m is a fuzzification factor. In this relation, the values of s and m can be altered (increased or decreased) to produce different shapes.

When considering a fuzzy representation, it is often instructive to represent the fuzzy set graphically and use that visualization to select the shape function that most closely resembles the graphic depiction can conceptualize.

5.2.2 Implementing Fuzzy-Based Strategies

Image classification based on the principles of fuzzy logic can be implemented following either a supervised or unsupervised classification procedure. Fuzzy supervised classification requires translating spectral response patterns for the land surface categories of interest into fuzzy representations and then applying those fuzzy patterns (signatures) in a classification algorithm. Perhaps, the most tractable algorithm for fuzzy supervised classification is fuzzy maximum likelihood. The maximum likelihood classifier is based on the probability theory where the mean and covariance summarize each class of interest. The algorithm solves for the probability function $P(A)$ such that

$$P(A) = \int H_A(s),$$

where s explains an element in class x, and H_A is the "hard" membership value that can only be 0 or 1. This basic relationship was extended to incorporate fuzziness by Wang (1990) and Maselli et al. (1995). In this extension, the probability function is replaced by a possibility of membership grade approximated by:

$$P(A) = \int \mu_A(s),$$

where μ_A denotes the membership function defining class A. The product of this algorithm is a membership value for each class that can be presented spatially as a sequence of individual map layers.

Unsupervised classification removes the analyst from direct manipulation of the spectral response patterns and relies instead on clustering techniques such as the K-means or ISODATA algorithms to discover natural groupings in the data. The fuzzy extension to these approaches is found in the fuzzy c-means algorithm and its derivatives (Bezdek 1981; Bezdek et al. 1984). Recalling that clustering is typically based on distance metrics to establish similarities and separate cluster centers, relaxing the "hard" representation of distance (whether expressed as Euclidean, Mahalanobis, or other) and replacing that with fuzziness, fuzzy cluster centers can be determined based on the function J_b:

$$J_b = \sum_{i=1}^{n} \sum_{j=1}^{n} (V_{ij})^b (d_{ij})^2,$$

where b is the weighting exponent that controls the degree of fuzziness (a value typically set between 1.5 and 0), d_{ij} defines a distance measure between each pixel x and a fuzzy cluster center V. Based on the above equation, the membership value of class j for the ith pixel becomes:

$$\mu_{ji} = \frac{1}{\sum_{k=1}^{m} (d_{ij}/d_{ki})^{2/(b-1)}},$$

where μ_{ij} must satisfy the following constraints:

$$\sum_{i=1}^{n} \mu_{ji} > 0 \quad j = 1, 2, \ldots, m,$$

$$\sum_{j=1}^{m} \mu_{ji} = 0 \quad j = 1, 2, \ldots, n.$$

The fuzzy cluster mean (V) is resolved iteratively from

$$V_j = \frac{\sum_k (V_{ij})^b X_n}{\sum_k (V_{ik})^b}.$$

Although widely used, the fuzzy c-means algorithm assumes that every information class has been specified, so that the membership for every pixel can be described independently of all other classes, a condition that may not be valid in all situations (Foody 2000; Tso and Mather 2001).

Fuzziness as a means to express imprecision in classification relies on the successful derivation of a membership function, which is not always a simple task. Furthermore, the membership functions do not necessarily stem from the data itself which can be problematic when the membership criteria do not sufficiently characterize uncertainty. While fuzzy classification is a useful approach when representing spatial objects that do not conform well to crisp delineation, there are examples in environmental analysis where uncertainty does not develop from the mixture of reflectance values within a pixel, but rather where the similar surface characteristics under investigation are explained by different spectral patterns. In these instances, there is a range of variation in reflectance within and between pixels, and the spectral patterns fail to model a perfect state (Lein 2003). Under these conditions, the problem is not one of fuzziness, but rather a situation underscored by "broadness" where several possibilities exist for placing a pixel into a class, and the evidence for deciding the membership may be incomplete or conflicting. In these examples, a method that can process the available evidence and narrow the possibilities of membership down to an objective set that best characterizes the informational categories offers an alternative solution to the problem of uncertainty.

5.3 Evidence-Based Classification

Evidence-based classification exploits the logic of evidential reasoning as defined according to Dempter–Shafer theory of evidence (Shafer 1976; Dempster 1967). Dempster–Shafer theory can be considered as a generalization of Bayesian theory of probability where probabilities are assigned to sets rather than to a single object (or a set containing a single member). As a variant of the Bayesian model, Dempster–Shafer theory replaces subjectivity with the concept of ignorance which can be loosely explained as a type of inconclusiveness given the evidence available to make a decision. Therefore, in contrast to probability theory where

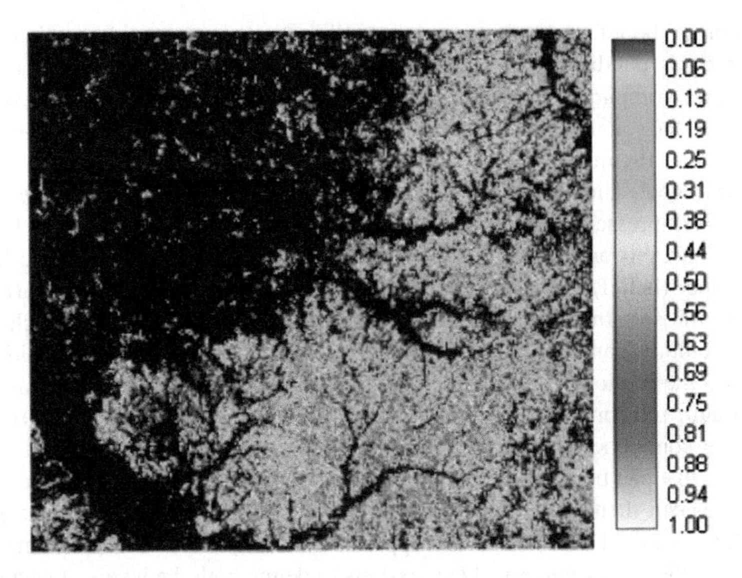

Fig. 5.8 Spatial representation of "belief"

evidence is associated with only one possible event or condition, Dempster–Shafer theory allows evidence to be associated with multiple possible events. This distinguishing aspect of this theory permits evidence to assume meaning at higher levels of abstraction without incorporating any underlying assumptions about the data. In remote sensing, the appeal of Dempster–Shafer theory rests in its ability to capture the natural behavior of reasoning by narrowing a hypothesis set down to a smaller number of possibilities as the evidence for making a selection increases (Mertikas and Zervakis 2001). In the example of image classification, this theory addresses the decision problem of selecting pixels to form one or more land cover categories under conditions where the evidence available to guide class assignment varies. For applications common to environmental sensing, by allowing for inconclusiveness in the evidence, Dempster–Shafer theory does not introduce bias into the probability assignment process. In a practical sense, the appeal of Dempster–Shafer as a method of classification relates to the common decision problem where a pixel must be assigned to one of the several categories; however, the evidence, as expressed by the set of spectral signatures, varies and conflicts. Labeling a pixel as a specific type of use or cover, for example "agricultural," proceeds by narrowing down this initial hypothesis based on the spectral response patterns to a level of agreement that can be supported by the spectral evidence (Fig. 5.8). When this method of reasoning is applied to guide spectral discrimination and classification uncertainty, image analysis based on the Dempter–Shafer model has been shown to produce more complete descriptions of surface characteristics (Le Hegarat-Mascle et al. 1997; Lein 2003, 2006; Cayuele et al. 2006).

Dempster–Shafer classification introduces the concept of belief as the device to examine the evidence surrounding class membership. As a decision criterion, the concept of belief is intuitive and compliments the manner by which human reasoning is conducted. In simple terms, if we were to observe the value of a pixel and compare its brightness number to those of the signatures we have collected, the statement "I believe this pixels is forest, based on the evidence before me" mimics the manner by which the Dempster–Shafer algorithm reasons toward the same goal. Unlike human reasoning, a belief according to Dempster–Shafer theory is expressed quantitatively based on the calculation of a belief function. Belief functions are based on a probability assignment that explains the level of acceptance (belief) that can be committed exactly to a specific hypothesis (i.e., that the spectral patterns exhibited by a pixel across its defining wavelengths characterize agricultural land). Measures of belief communicate the level of confidence in the hypothesis along the familiar range of numbers from 0.0 to 1.0. Along this continuum, 0.0 communicates "no confidence" in the hypothesis while 1.0 explains absolute confidence in the decision. Intermediate values express the degrees of belief in that can be ascribed to the hypothesis given the available evidence. Belief measures can then be summed up to produce an overall expression of certainty, which makes the Dempster–Shafer approach a useful method for qualifying classification belief using numerical expressions to define both the levels of support and the degrees of uncertainty.

There are three important mathematical functions that form the foundation of Dempster–Shafer theory and direct its use as a classification algorithm: the basic probability assignment function (BPA), the belief function (BEL), and the plausibility function (PLS). In the language of Dempter–Shafer theory, the BPA represents a mapping of the power set to the interval between 0 and 1, where its null set is 0, and the summation of the PBA's of all the subsets of the power set is 1. Although expressed as a value between 0 and 1, BPA is not a value of probability in the classic sense, but rather the proportion of all relevant and available evidence that supports the allocation of a pixel (X) to the information class (D). From this basic probability assignment, the upper and lower bounds of an interval, expressing class membership, can be determined. It is this interval that contains the actual probability of the set under evaluation, which is then bounded by two nonadditive continuous measures describing belief and plausibility. Therefore, given a set of mutually exclusive alternatives, such as three distinct land cover types, a basic probability assignment (m) is determined for each subset (A), which is written as $m(A)$. The lower boundary of belief for a set (A) is the sum of all the basic probability assignments for the set such that

$$\mathrm{BEL}(D) = \sum_{A \cap X} m(A)$$

and plausibility develops as the sum of all BPA's determined by

$$\mathrm{PLS}(D) = 1 - \mathrm{BEL}(D),$$

a term that expresses the degree to which an assignment to a class (hypothesis) cannot be disbelieved. When the two equations given above are compared, belief (BEL) quantifies the degree of evidence in support of the hypothesis (that pixel X is agricultural), while plausibility indicated the degree to which the condition appear to be correct for that hypothesis, although the evidence to support that conclusion is incomplete (Lein 2003). To illustrate this relationship, suppose we assume that three training sites have been generated for three land cover classes (e.g., urban, forest, and water), a pixel that shows some similarity to the urban signature, but not to any other would be typically named urban. However, if the evidence supports the hypothesis that the pixel is urban to the degree 0.40, the Dempster–Shafer algorithm would assign 0.40 as its value of belief to urban with a plausibility of 1.0. The result would produce a belief interval of 0.60, and the algorithm would assign all other classes a belief of 0.0 and a plausibility of 0.60. This assignment creates a measure (degree of belief) that can be employed to represent the cover type as a continuous surface where both belief and uncertainty become spatial variables that express the geographic pattern of class membership over the study area (Lein 2003, 2006). Following the combination of probability assignment, a decision rule must be selected in order to assemble pixels into classes and form representative informational classes. Decision rules are based on the Dempster evidence function for belief and plausibility (Mantara 1990). This function is essentially an evidence gathering process that requires combining the support for a hypothesis based on multiple observations. The basic rule guiding the algorithm states that the BPA representing the combination of m and m_2 apportions the total amount of belief among the subsets by assigning m and m_2 to the intersection of the set.

Applying Dempster–Shafer theory as a soft classification strategy follows the general format of the supervised image classification procedure. The main point of departure is simply the idea that the Dempster–Shafer algorithm would provide a more correct or logically appropriate characterization of the remotely sensed data. There are four main phases of the procedure:

1. *Signature development and training* – involving the creation of spectral response patterns for the informational categories of interest. However, unlike traditional supervised classification, there is little need to be concerned over the statistical parameters that summarize the land surface objects or their potential overlap.
2. *Soft classification* – employing the Dempster–Shafer algorithm to generate belief and plausibility surfaces for the thematic classes of interest.
3. *Class hardening* – requiring the selection of belief thresholds for each thematic class to produce a finished map product. The hardening algorithm evaluates the degree of support for the assignment of a pixel to a specific informational class and involves detailed consideration of the hierarchy of classes and all possible combinations (mixtures).
4. *Assessment and evaluation* – focusing on "face validity," this final step examines how well the belief surfaces communicate the uncertainty, the overall correctness of the hardened classification, and the level of agreement achieved between surface arrangements and their spatial representation. Because belief surfaces

serve as a basis for quantifying class membership, reference to the spatial patterns of belief can help to refine optimal thresholds for forming conventional class boundaries.

5.4 Bayesian Classification

The Bayesian approach to representing classification uncertainty is based on the supposition that the prior probability of an event should be incorporated into the interpretation of a current situation (Lein 1997). As a classification strategy, Bayesian methods define class membership as a "probability-to-feature" problem as opposed to developing decision rules that explain class membership according to a "distance-to-feature" logic. The underlying assumption guiding Bayesian classification is that the decision problem surrounding the question of assigning pixels to classes can be expressed in probabilistic terms and that all of the relevant probability values are known a priori. Under this assumption, Bayesian classification performs in a manner sensitive to the selection of prior probabilities and introduces the use of subjective probability into the classification process. When compared to hard classification techniques, Bayesian classification requires the analyst to draw from theoretical or heuristic knowledge the nature of the expected probability relationships rather than from strict statistical assumptions stemming from the data (Wither 2002). Because there are no prior assumptions about class membership, the analyst must carefully consider what the expectations are, given the nature of the classification problem. The use of subjective prior knowledge, in this context, while permitting greater flexibility to unique situations, requires the explicit consideration of uncertainty. The application of subjective judgment makes the Bayesian approach well suited to situations where the data are inexact and the general rules of thumb are called upon to contend with the unknowns.

As a method of classification, Bayesian theory is predicated on three defining principles: (1) a set of mutually exclusive hypotheses, (2) conditionally independent evidence, and (3) complete enumeration of the hypothesis set. Working with these principles, the goal of Bayesian classification is to assign objects to categories such that

$$P\left(\frac{H}{E}\right) = P(H)\frac{P(E/H)}{P(E)},$$

which describes the conditional probability of E (evidence) given a hypothesis (H), where E is typically expressed as a subject probability value. In the example of land cover analysis, we might assume that there is a prior probability $p(w^*)$ that a parcel (pixel) of vegetated land is chaparral, and a probability $p(w^\wedge)$ that it is not chaparral. The prior probability terms reflect the subjective judgment on the analyst as to how likely chaparral serves as a reasonable estimate of any feature descriptive of the parcel (pixel). Without any additional information, the decision about how the parcel (pixel) should be labeled is simply a comparison of $p(w^*)$ and $p(w^\wedge)$. However, the conditional probability for vegetated cover can be explained more completely by other variables such as leaf reflectance of by a vegetation index (VI) that establish a cleared decision boundary and threshold for evaluating the condition. In this example, a vegetation index value from chaparral pixels and nonchaparral pixels can be used to define a relationship:

$$p\left(\frac{w_1}{\mathrm{VI}}\right) \quad \text{and} \quad p\left(\frac{w_2}{\mathrm{VI}}\right),$$

where a pixel that is chaparral can be defined and classified. By determining appropriate thresholds for the VI values, a relationship can be evaluated of the form:

$$\begin{gathered} \text{If } (p(w_1/\mathrm{VI}) > p(w_2/\mathrm{VI})) \\ \text{Then Class} = \text{chaparral} \\ \text{Else Class} = \text{nonchaparral} \end{gathered}$$

According to this simple example, if both the prior probabilities $p(w)$, conditional probabilities $p(w/\mathrm{VI})$, and the VI values can be estimated from an image, we can employ Bayes rule to formulate the decision taking the general form:

$$P\left(\frac{w_i}{\mathrm{VI}}\right) = p\left(\frac{\mathrm{VI}}{w_i}\right)\left(\frac{p(w_i)}{p(\mathrm{VI})}\right),$$

where $p(\mathrm{VI})$ is the prior probability of VI, which develops out of the total probably given as

$$P\left(\frac{w_i}{\mathrm{VI}}\right) = p\left(\frac{\mathrm{VI}}{w_i}\right)\frac{p(w_i)}{\sum[p(\mathrm{VI}/w_i)p(w_i)]}.$$

Pixel classification is performed by maximizing the a posteriori probability obtained from this solution. As a classification algorithm, the Bayesian approach searches for the best class descriptions which can be defined as the solution that optimally trades off predictive accuracy against class complexity without overfitting the data. The resulting classes are "approximate," and pixel membership in a class is expressed in terms of a probability estimate in the range 0.0–1.0. When classification results are presented spatially, membership explains a probability surface, which can be hardened or manipulated further depending on the purpose (Rocchini and Ricotta 2007; Binaghi et al. 1999).

Soft classification offers many possibilities when confronted with the uncertainty stemming from conceptual imprecision and vagueness, and the use of these methods can support a range of applications' problems. One common problem encountered in environmental remote sensing relates to the presence of mixed pixels and the uncertainty mixtures introduce into thematic extraction. In the next section, we will examine the issues surrounding mixed pixels.

5.5 Mixtures and Subpixel Analysis

To this point in our discussion, uncertainty was characterized as a product of the confusion resulting from the attempts to organize spectral measures into informational classes based on the terminology that can be vague or poorly defined in language. Interrelated to this important source of confusion are those uncertainties that grow out of the complex arrangements of objects at the surface that cannot be aptly described by a single "unique" spectral pattern. Implicit in this discussion are the constraints introduced by the spatial resolution of the sensors, which, due to varying dimensions, impose an artificial gridding on real and highly varied landscapes. The result of superimposing this ridged, stratified sampling design on a surface that is likely to be a heterogeneous complex of natural features and human-made objects is a form of uncertainty well recognized as the mixed pixel problem (Fisher 1997; Lewis et al. 2000). The production of mixed or composite pixels originates from the observation that at some level of spatial resolution, natural surfaces are not uniform, and whereas human-created surfaces might conceivably be uniform at a given scale, homogeneity is not typically the case (Schowengerdt 1996). Rather, composite pixels in a remotely sensed image represent a spatial average of the spectral response patterns from at least two or more surface conditions. Composites are also unique to the manner in which features occupy the pixel, displaying variations resulting from subtle differences in the material, chemical or biological fabric descriptive of that surface. The mixing of spectral response patterns arise from four main sources:

1. The intrinsic, spatially mixed nature of most land cover categories
2. The physical continuum that may exist between discrete categorical labels
3. Resampling for geometric rectification during preprocessing operations
4. The spatial integration induced by the sensor's point spread function

Only in controlled and limited situations can the mixing induced by these contrasting sources be distinguished (Schowengerdt 1996). Consequently, the categories that describe our thematic (informational) classes are often spectrally ambiguous in addition to being linguistically vague. Unraveling the mixing with a single pixel resulting from spectral ambiguity is a challenge, and the soft classification strategies introduced previously have been used to address this problem. However, estimating subpixel composition and quantifying the land cover/land use fractions contained with the pixel are better addressed by mapping the subpixel components of the surface using spectral mixture analysis (SMA) (Powell et al. 2007; Song 2005; Small 2001).

5.5.1 Spectral Unmixing and Spectral Mixture Analysis

The process of unmixing the spectral composition of a pixel and labeling the cover types that comprise the observed mixture introduces the concept of an endmember. In simple terms, an endmember is a pure reflectance spectra that has been derived from a specific target material or precise land cover arrangement where no mixing with other objects or features has taken place. Endmembers, are therefore, features recognizable in the scene as being abstractions of land cover materials with uniform properties (Rashed et al. 2001). The material nature of the pixel is a central idea in the definition of an endmember. Based on this material definition, an endmember is a pattern of reflectance that describes the characteristic properties of an object or feature. For example, if the target of interest were asphalt, the asphalt endmember would be similar in concept to its pure spectral signature. As a characteristic pattern of reflectance, endmembers serve as reference spectra for objects. Also because they are free of any mixtures, they can be used to separate any mixing of reflectance patterns within a pixel and determine the combination of composite reflectance patterns exhibited by an observed mixture in the scene.

Quantifying the spectral composition of a pixel is accomplished using SMA. As a method of subpixel classification, SMA assume that the net radiance at the sensor (without the effects introduced by atmospheric attenuation) is a linear combination of the spectral patterns of all cover types or objects that fall within a given pixel. According to this technique, each pattern explains one component of the mixture and contributed in some way to the observed value of reflectance measured by the sensor. The process of unmixing is comparatively straight forward. If the properties of each object or cover type's contribution to the observed brightness value can be determined, then the relative proportion of each patterns contribution to the overall spectral response can be estimated (Small 2001; Foody 2000). Expressing the idea of unmixing in more quantitative terms, a composite pattern of reflectance of a pixel (P) for a specific spectral band (i) can be explained as (after Settle and Drake 1993; Wu and Murray 2003)

$$P_i = \sum_{k=1}^{N} F_k P_{ik} + E_i,$$

where N is the number of spectral bands, P_i is the reflectance of a pixel in the ith band, P_{ik} is the reflectance of the kth component of the pixel in the ith band, F_k is the weight of the mixing or fraction of endmember k, and E_i is the error.

The fit of this relationship is typically evaluated by the residual (error) term E_i of the root mean square (RMS) error over all of the bands in the image (n) such that

$$RMS = \left(\frac{\sum_{i=1}^{n} e_i^*}{n} \right)^{1/2}.$$

The fraction of each endmember can be estimated by applying the least square technique to the data in order to minimize the unmodeled residual error (e_i) with the assumption that selected endmembers are independent of each other, the number of endmembers is less than or equal to the available spectral bands, and the selected spectral bands are not highly correlated (Lu and Weng 2006). The products of the SMA algorithm are a series of images that depict the spatial distribution of each endmember's abundance across the scene. Fractional abundance surfaces, therefore, communicate a continuum for each endmember and provide a means to visualize both the unmixing of pixels and the relative material abundance of the features that compose the land surface.

Although a useful way to decompose the mixtures produced by complex land surface arrangements, the linear mixture model may not be the appropriate choice for applications where only subtle spectral differences exist across the selected bands. Validity is also an important consideration in the use of this method. The validity of the SMA technique depends on the selection of endmembers. Generally, the rule of thumb is that more endmembers can explain more spectral variation, which improves the fit of the SMA model. However, there is a trade-off between the number of endmembers and model fit. In a typical application of SMA, a fixed number of representative endmembers usually between 2 and 5 are chosen, and the image is modeled relative to those spectral components. There are limitations to this rule of thumb, particularly in situations when the selected endmember patterns do not effectively explain all features present in the scene or in examples where a pixel may be characterized by endmembers that do not correspond to the materials found within its spatial dimensions (Powell et al. 2007).

As a method for sensing uncertainty, SMA has traditionally be a technique reserved for use with hyperspectral imagery. Recently, some success with mixture uncertainty and land cover classification has been achieved using multispectral data such as Landsat, AVHRR, and ASTER (Wu 2004; Quarmby et al. 1992; Sabol et al. 1992). In either case, the procedure followed while applying the SMA algorithm consists of five main steps:

1. Selection of endmembers
2. Application of the SMA algorithm
3. Evaluation of the SMA fractional surfaces and RMS error
4. Classification of the SMA fractions
5. Assessment of classification accuracy

Each step in the procedure requires detailed consideration to produce useful end results. Perhaps the most critical aspect of the SMA approach centers on the final classification of the SMA fractions into informational classes. Classification focuses on the fractional surfaces produced by the SMA algorithm. A fractional surface shares properties similar to those obtained through fuzzy or evidence-based classification. As the SMA algorithm decomposes the mixture, it produces a surface for each endmember that depicts the abundance of that endmember for every pixel in the scene. Using values ranging from 0.0 to 1.0 (abundance values of <0.0 or >1.0 suggest missing or incorrect endmembers), the per-pixel fraction of that endmember is given a spatial expression that can be used to form a decision regarding the labeling of pixels. Depending on purpose, the spatial distribution of fractional abundance can be treated as a continuous expression used to characterize the representation uncertain of the endmember across the scene, or it can be subjected to "hardening" in order to produce a standard thematic map. Combining fractional solutions into a crisp thematic presentation involves visualization of the fractional images, analysis of the fraction characteristics of representative land cover/feature types, and the evaluation of error images. Each of these activities helps to establish thresholds that can be used to select the abundance value that best categorizes the pixel. Through subject assessment based on the fractional surfaces, the analyst can determine which combinations or thresholds provide the best description of the thematic categories. Class boundaries can be established using operations such as simple Boolean logic to reclassify the fractions into a 0 or 1 representation of the endmember, or combined using a variety of other strategies including

- Derived exogenous groupings based on application-specific criteria
- Statistical-based classification
- Clustering
- Majority analysis
- Multiattribute overlay

5.6 Summary

Uncertainty remains a vexing issue in remote sensing and has numerous sources which conspire to degrade the fidelity of the thematic information taken from the image. In this chapter, a selection of methods for managing uncertainty as it applied to the image classification problem were examined. Each has its value depending on the nature of the problem under investigation, and each offers a unique way to present and explain thematic content as a continuous phenomenon that allows the end user to make the final decision regarding the precise allocation of pixels into informational categories. Through the logic of fuzziness, the explication of belief, the introduction of prior probability, and the notion of pixel unmixing, more adaptive approaches to image classification are encouraged, and useful information that might otherwise have been lost due to the inflexibility of statistical-based classification algorithms is preserved. When applied to the questions that envelop environmental analysis, the method of sensing with uncertainty extends the utility of remotely sensed data and facilitates their use where the environmental process is poorly defined and our understanding is tempered by inexactness.

The process of image classification based on pattern recognition methods assumes that pixels can be organized into mutually exclusive and exhaustive categories. In some cases, the prospects for a perfect delineation of surface features into classes fail, and information that might otherwise be useful is lost into the class of "unknowns." In this chapter, alternative strategies based on approximate reasoning logics such as fuzzy set theory and the Dempster–Shafer theory of evidence were explained. Both models offer useful advantages to the task of information extraction, particularly where classification suffers from practical and semantic imprecision and inexactness, and both models represent a form of many-valued logic to deal with reasoning approximate, rather than fixed and exact. Here, we examined the usefulness of explaining surface categorizations expressed along a continuum that ranges in degree between 0 and 1, and explored situations where partial truth can describe membership in land cover categories producing surface patterns that defer judgment regarding class membership to the analyst's subjective technical judgment. Recognizing that human–environmental processes, such as land degradation, deforestation, drought, and urban sprawl, produce observable patterns on the land surface, which also suggest a continuum of states that transition, encourages using the logic of "maybe" as a means of revealing environmental uncertainties and capturing complex environmental patterns in a more realistic way.

References

Bennett, B. (2001) What is a Forest? On the Vagueness of Certain Geographic Concepts, TOPOI, 20, 189–201.

Bezdek, J. (1981) Pattern Recognition with Fuzzy Objective Function Algorithms, Kluwer Academic Publishers, 272p.

Bezdek, J., Ehrlich, R. and Full, W. (1984) FCM: The Fuzzy c-Mens Clustering Algorithm, Computers and Geosciences, 10, 191–203.

Binaghi, E., Brivio, P., Ghezzi, P. and Rampini, A. (1999) A Fuzzy set-based Accuracy Assessment of Soft Classification, Pattern Recognition Letter, 20, 935–948.

Cayuele, L., Golicher, J., Salas Rey, J., Rey Benayas, J. (2006) Classification of a Complex Landscape using Dempster-Shafer Theory of Evidence, International Journal of Remote Sensing, 27, 1951–1971.

Dempster, A. (1967) Upper and Lower Probabilities Induced by a Mutlivalues Mapping, The Annals of Statistics, 28, 325–339.

Eastman, R. and Laney, R. (2002) Bayesian Soft Classification for Sub-pixel Analysis: a critical evaluation, Photogrammetric Engineering and Remote Sensing, 68, 1149–1154.

Fayyad, U., Piatetsky-Shapiro, G., Smyth, P., and Uthurusamy, R. (1996) Advances in Knowledge Discovery and Data Mining, The MIT Press, 560p.

Fisher, P. (1997) The Pixel, A Snare and a Delusion, International Journal of Remote Sensing, 18, 679–685.

Fisher, P. (2010) Remote Sensing of Land Cover as Type 2 Fuzzy Set, Remote Sensing of Environment, 114, 309–321.

Foody, G. (1992) A Fuzzy Sets Approach to the Representation of Vegetation Continua from Remotely Sensed Data: an example of lowland heath, Photogrammetric Engineering and Remote Sensing, 58, 221–225.

Foody, G. (1999) The Continuum of Classification Fuzziness in Thematic Mapping, Photogrammetric Engineering and Remote Sensing, 65, 443–451.

Foody, G. (2000) Estimation of sub-pixel land cover composition in the presence of untrained classes, Computers and Geosciences, 26, 469–478.

Gomez, D. and Montero, J. (2008) Fuzzy Sets in Remote Sensing Classification, Soft Computing, 12, 243–249.

Ibrahim, M., Arora, M. and Ghosh, S. (2005) Estimating and accommodating uncertainty through the soft classification of remote sensing data. International Journal of Remote Sensing, 26, 2995–3007.

Kempton, W. (1984) Interview Methods for Eliciting Fuzzy Categories, Fuzzy Sets and Systems, 14, 43–64.

Klir, G. and Yuan, B. (1995) Fuzzy Sets and Fuzzy Logic-Theory and Application Prentice-Hall, 592p.

Le Hegarat-Mascle, S., Bloch, I., Vidal-Madjar, D., Taconet, O., Normand, M. and Loumagne, C. (1997) Application of Demster-Shafer Evidence Theory to Unsupervised Classification in Multisource Remote Sensing, IEEE Transactions on Geosciences and Remote Sensing, 35, 1018–1031.

Lein, J. (1997) Environmental Decision Making: an information technology approach, Blackwell Science, Malden, MA. 213p.

Lein, J. (2003) Applying Evidential Reasoning Methods to Agricultural Land Cover Classification, International Journal of Remote Sensing, 24, 4161–4180.

Lein, J. (2006) Toward the Rapid Characterization of the Built Environment within the Wildland-Urban Interface: A Soft Classification Strategy, GIScience & Remote Sensing, 43, 179–196.

Leung, Y. (1988) Spatial Analysis and Planning Under Imprecision. North-Holland, Amsterdam, 375p.

Lewis. H., Brown, M., Tatnall, A. (2000) Incorporating Uncertainty in Land Cover Classification from Remote Sensing Imagery, Advances in Space Research, 26, 1123–1126.

Lu, D. and Wend, Q. (2006) Specral mixture analysis of ASTER images for examining the relationship between urban thermal features and biophysical descriptors in Indianapolis, Indiana, USA, Remote Sensing of Environment, 104, 157–167.

Mantara, R. (1990) Approximate Reasoning Models, Ellis Horwood, Chichester, 287p.

Maselli F, Conese C, De Filippis T, Norcini S. (1995) Estimation of forest parameters through fuzzy classification of TM data. IEEE Trans Geoscience Remote Sensing 33, 77–84.

Melgani, F., Al Hasemy, B. and Taha, S. (2000) An Explicit Fuzzy Supervised Classification Method for Multispectral remote Sensing Images, IEEE Trans Geoscience and Remote Sensing, 38, 287–295.

Mertikas, P. and Zervakis, M. (2001) Exemplifying the Theory of Evidence in Remote Sensing Image Classification, International Journal of Remote Sensing, 22, 1081–1095.

Metternicht, G. (2003) Categorical Fuzziness: a comparison between crisp and fuzzy set boundaries for Mapping salt-affected Soils, Ecological Modeling 168, 371–389.

Nachtegael, M., Van der Werken, D., Kerre, E. and Philips, W. (2007) Soft Computing in Image Processing, Springer, 497p.

Ozesmi, S. and Bauer, M. (2002) Satellite remote sensing of wetlands Wetlands Ecology and Management, 10, Number 5, 381–402.

Powell, R. Roberts, D., Dennison, P. and Hess, L. (2007) Sub-pixel Mapping of Urban Land Cover Using Multiple Endmember spectral mixture Analysis, Remote Sensing of Environment, 106, 253–267.

Quarmby, N., Townshend, J., Settle, J., White, K., Milnes, M., Hindle, T., Silleos, N. (1992) Linear mixture modelling applied to AVHRR data for crop area estimation, International Journal of Remote Sensing, vol. 13, issue 3, pp. 415–425.

Robinson, V. (2003) A Perspective on the Fundamentals of Fuzzy Sets and their Use in Geographic Information Systems, Transactions in GIS. 7, 3–30.

Rashed, T., Weeks, J., Gadalla, M. (2001) Revealing the Anatomy of Cities through Spectral Mixture Analysis of Multispectral Satellite Imagery: A Case Study of the Greater Cairo Region, Egypt., Geocarto International, 16, 5–16.

Rocchini, D. and Ricotta, C. (2007) Are Landscape as Crisp as We May Think? Ecological Modelling, 204, 535–539.

Sabol, D., Adams, J., Smith, M. (1992) Quantitative subpixel spectral detection of targets in multispectral images, Journal of Geophysical Research, 97, 2659–2672.

Schowengerdt, R. (1996) Soft Classification and Spatial-spectral mixing in Binaghi, E., Brivio, P. and Rampini, A. (eds) Soft Computing in Remote Sensing Data Analysis, World Scientific, Singapore, 1–6.

Settle, J. and Drake, N. (1993) Linear mixing and the estimation of ground cover proportions, International Journal of Remote Sensing, 14, 1159–1177.

Shafer, G. (1976) A Mathematical Theory of Evidence, Princeton University Press, Princeton, NJ, 314p.

Small, C. (2001) Estimation of urban vegetation abundance by spectral mixture analysis, Intenaional Journal of Remote Sensing, 22, 1305–1334.

Small, C. (2002) Mutlitemporal analysis of of urban reflectance, Remote Sensing of Environment, 81, 427–442.

Song, C. (2005) Spectral Mixture Analysis for Subpixel Vegetation Fractions in the Urban Environment, Remote Sensing of Environment, 95, 248–263.

Townsend, P. (2000) A Quantitative Fuzzy Approach to Assess Mapped Vegetation Classifications for Ecological Applications, Remote Sensing of Environment, 72, 253–267.

Tso, B. and Mather, P. (2001) Classification Methods for Remotely Sensed Data, Taylor and Francis, London, 332p.

Turksen, I. (1991) Measurement of Membership Functions and Their Acquisition, Fuzzy Sets and Systems, 40, 5–38.

Watanabe, N. (1979) Statistical Methods for Estimating Membership Functions, Japanese Journal of Fuzzy Theory and Systems 5, 17–25.

Wang, F. (1990) Fuzzy Supervised Classification of Remote Sensing Images. IEEE Trans Geoscience Remote Sensing, 28, 194–201.

Wither, S. (2002) Quantitative Methods: Bayesian Inference, Bayesian Thinking, Progress in Human geography, 26, 553–566.

Wu, C. and Murray, A. (2003) Estimating impervious surface distribution by spectral mixture analysis, Remote sensing of Environment, 84, 493–505.

Wu, G. (2004) Normalized spectral mixture analysis for monitoring urban composition using EMT + imager, Remote Sensing of Environment, 93, 480–492.

Zaheh, L. (1988) Fuzzy Logic, Computer, 21, 83–92.

Zadeh, L. (1994) Fuzzy Logic, Neural Networks and Soft Computing, Communications of the ACM, 37, 77–84.

Chapter 6
Environmental Characterization

Environmental process and the products of human decision making that assume discernable geographic patterns are actors on a landscape that exhibit behaviors expressed as a function of time. The temporal dimension, while implied in most remote sensing investigations, becomes explicit when concern is focused on the question of change. Change is a familiar theme in the context of applied remote sensing whether couched in terms of land cover, land use, or more specific environmental conditions (Walker and Peters 2007; Berberoglu and Akin 2009). As a concept, change is perhaps on the more intriguing subjects in science, if only because its presence illustrates the fact that events do not simply occur in the here and now, but along a continuum of constant flux and adjustment. The challenge to us is recognizing when change has occurred and connecting together the myriad pathways and processes that conspired to produce a difference in the status of the landscape that we now observe. Tracing the pathways followed to a changed system is fundamentally more involved than the present looking backward into the past with perfect knowledge, but the more complicated problem of projecting the present into tomorrow.

At present, there is a sense of urgency regarding our environmental future. However, the study of change begins much earlier with a detailed understanding of the initial state of our environmental system. From this perspective, the status and the active processes that define our environment are described and organized into a framework that explains its present disposition and captures the dynamic nature of those processes active in the landscape. Although the concept of change is important to this discussion, the focus of this chapter is directed at this beginning phase of our appreciation of change. For the purposes of our discussion, we can call this point of beginning as the baseline phase in which the attention is restricted to the current status of the environment and the identification of those attribute that characterize the system. The questions we ask of our data are twofold and interrelated:

1. How do we describe the qualities or peculiarities of the environmental system?
2. Which attributes mark or distinguish this system?

J.K. Lein, *Environmental Sensing: Analytical Techniques for Earth Observation*,
DOI 10.1007/978-1-4614-0143-8_6,

In this chapter, we will examine the role satellite remote sensing plays in answering these questions; exploring the methods and techniques that contribute information to assist in the characterization of our environment, from the derivation of environmental indicators to the incorporation of remotely sensed technology in the implementation of environmental monitoring systems.

6.1 The Characterization Question

Environmental characterization may be defined as the systematic collection of attributes describing the qualities that give distinction to a specific geographic area. Measurement is an essential aspect of the characterization process and methods that simplify procedures and enable rapid appraisal of the environment can greatly benefit the characterization activities. The term "site characterization" was introduced by the nuclear waste industry but has expanded over time to describe a range of pursuits undertaken to gather hydrogeologic information, guide design remediation, evaluation site restoration strategies, and support the environmental impact assessment process (Cook 2006; Artiola et al. 2004). As an investigative activity, characterization involves the collection, analysis, and interpretation of data together with the application of the knowledge gained from that data to evaluate the support of resource potential of a site. Typically, site characterization examines natural phenomena and those human-induced conditions important to the definition of the local environment through a combination of sampling and monitoring activities (Kays 2000). The goal of characterization procedures is to develop an operational concept of the site and to identify those aspects of the area that influence both environmental functioning and the performance of human constructions committed to that location.

The data needed to guide environmental characterization will vary, but tend to explain the controlling variables that given the landscape its definition. These natural factors are selected to describe the baseline against which change can be evaluated. The idea of a "baseline" or beginning point is a central concept in environmental characterization. We can think of baseline information addressing two fundamental questions: (1) what is there, and (2) what does it mean. From a basic accounting of the locale under consideration whose present disposition may be poorly understood, characterization structures the salient environmental variables into a "context" (Lein 2006). Therefore, to be useful, these natural factors should aptly typify the environment, providing sufficient detail to inform and demonstrate relevance to the motivating problem. The environmental attributes that are critical to a site characterization may be placed into two broad categories (Table 6.1). When viewed collectively the elements listed can be conceptualized as a series of themes that relate and interact over time and area. A careful review of Table 6.1 will reveal that some aspect of each of the attributes listed can be acquired through remote sensing either directly as the product of image classification or through the application of one or more environmental indicators designed to identify the phenomena or explain trends that may not be readily perceptible.

Table 6.1 Environmental factors relevant to site planning and inventory

Natural elements
Physiography: slope, unique features
Geology: bedrock formations, faults, fractures, slumps, slips
Soils: type, composition, permeability, erosion potential
Hydrology: drainage systems, springs, seeps, wetlands
Vegetation: plant associations, unique communities
Wildlife: habitats
Climate: temperature, precipitation, wind flow, humidity, evaporation
Cultural elements
Transportation: roads, rail lines, airports
Utilities: oil, gas, electric, water networks
Structures and excavations: buildings, mines, dumps
Historical/archeological sites

In some respects, we can think of environmental characterization as a type of landscape inventory with a more directed and rigorous focus, where we wish to document the opportunities, constraints, and sensitivities inherent to the landscape to form a systematic understanding of the environmental system at both a synoptic scale and a local scale.

From a remote sensing perspective, effective environmental characterization depends on the selection of what to characterize and the methods required in order achieving environmental understanding (Herold et al. 2006). Although the selection problem will vary according to purpose, and in some instances may be predetermined by policy directives, typically characterization is built around the dominant or preeminent features of the landscape system. Given the multidimensional nature of the environment, selection is no simple task; however, detailed consideration is often shown to:

- Habitats found within or adjacent to the focus of interest
- Major cultural resources
- Major land uses and related land use activities
- Major geologic and geomorphic features

A more comprehensive listing is suggested in Table 6.2. Characterizing each of the factors presented in the table usually demands a combination of mapping and field data collection, which can limit the applicability of remote sensing technology in some cases. However, for those variables that can be measured remotely, how measurement proceeds becomes a critical next step in the process. The answer to this procedural issue is closely related to the level of detail needed based on the nature of the characterization problem. This level of detail in large part determines the degree of spatial and spectral resolution that must be available to match this fundamental data requirement. With specific reference to the required level of

Table 6.2 Comprehensive listing of environmental characterization targets

A. *Physical and chemical characteristics*	
1. Earth	3. Atmosphere
(a) Mineral resources	(a) Quality (gases, particulates)
(b) Construction material	(b) Climate (micro, macro)
(c) Soils	(c) Temperature
(d) Landform	4. Processes
(e) Force fields and background radiation	(a) Floods
(f) Unique physical features	(b) Erosion
2. Water	(c) Deposition (sedimentation, precipitation)
(a) Surface	(d) Solution
(b) Ocean	(e) Sorption (ion exchange, complexing)
(c) Underground	(f) Compaction and settling
(d) Quality	(g) Stability (slides, slumps)
(e) Temperature	(h) Stress–strain (earthquake)
(g) Snow, ice, and permafrost	(f) Recharge
	(i) Air movements
B. *Biological conditions*	
1. Flora	2. Fauna
(a) Trees	(a) Birds
(b) Shrubs	(b) Land animals including reptiles
(c) Grass	(c) Fish and shellfish
(d) Crops	(d) Benthic organisms
(e) Microflora	(e) Insects
(f) Aquatic plants	(f) Microfauna
(g) Endangered species	(g) Endangered species
(h) Barriers	(h) Barriers
(i) Corridors	(i) Corridors
C. *Cultural factors*	
1. Land use	3. Esthetics and human interest
(a) Wilderness and open spaces	(a) Scenic views and vistas
(b) Wetlands	(b) Wilderness qualities
(c) Forestry	(c) Open space qualities
(d) Grazing	(d) Landscape design
(e) Agriculture	(e) Unique physical features
(f) Residential	(f) Parks and reserves
(g) Commercial	(g) Monuments
(h) Industrial	(h) Rare and unique species or ecosystems
(i) Mining and quarrying	(i) Historical or archeological sites and objects
2. Recreation	(j) Presence of misfits
(a) Hunting	4. Cultural status
(b) Fishing	(a) Cultural patterns (life style)
(c) Boating	(b) Health and safety
(d) Swimming	(c) Employment
(e) Camping and hiking	(d) Population density
(f) Picnicking	
(g) Resorts	

(continued)

Table 6.2 (continued)

5. Man-made facilities and activities	
(a) Structures	
(b) Transportation network (movement, access)	
(c) Utility networks	
(d) Waste disposal	
(e) Barriers	
(f) Corridors	
D. *Ecological relationships*	
(a) Salinization of water resources	(e) Salinization of surfacial material
(b) Eutrophication	(f) Brush encroachment (g) Other
(c) Disease–insect vectors	(g) Other
(d) Food chains	
E. *Modification regimes*	
1. Alterations	5. Land alteration
(a) Exotic flora or fauna introduction	(a) Erosion control and terracing
(b) Biological controls	(b) Mine sealing and waste control
(c) Modification of habitat	(c) Strip mining rehabilitation
(d) Alteration of ground cover	(d) Landscaping
(e) Alteration of ground-water hydrology	(e) Harbor dredging
(f) Alteration of drainage	(f) Marsh fill and drainage
(g) River control and flow codification	6. Resource renewal
(h) Canalization	(a) Reforestation
(i) Irrigation	(b) Wildlife stocking and management
(j) Weather modification	(c) Ground-water recharge
(k) Burning	(d) Fertilization application
(l) Surface or paving	(e) Waste recycling
(m) Noise and vibration	7. Changes in traffic
2. Land transformation	(a) Railway
(a) Urbanization	(b) Automobile
(b) Industrial sites and buildings	(c) Trucking
(c) Airports	(d) Shipping
(d) Highways and bridges	(e) Aircraft
(e) Roads and trails	(f) River and canal traffic
(f) Railroads	(g) Pleasure boating
(g) Cables and lifts	(h) Trails
(h) Transmission lines, pipelines, and corridors	(i) Cables and lifts
(i) Barriers, including fencing	(j) Communication
(j) Channel dredging and straightening	(k) Pipeline
(k) Channel revetments	8. Waste emplacement and treatment
(l) Canals	(a) Ocean dumping
(m) Dams and impoundments	(b) Landfill
(n) Piers, seawalls, marinas, and sea terminals	(c) Emplacement of tailings, spoil, and overburden

(continued)

Table 6.2 (continued)

(o) Offshore structures
(p) Recreational structures
(q) Blasting and drilling
(r) Cut and fill
(s) Tunnels and underground structures

3. Resource extraction
   (a) Blasting and drilling
   (b) Surface excavation
   (c) Subsurface excavation and retorting
   (d) Well drilling and fluid removal
   (e) Dredging
   (f) Clear cutting and other lumbering
   (g) Commercial fishing and hunting
4. Processing
   (a) Farming
   (b) Ranching and grazing
   (c) Feed lots
   (d) Dairying
   (e) Energy generation
   (f) Mineral processing
   (g) Metallurgical industry
   (h) Chemical industry
   (i) Textile industry
   (j) Automobile and aircraft
   (k) Oil refining
   (l) Food
   (m) Lumbering
   (n) Pulp and paper
   (o) Product storage

(d) Underground storage
(e) Junk disposal
(f) Oil-well flooding
(g) Deep-well emplacement
(h) Cooling-water discharge
(i) Municipal waste discharge including spray irrigation
(j) Liquid effluent discharge
(k) Stabilization and oxidation ponds
(l) Septic tanks: commercial and domestic
(m) Stack and exhaust emission
(n) Spent lubricants

9. Chemical treatment
   (a) Fertilization
   (b) Chemical deicing of highways, etc.
   (c) Chemical stabilization of soil
   (d) Weed control
   (e) Insect control (pesticides)
10. Accidents
   (a) Explosions
   (b) Spills and leaks
   (c) Operational failure

information content, remote sensing data collection and image processing can be targeted at one of three geographic of details:

1. Reconnaissance level appropriate for describing regional scale patterns or factors that explain "large-scale" regional trends
2. Mesoscale definitions oriented toward more generic inventory of variable at the local level
3. Site-specific descriptions providing detailed analysis needed to guide citing decisions and environmental impact concerns

The necessary features of the sensor to deliver data follow based on its ability to supply measurement at the requisite level of discreteness (Table 6.3). A careful review of Table 6.3 shows that sensor systems can be indentified that collect measurements with a spatial precision sufficient in most instances to support characterization objectives (Melesse et al. 2007). Transforming these raw spectral measures into information that is meaningful to the characterization problem, calls on the image analyst's skills to generate customized information products from the

Table 6.3 Sensor resolution requirements for selected classification level discreteness

Land use/cover	Temporal (years)	Spatial (m)	Spectral
USGS Level 1	5–10	20–100	VIS–NIR
USGS Level 2	5–10	5–20	VIS–NIR
USGS Level 3	3–5	1–5	Pan–VIS–NIR
USGS Level 4	1–3	0.25–1	Pan

imagery. In the sections to follow several of the more commonly used remote sensing techniques for deriving specialized information products germane to environmental characterization are examined.

6.2 Classification as Characterization

Classification is an operation that seeks to organize observations into a set of mutually exclusive and exhaustive categories based on one or more unifying criteria. As a logical procedure the complexities of the landscape is simplified and structured into new representations that improve communication and enhances understanding. As we recall from Chap. 3, observations once arranged according to a system of classification reveal patterns in the data, which can be expressed geographically and compared over time. Within the context of environmental characterization, the structure imposed by a classification system serves to identify the presence of specific environmental phenomena, documenting their geographic consequence and selecting their locations from the background clutter of the landscape. The essential ingredient in this formula is the classification system we select to use and its relationship between spectral measurement and the typology of the constructs it embodies that organizes environmental features into categories.

Perhaps the best known example of a classification system designed exclusively to organize spectral measurements into nominal information categories is the Anderson–Hardy system developed for the United States Geological Survey (Anderson et al. 1976). This method of land cover/land use classification serves to illustrate a logic that translates spectral data obtained via supervised or unsupervised image analysis into carefully defined classes. Based on the underlying principle of descending levels of discreteness, this system moves from the general categorization of broad land cover themes to increasingly more specific and detailed definitions of land use. As a resolution-depended method, the spectral data fall into informational classes that facilitate the representation of the biophysical qualities that explain the land surface (Table 6.4). Since its introduction over four decades ago, other classification systems have been created to support general land cover characterization needs. Representative examples include the national vegetation classification system (NVCS) and the national land cover characterization (NLCC) system (The Nature Conservancy 1994; Homer et al. 2004).

Table 6.4 General examples of land cover classes and their definitions

Class	Definition
Developed land	This class is composed of areas of intensive anthropogenic use. Much of the land is covered by structures and impervious surfaces. Anderson et al. (1976) called these areas "Urban or Built-up Land" although the definition clearly included suburban and rural areas. Included in this category are cities; towns; villages; strip developments along highways; transportation, power, and communications facilities; and areas such as those occupied by mills, shopping centers, industrial and commercial complexes, and institutions that may, in some instances, be isolated from urban areas
Cultivated land	Agricultural land used primarily for production of food and fiber. On high-altitude imagery, the chief indications of agricultural activity will be distinctive geometric field and road patterns on the landscape and the traces produced by livestock or mechanized equipment
Grassland	The grassland category includes lands covered by natural and managed herbaceous cover. Grassland historically has been defined as land where the potential natural vegetation is predominantly grasses, grass-like plants, and forbs, and where natural herbivore was an important influence in its precivilization state. Some grasslands have been or may be seeded to introduce or domesticate plant species
Woody land	The woody land class includes any species with an aerial stem that persists for more than one season. The woody class is divided into three subclasses: 1.41, deciduous; 1.42, evergreen; and 1.43, mixed
Bare land	Bare land is composed of bare rock, sand, silt, gravel, or other earthen material with little or no vegetation regardless of its inherent ability to support life. Vegetation, if present, is more widely spaced and scrubby than that in the vegetated categories. Unusual conditions, such as a heavy rainfall, occasionally may result in a short-lived, luxuriant plant cover. Wet, nonvegetated exposed lands are included in the wetland categories
Tundra	Tundra is the term applied to the treeless cover beyond the latitudinal limit of the boreal forest in poleward regions and above the elevation range of the boreal forest in high mountains. The vegetative cover of the tundra is low and dwarfed and often forms a continuous mat. Plant characteristics are an adaptation to an extreme physical environment in which temperatures may average above freezing only 1 or 2 months out of each year, strong desiccating winds may occur, great variation exists in solar energy, and permafrost is ubiquitous beneath the surface

While the nature of these classification systems differs, each recognizes the important role land cover enjoys as an "index" of environmental conditions. Land cover, as a focus of characterization, defines the observable, biophysical cover of the Earth's surface (Herold et al. 2006). Included in this definition are the vegetative and human-made features that populate the landscape as well as the bare rock, exposed soil, inland water surface, and other elements forming the landscape's skin. Taken in total, the status and arrangement of a land cover category becomes an important indicator of underlying environmental process and existing ecological conditions. Therefore, at this fundamental level, land cover is not just an observable element in a characterization study, but an obvious and detectable signal of both natural process and human interventions at the surface. At present, land cover characterization is hampered by the absence of mapping standards, and none of the current systems of land cover classification have gained international acceptance (Di Gregorio and Jansen 2000). The emergence of the land cover characterization system (LCCS) method may move the characterization issue closer toward a standard solution. The LCCS is a comprehensive a-priori classification system containing systematic and strict class boundary definitions, yet it maintains important flexibility to meet a variety of user needs (DiGregorio 2005).

As noted in Di Gregorio and Jansen (2000), the classification uses a set of independent diagnostic criteria that allow correlation with existing classifications and legends. Land cover classes are defined by a combination of a set of independent diagnostic criteria that are hierarchically arranged to assure a high degree of geographical accuracy. Because of the heterogeneity of land cover, the same set of classifiers cannot be used to define all land cover types. The hierarchical structure of the classifiers may differ from one land cover type to another. Therefore, the classification has two main phases:

- An initial *dichotomous phase*, where eight major land cover types are distinguished
- A subsequent *modular-hierarchical phase* where the set of classifiers and their hierarchical arrangement are tailored to the major land cover type

The classification system leads to mutually exclusive land cover classes, which comprise of: (1) a unique Boolean formula (a coded string of classifiers used); (2) a standard name; and (3) a unique numerical code. Both the numerical code and standard name can be used to build an automatically generated legend, with the classes created grouped according to the main land cover categories and their domains according to the level of detail. The nomenclature can be linked to a user-defined name in any language.

Further, the definition of the land cover class can be achieved by adding attributes. Two types of attributes, which form separate levels in the classification, are distinguished:

- *Environmental attributes*: These are attributes (e.g., climate, landform, altitude, soil, lithology, and erosion) which influence land cover but are not inherent features of it and should not be mixed with "pure" land cover classifiers.
- *Specific technical attributes*: These are associated with specific technical disciplines (e.g., for (semi-)natural vegetation, the floristic aspect can be added; for cultivated areas, the crop type; and for bare Soil, the soil type).

The advantages of the classifier approach are found in its highly flexible approach in which each land cover class is clearly and systematically defined, providing essential internal consistency. In addition, the system is truly hierarchical and applicable at a variety of scales. Rearrangement of the classes based on regrouping of the classifiers used facilitates extensive use of the outputs by a wide variety of end users. Also, accuracy assessment of the end product can be generated by a class or by the individual classifiers forming the class, and all land covers can be accommodated in this highly flexible system. An overview of the LCCS (DiGregorio 2005) is given in Fig. 6.1. Applying this system produces land

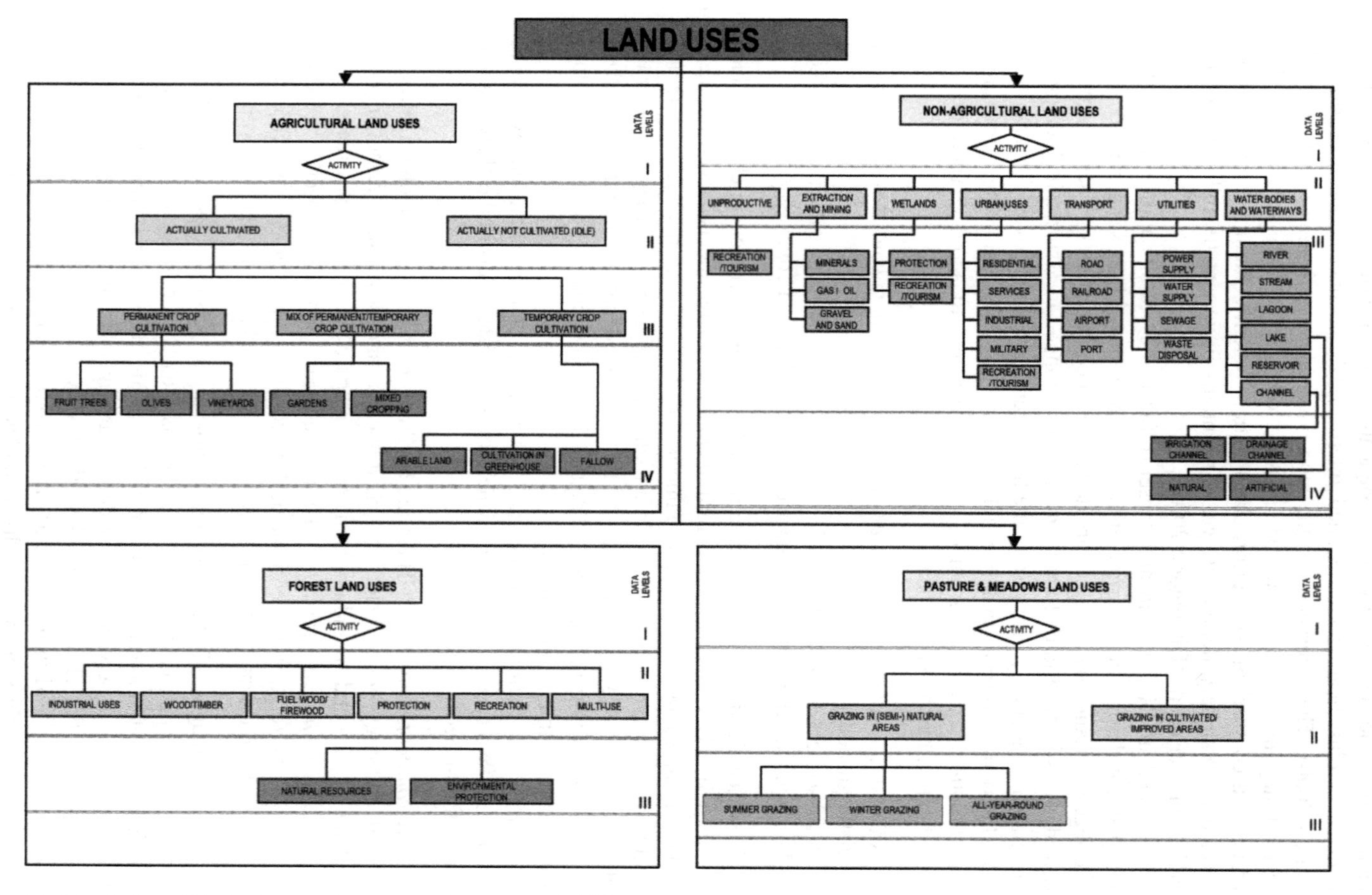

Fig. 6.1 The design of the land cover characterization system (LCCS)

cover mapping results that identify biophysical surface types as measurement endpoints whose values quantify key environmental relationships moving land cover into the status of an indicator.

6.3 Indicator Science

Environmental indicators are instruments designed to quickly and easily inform a target audience about the status of an object of interest (Lein 2004). When abstracted from remotely sensed data, indicators serve to communicate information about environmental conditions and, over time, about significant changes and trends that are actively reshaping the landscape (National Research Council 2000). Communication is perhaps the most important function of an indicator, and to be effective in this role, an indicator should enable or promote information exchange regarding the characteristics it has been designed to address (Smeets and Weterings 1999). Because the effective communication requires simplicity, useful indicators strive to reduce the complexity of a specific condition or situation by focusing attention on those aspects of the environment that are considered to be the most relevant. In relation to the goals of environmental sensing, indicators serve three critical functions:

1. They supply information on the status and condition of the environment.
2. They support management and policy decision making.
3. They facilitate monitoring of critical environmental thresholds.

Therefore, by developing indicators and observing their spatio-temporal variation, our ability to identify environmental problems as they evolve is greatly improved. Acquiring the data to capture and clearly illuminate changes in environmental functioning can be problematic (Noss and Cooprrider 1994). This reality emphasizes the importance of selecting the appropriate indicator to meet the challenges of the environmental problem under investigation. To be effective as means of analysis and inquiry, an environmental indicator must possess certain traits; that not only determine its applicability to a problem, but also its scientific validity (Lein 2004; Hammond 1995; Kogan 1995). Several of the more critical attributes that influence the selection of an indicator include (Riley 2000; Dale and Beyeler 2001):

- *General relevance* – Consideration of relevance helps to determine how well the indicator characterizes the environment and facilitates the definition of process and change.
- *Conceptual integrity* – Integrity speaks to the overarching rationale, which may be theoretical or practical, that supports and justifies the use of the indicator.
- *Reliability* – Focus of reliability centers around the question of how successful using the indicator will be now and over time, and the level of explanation that can be delivered based on its use.

Table 6.5 Criteria to guide indicator selection

A clear representation of the indicandum (the indicated issue) by the indicator
A clear proof of relevant cause–effect relations
An optimal sensitivity of the representation
Information for adequate spatio-temporal scales
A very high transparency of the derivation strategy
A high degree of validity and representativeness
A high degree of comparability in and with indicator sets
An optimal degree of aggregation
A good fulfillment of statistical requirements concerning verification, reproduction, and validity
Information and estimations of the normative loadings
High political relevance concerning the decision process
High comprehensibility and public transparency
Direct relations to management actions
An orientation towards environmental targets
A high utility for early warning purposes
A satisfying measurability
A high degree of data availability
Information on long-term trends of development

- *Scale appropriateness* – Scale directs our attention to the ability of the indicator to detect the desired environmental quality at the appropriate temporal and spatial scale as dictated by the problem or purpose.
- *Statistical sensitivity* – Sensitivity relates to the level of measurement precision and accuracy that can obtained from the indicator as well as the level of confidence that can be ascribed to the results it produces when applied.
- *Robustness* – Focusing on the potential of the indicator to produce consistent results under a range of external conditions and environmental perturbations, robustness directs a selection to look critically at those factors that influence its capacity to deliver useful measures of the environment.

In a purely pragmatic sense, relevance ultimately narrows down to the indicator's ability to quantify and simplify information in a cost-effective manner (Babu and Reidhead 2000; Landres 1992). With these basic considerations kept in mind, criteria can be offered to guide indicator selection in a systematic way (Niemeijer and deGroot 2008). Several of these criteria more pertinent to environmental remote sensing are given in Table 6.5.

Once an indicator has been chosen, the main concern surround its use within the context of environmental analysis tends to be conceptual in nature. As noted in our discussion previously, an environmental indicator must be sufficiently complete to capture key expressions of process without becoming overly complex. We can draw on a conceptual model to enhance our understanding of the environmental system in which indicators, as surrogates for more complicated actors, can be connected to a process-oriented view of the landscape. One useful model that offers a lucid framework to achieve this linkage is expressed in the DPSIR model (Fig. 6.2). The DPSIR model is a causal framework for describing the interactions between

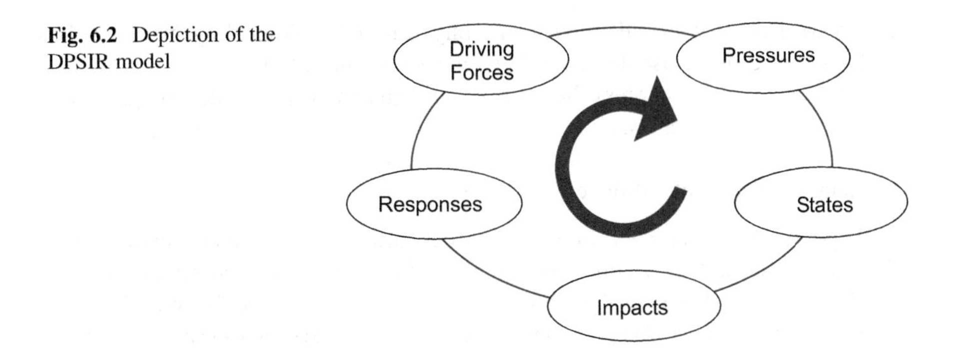

Fig. 6.2 Depiction of the DPSIR model

society and the environment. This framework, developed by the European Environment Agency adopts a systems view of environmental process, simplifying process and change as a function of the interaction among five main components:

- *D*riving forces
- *P*ressures
- *S*tates
- *I*mpacts
- *R*esponses

Applying this model in an environmental sensing application structures the need for information and concentrates the selection of indicators on:

- The identification and monitoring of driving forces
- The resulting environmental pressures
- The baseline state of the environmental system
- The impacts resulting from changes in the environmental functioning and quality
- The societal response to environmental change

Ideally environmental indicators are called on to capture each element of this casual network. In this manner, the DPSIR framework becomes a useful device for both describing the relationships between the origins of an environmental problem and its consequences.

A typology of indicators has been offered by Smeets and Weterings (1999) that can assist both the selection problem and integrate chosen indicator into the DPSIR framework. Although simple, the typology is appealing because it relates the indicator to type of information it provides for a given environmental situation. According to this typology the initial class of indicators, Type A, is purely descriptive in nature. Information obtained from these metrics address the general questions of "what is happening to the environment and human populations." Descriptive indicators of this type can be refined further into four subcategories:

1. *Driving force indicators* – describing social/demographic forces that provoke changes in the production or consumption of resources that exert pressure on the environment

2. *Pressure indicators* – define specific physical and biological agents as well as land and resource use that contribute emissions and releases
3. *State indicators* – express the quantity and quality of physical, biological, and chemical phenomena such as temperature or carbon dioxide concentrations
4. *Response indicators* – identify how groups and individuals respond to or adapt to changes in the state of the environment

Type B indicators refer to those metrics designed to evaluate performance. A performance indicator compares actual conditions to a specific set of reference conditions. Through the use of performance indicators the difference between the actual condition, the reference state, and the desired state can be measured and assessed. In this way, progress toward specific goals or policy objectives can be evaluated. Performance indicators can be based on a variety of reference conditions, but typically involve policy targets established by national or international instruments. Type C indicators define metrics that measure effectiveness. Efficiency indicators address the broad question: "are we improving," while the last category in this typology, Type D indicators, explains metrics designed to measure improvements in total welfare.

6.4 Environmental Indicators and Transforms

Recognizing that many physical and human elements of the environment can be measured using remote sensing, satellite technology plays a vital role in the development and application of indicators. The range of indices derived from the remotely sensed data is impressive and continue to expand as new sensor systems are introduced and as indicator science progresses. Satellite-derived indicators that support the environmental characterization problem fall into four general categories: (1) vegetation transforms, (2) landscape metrics, (3) customized band ratios, and (4) statistical approximations.

6.4.1 Vegetation Transform

A vegetation transform is an index derived from important spectral characteristics of vegetation. The index is a quantitative measure generated by an algebraic combination of specific spectral bands that define a relationship to the status of vegetation in a pixel. This summary value forms an expression of vegetation biomass or plant vigor, serving as an indicator of those environmental qualities that influence ecosystem vitality, and is often used as a surrogate of climate and other environmental factors that affect ecosystem conditions. Vegetation transforms are supported by empirical evidence that suggest that the cells in plant leaves are effective at scattering electromagnetic radiation. Scattering at the level of the leaf is due to the high contrasts in refraction between water-rich cells and intercellular air spaces. When viewed spectrally vegetation appears dark in the visible portion of

Table 6.6 Selected vegetation transforms

Vegetation index	Defining equation
Band ratio (RATIO)	(NIR/RED)
Normalized difference vegetation index (NDVI)	(NIR − RED)/(NIR + RED)
Transformed vegetation index (TVI)	(NDVI + 0.5)0.5
Infrared percentage vegetation index (IPVI)	NIR/(NIR + RED)
Perpendicular vegetation index (PVI)	[NIR − a_1(RED) − a_0]/[1 + (−a_1)(0.5)]
Soil-adjusted vegetation index (SAVI)	[(NIR − RED)/(NIR + RED + L)] × (1 + L)
Transformed soil-adjusted vegetation index (TSAVI)	a_1[NIR − a_1(RED) − a_0]/ [RED + a_1(NIR) − $a_1 \times a_0$]
Modified soil-adjusted vegetation index (MSAVI)	(2NIR + 1 − [(2NIR + 1)2 − 8(NIR − RED)] 0.5)/2
Difference vegetation index (DVI)	NIR − RED
Renormalized difference vegetation index (RDVI)	(NDVI × DVI)0.5
Weighted difference vegetation index (WDVI)	NIR − a_1 × RED

NIR, crop reflectance in the near infrared band (TM4); RED, crop reflectance in the red band (TM3); L, constant (taken as 0.5), and a_0, a_1 are the intercept and slope of the soil line, respectively

the spectrum (0.4–0.7 μm) resulting from the high absorption of energy by leaf pigments, such as chlorophyll, proto-chlorophyll, and xanthophylls. In the green portion of the spectrum, reflectivity increases since these pigments are less absorptive. Consequently, when examined within the 0.7–1.3 μm portion of the spectrum, plant material appears bright since absorption is weaker and scattering increases. Beyond 1.3 μm to approximately 2.5 μm vegetation appears dark once again primarily the result of leaf water, cellulose, and lignin absorption. From this comparatively simple relationship, one can infer that any internal or external influence that modifies water in the leaf, cellulose production, and the presence of plant pigments will be detectable owing to changes in the patterns of absorption and reflectance they produce. Therefore, vegetation transforms are proportional to the value of key biophysical parameters, such as leaf area (as quantified by the leaf area index, LAI), green vegetation fraction, net primary productivity, and the fraction of absorbed photosynthetically active radiation. Drawing on these parameters, vegetation transforms poses the capacity to characterize the structure and dynamics of ecosystem; a capability that has been well documented in the remote sensing literature (Xie et al. 2008; Nagler et al. 2001). A selection of commonly applied vegetation transforms is given in Table 6.6.

Although this topic remains an active area of research, a selective review of vegetation transforms permits us to examine their utility to the task of environmental characterization is more detail. This cursory review focuses on the normalized difference vegetation index (NDVI), the Kauth–Thomas transform, and the leaf area index (LAI).

1. *The Normalized Difference Vegetation Index (NDVI)* – The NDVI dates to the early work introduced by Kriegler et al. (1969) and Rouse et al. (1974).

Fig. 6.3 An NDVI image

This index is calculated based on the interaction between the visible and near-infrared portion of the electromagnetic spectrum as reflected by vegetation (Fig. 6.3). According to this relationship, healthy vegetation absorbs most of the incident energy in the visible wavelengths and reflects a large portion of energy in the near-infrared bands. The NDVI captures this relationship and, because of the ratioing involved in its calculation, the index tends to reduce several sources of noise in the imagery that originates from variations in solar illumination, cloud shadow, and topography. The index is calculated based on the formula:

$$\mathrm{NDVI} = (\mathrm{NIR} - \mathrm{RED})/(\mathrm{NIR} + \mathrm{RED}),$$

where NIR stands for spectral measures acquired in the near-infrared region of the spectrum, and RED explains spectral measures taken within wavelengths defining the red portion of the spectrum. The spectral reflectances are ratios of the reflected over incoming radiation in each spectral band. Typically values of this ratio assume values between −1.0 and +1.0. Although the index is dimensionless, NDVI values for dense vegetation tend to migrate toward positive values from 0.3 to 0.8 whereas barren or snow covered surfaces display NDVI values in the negative range below 0.0.

There are several noteworthy derivations of the NDVI formula. Three of the more common ones include:

(a) *Transformed Vegetation Index (TVI)* – Introduced by Deering et al. 1975, the TVI was created by adding a constant to the NDVI and taking the square root such that:

$$\mathrm{TVI} = \sqrt{(\mathrm{NDVI} + 0.5)}$$

Which avoids problems related to negative values, even though negative values may still result at NDVI $= -0.5$. In general, a TVI ≤ 0.71 indicates a nonvegetated surface.

(b) *Soil-Adjusted Vegetation Index (SAVI)* – Developed by Huete 1988, the SAVI is designed to minimize the effects of soil background noise by introducing a soil adjustment factor (L) such that:

$$\mathrm{SAVI} = [(\mathrm{NIR} - \mathrm{Red})/(\mathrm{NIR} + \mathrm{Red}) + \mathrm{I}] \times (1 + L),$$

where L ranges from 1.0 for low density vegetative cover to values of 0.25 for densely vegetated surfaces.

(c) *Enhanced Vegetation Index (EVI)* – An optimized index designed to enhance the vegetation signal by improving sensitivity to high biomass, reducing canopy background noise, and atmospheric influences. The EVI is computed from the formula:

$$\mathrm{EVI} = 2.5 \times [(\mathrm{NIR} - \mathrm{red})/\mathrm{NIR} + C_1 \times \mathrm{Red} - C_2 \times \mathrm{Blue} + L)],$$

where NIR, Red and Blue are the atmospherically correct surfaces reflectance in the near-infrared, red and blue bands, respectively, L is the canopy background adjustment factor, and C_1 and C_2 are aerosol resistance coefficients. In most applications of the EVI, the parameters used for calculating the index are: $L = 1$, $C_1 = 6$, $C_2 = 7.5$, and G, a gain factor is usually set to 2.5.

2. *Kauth–Thomas Transform* – It is commonly referred to as the Tassle-cap transformation (Kauth and Thomas 1976). The Tassle-cap transformation converts multispectral data into a new set of characteristics that correspond to physical conditions at the surface (Fig. 6.4). Specifically, this transform produces four vegetation indices:

(a) *The Brightness Indicator* – a metric forming out of the weighted sum of all bands and describes the principal variation in soil reflectance.

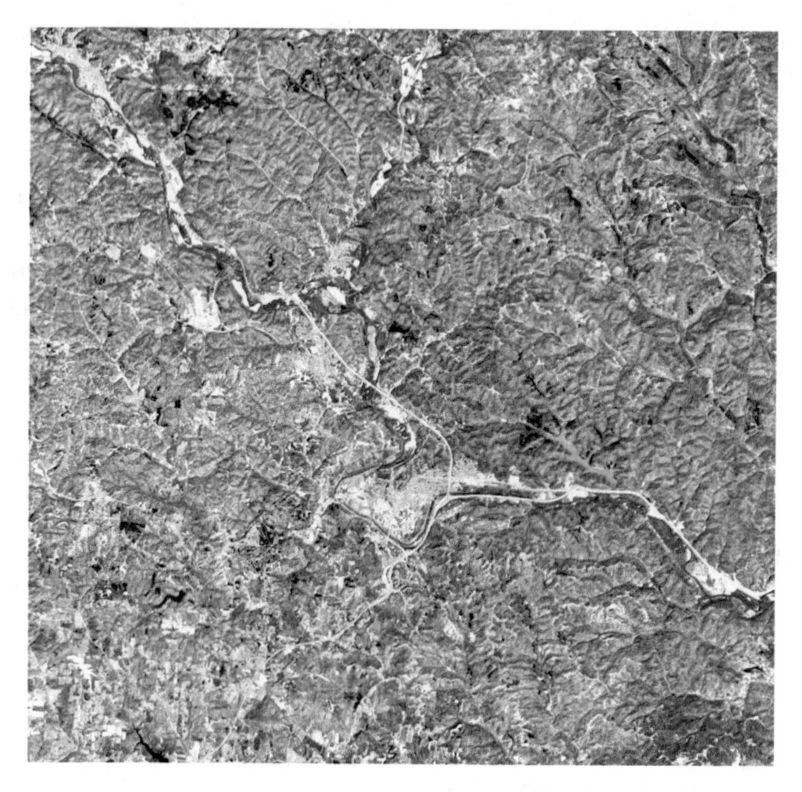

Fig. 6.4 A Tassle-cap greenness surface

(b) *The Greenness Indicator* – an index that exploits the contrast between the NIR and the visible bands, serving as a measure of the density of green vegetation.
(c) *The Wetness Indicator* – a measure that explains the pattern of canopy and soil moisture due to contrasts between short-wave infrared and visible/near infrared reflectance.
(d) *Haze* – an index that essentially expresses noise.

Although originally developed for use with Landsat MSS data, the Kauth–Thomas transform has been extended to other satellite systems including Landsat TM, ASTER and IKONOS. Regardless of platform, this transform projects soil and vegetation through a linear combination of bands. Successful application, however, depends on careful calibration information for the sensor (Kaufman and Tanre 1992).

3. *Leaf Area Index* (*LAI*) – The LAI expresses an important structural property of a plant canopy; the number of equivalent layers of leaves the vegetated surface displays relative to a unit ground area. Because leaves are the fundamental

photosynthetic organs of vegetation, LAI is a useful indicator of carbon dioxide release, interception, net photosynthesis, volume growth, and light penetration through the canopy (Jensen and Binford 2004; Zheng and Moskel 2009). In the majority of applications, LAI is estimated based on a combination of field measurement and statistical methodologies using regression analysis to establish the statistical fit between field-collected samples of LAI and selected spectral bands from the sensor system. For example, in a study of urban LAI for Terra Haute, Indiana, Jensen and Hardin (2005) found the relation:

$$\mathrm{LAI} = 3.99 \times (\mathrm{Green}/\mathrm{Red}) + 0.02 \times \mathrm{IR} - 7.10,$$

which produced a suitable estimator using the Aster SWIR imagery. Other approaches involved deriving LAI from vegetation indices. One example of this alternative solution uses the soil-adjusted vegetation index (SAVI) (Schultz and Engman 2000). According to this approach, LAI is found using the formula:

$$\mathrm{LAI} = -\ln(\mathrm{SAVI} = 0.371)/0.48.$$

6.4.2 Landscape Metrics

Environmental characterization is often concerned with understanding the structure and function of landscapes and how these two qualities change in response to internal and external forces. As fundamental descriptions of landscape pattern, the relationship between structure and function strongly influence ecological processes, biotic abundance, and diversity (Herold et al. 2002). From an analytical perspective, changes in landscape structure alter landscape function and vice-versa. Therefore, a fundamental understanding of the dynamic interactions that form between their two attributes of the environmental system is an essential ingredient of effective environmental planning and management strategies. Characterizing the main structural elements and the fluxes or processes that dominate is central to the definition of landscape function. In practical terms, establishing this relationship also helps illuminate the ecological consequences of human actions on the land surface, serving as integrative indicators of ecological sustainability (Helming et al. 2008; Renetzeder et al. 2010).

The concept of a landscape metric refers exclusively to indicators developed from categorical map patterns. Metrics, taken from the spatial arrangement of thematic data, are essentially algorithms that quantify geographic characteristics of pattern, focusing on either the composition of the map without reference to spatial attributes or the spatial configuration of the map based on the specific arrangements and juxtaposition of map categories. As a characterization device, landscape metrics concentrate attention on the geometric and geographic properties of categorical map patterns at a specific scale of reference (Frohn and Hao 2006). Three important levels of metric quantification can be noted:

1. *Patch-level Metrics* – which are defined for individual patches, and characterize the spatial character and context of the landscape patches
2. *Class-level Metrics* – which are integrated over all the patches of a given type or category
3. *Landscape-level Metrics* – which are integrated over all patch types or categories across the full extent of the landscape (scene). This can be further subdivided into eight measurement groupings:

   (a) Area/density/edge metrics
   (b) Shape metrics
   (c) Core area metrics
   (d) Isolation/proximity metrics
   (e) Contrast metrics
   (f) Contagion/interspersion metrics
   (g) Connectivity metrics
   (h) Diversity metrics

A detailed listing of the available metrics arranged according to measurement group is given in Table 6.7. These metrics are carefully reviewed and extensively described by McGarigal and Marks (1995). A suite of improved landscape metrics that are independent of the characteristic variations inherent to remotely sensed imagery has been introduced by Frohn (1998). These together with the landscape metrics listed in Table 6.7 demonstrate that the list of available metrics is extensive. However, studies have shown that many of these measures are highly correlated with one another, and therefore, their use becomes redundant (Riitters et al. 1995; Cushman et al. 2008). A core set of landscape metrics useful for landscape analysis and environmental monitoring/characterization can be offered (Boteguilha and Ahern 2002; Schindler et al. 2008). This core selection includes the following:

- *Diversity* – expressed according to

$$H = -\text{sum}[p \times \ln(p)],$$

  where p is the proportion of each class in the kernel.
- *Dominance* – a metric calculated according to the formula:

$$D = H_{\max} - H,$$

  where H is a measure of diversity, and $H_{\max}$ is the maximum diversity based on the natural logarithm of the number of different classes present in the kernel.
- *Fragmentation* – expressed according to:

$$F = (n - 1)/(c - 1)$$

Table 6.7 Comprehensive listing of available landscape metrics

Type	Metric
Area/density/edge	Patch area
	Patch perimeter
	Radius of gyration (patch)
	Total area (class)
	Percentage of landscape
	Number of patches
	Patch density
	Total edge
	Edge density
	Landscape shape index
	Normalized landscape shape index
	Largest patch index
	Patch area distribution
	Radius of gyration distribution (class)
	Total area (landscape)
	Number of patches
	Patch density
	Total edge
	Edge density
	Landscape shape index
	Largest patch index
	Patch area distribution
	Radius of gyration distribution
Shape	Perimeter–area ratio (patch)
	Shape index
	Fractal dimension index
	Linearity index
	Related circumscribing circle
	Contiguity index (patch)
	Perimeter–area fractal dimension (class)
	Perimeter–area ratio distribution
	Shape index distribution
	Fractal index distribution
	Linearity index distribution
	Related circumscribing circle distribution
	Contiguity index distribution (class)
	Perimeter–area fractal dimension (landscape)
	Perimeter–area ratio distribution
	Shape index distribution
	Fractal index distribution
	Linearity index distribution
	Related circumscribing circle distribution
	Contiguity index distribution (landscape)
Core area	Core area (patch)
	Number of core areas
	Core area index

(continued)

Table 6.7 (continued)

Type	Metric
	Average depth index
	Maximum depth index (patch)
	Total core area (class)
	Core area percentage of landscape
	Number of disjunct core areas
	Disjunct core area density
	Core area distribution
	Disjunct core area distribution
	Core area index distribution (class)
	Total core area (landscape)
	Number of disjunct core areas
	Disjunct core area density
	Core area distribution
	Disjunct core area distribution
	Core area index distribution (landscape)
Isolation/proximity	Proximity index (patch)
	Similarity index
	Euclidean nearest-neighbor distance
	Functional nearest-neighbor distance (patch)
	Proximity index distribution (class)
	Similarity index distribution
	Euclidean nearest-neighbor distance distribution
	Functional nearest-neighbor distance distribution (class)
	Proximity index distribution (landscape)
	Similarity index distribution
	Euclidean nearest-neighbor distance distribution
	Functional nearest-neighbor distance distribution (landscape)
Contrast	Edge contrast index (patch)
	Contrast-weighted edge density (class)
	Total edge contrast index
	Edge contrast index distribution (class)
	Contrast-weighted edge density (landscape)
	Total edge contrast index
	Edge contrast index distribution (landscape)
Contagion/interspersion	Percentage of like adjacencies (class)
	Clumpiness index
	Aggregation index
	Interspersion and juxtaposition index
	Mass fractal dimension
	Landscape division index
	Splitting index
	Effective mesh size (class)
	Percentage of like adjacencies (landscape)
	Contagion
	Aggregation index
	Interspersion and juxtaposition index

(continued)

Table 6.7 (continued)

Type	Metric
	Landscape division index
	Splitting index
	Effective mesh size (landscape)
Connectivity	Patch cohesion index (class)
	Connectance index
	Traversability index (class)
	Patch cohesion index (landscape)
	Connectance index
	Traversability index (landscape)
Diversity	Patch richness (landscape)
	Patch richness density
	Relative patch richness
	Shannon's diversity index
	Simpson's diversity index
	Modified Simpson's diversity index
	Shannon's evenness index
	Simpson's evenness index
	Modified Simpson's evenness index (landscape)

with n representing the number of classes present in the kernel, and c the number of cells directing the calculation (9, 25, or 49).

- *Relative richness* – determined by the equation:

$$R = n/n_{\max} \times 100,$$

where n is the number of different classes present in the kernel, and $n_{\max}$ is the maximum number of classes in the scene.

- *Fractal dimension* – computed over the entire surface using a moving window operation typically set to a 3 row by 3 column design, fractal dimension is found using the formula:

$$\begin{array}{lll} \boxed{\text{Line}} & \boxed{\text{Square}} & \boxed{\text{Cube}} \\ = \log N\text{^}1/\log N & = \log N\text{^}2/\log N & = \log N\text{^}3/\log N \\ = 1 \times \log N/\log N & = 2 \times \log N/\log N & = 3 \times \log N/\log N \\ D = 1 & D = 2 & D = 3 \end{array}$$

As a general rule of thumb, it is often necessary to employ more than one metric to characterize the landscape. The rational for this is simple; the observation that one number does not adequately explain the pattern in sufficient detail. When selecting metrics for the purposes of environmental characterization, choice should be given to those that are relatively independent of one another and center on those measures that facilitate detection of ecologically meaningful landscape properties. Detection will also be enhanced by insuring that: (1) the resolution of the image is at least two times smaller than the spatial features to be analyzed, and (2) the extent of the image exceeds the largest patch visible in the scene.

6.4.3 Customized Band Ratios

Although frequently used as an image enhancement technique, band ratios can also be employed to produce specialized results that give prominence to unique features or materials in a remotely sensed image. Through carefully selected ratios distinctive gray tones can be produced that highlight desired characteristics at the surface preserving unique information and subtle spectral-reflectance or color differences between surface materials that are often difficult to detect in a standard image. The general form of a customized band ratio follows the formula:

$$\mathrm{BV}_{i,j,\mathrm{ratio}} = \frac{\mathrm{BV}_{i,j,k}}{\mathrm{BV}_{i,j,l}},$$

where

$\mathrm{BV}_{i,j,k}$ is the original input brightness value in band k,
$\mathrm{BV}_{i,j,l}$ is the original input brightness value in band l, and
$\mathrm{BV}_{i,j,ratio}$ is the ratio output brightness value.

The number of possible ratio combinations for a multispectral sensor with P bands can be determined from:

$$n = P(P-1).$$

For example, using the six reflectance bands of the Landsat TM, there are 30 different ratio combinations, 15 original and 15 reciprocal; and based on these possible combination, certain surface characteristics can be discovered. In a study based on the use of Landsat MSS data, Avery and Berlin (1992) identified ratios for depicting soil and rock units (shown by band 1/band 2, band 1/band 4, band 2/band 4, and band 3/band 4), vegetation (shown by band 3/band 1, band 3/band 2, band 4/band 1, and band 4/band 2), and the ratio band 2/band 3 and band 3/band 2 to differentiate between vegetated areas and rock and soil units, whereas a band 3/band 1 ratio derived from Landsat TM produces a resulting image that emphasizes features that are red or orange in color. These examples illustrate that ratios can be discovered that are useful for maximizing the visibility of unique spectral difference. However, the utility of a spectral ratio depends on the reflectance characteristics of the features under investigation and their abundance relative to other surface cover types in the scene.

A ratio can discriminate subtle spectral variations that might otherwise be masked by brightness differences. The two factors that influence a ratio's power of discrimination are the spectral properties of the surface and the material abundance of the object or feature composing the scene. Since any ratio image will portray the variation in the slopes of the spectral reflectance curves between two bands, regardless of their absolute reflectance values, an effective ratio will exploit the difference in the spectral curves for the materials or surface arrangements we wish to characterize. When crafting a band ratio the numerator is typically the band where the material of interest is highly reflective and the denominator defines the band that describes an absorption feature of that material. For example, surface types with a high abundance of iron oxide tend to reflect more strongly in band 3 of the Landsat TM than in band 1; therefore, a band 3/band 1 ratio provides a means to identify iron oxides in an image. Similarly, other ferrous minerals can be detected using a Landsat TM band 5/band 4 ratio. Surfaces with a high clay mineral abundance display strong reflectance in band 5 than in band 7, which enables a Landsat TM band 5/band 7 ratio to highlight their occurrence in the image. A variety of band ratio combinations have been introduced for applications ranging from water quality assessment to geologic prospecting. Consulting ratios published in the remote sensing literature can simplify the band ratio selection problem and, with some modification, support environmental characterization requirements. However, caution should be exercised when interpreting ratioed image. In some circumstances materials with different radiances, but similar spectral reflectance curves are encountered. When ratios are produced these instances, surface types appear identical even though their constituent materials are not the same. Noise can also degrade the quality of a ratio which will require careful noise removal before any computations are attempted.

6.4.4 Statistical Estimation

Statistical estimation describes the general procedure whereby a material or surface feature is identified as a function of a set of spectral patterns or measurements. This functional relationship can be expressed as

$$S_i = f(R_1, R_2, R_3 \ldots, R_n)$$

Where S_i is the surface type or condition of interest and R is a reflectance value or ration combination from the sensor. Estimating the functional form of this relationship implies either a type of prediction where a known characteristic of (S) helps to fit a model to explain the observed pattern or where quality (S) forms out of the linear combination of reflectance values. In the first example, characterization follows the logic of regression analysis and its multivariate extension whereas in the second case we can employ factor analysis or principle component analysis to identify a construct from the data that explains the observed condition (S).

1. *Regression estimation* – Baseline characterization based on regression analysis follows a multivariate logic that seeks to model the pattern or status of a landscape variable according to a series of predictor variables measured over the range of wavelengths that define the image. These wavelengths can be individual reflectance bands and/or band ratios selected to highlight specific surface features. By acquiring a value from field collected samples of the target condition, the attempt is made to fit a model that explains the variance exhibited by that sample. The statistical model is then used to characterize the pattern of the condition over the remainder of the scene or to derive a predictive algorithm that can be employed to identify and quantify the expressed condition over time. To illustrate this approach to remote environmental appraisal, consider the characterization problem where the goal is not to simply identify a material in the image, but to relate spectral measures of that material to a quantified estimate of its value or concentration in a medium (i.e., soil or water). Concentration could be focused on algae in a lake, pH levels in a soil, or a contaminant introduced by human activities on the land surface or in a body of water. In each instance, the underlying hypothesis and defining assumption is that there are detectable differences in the reflectance values for that medium that may be attributed to the presence of or variation in the concentration or amount of a given substance. Thus, the observed difference in the quality of a rangeland, as evidenced by brightness contrasts in the image, is a function of differential values of pH in the soil, or in the example of water quality, the heightened presence of an invasive aquatic plant is an assumed function of elevated water temperature. Testing either of the two hypotheses using the regression estimation approach would follow a carefully scripted seven phase procedure (Fig. 6.5).

   (a) *Field data collection* – A field campaign is conducted to sample the medium of interest to derive concentrations or values of the target variable of interest. Field data collection phase concentrates on the sampling design and timing of the field campaign in a manner that is sensitive to the variable of interest. Seasonal and logistical considerations drive this phase as does the larger issue of selecting the appropriate sensor system to insure that field data collection is coordinated to correspond with sensor overpass dates.

   (b) *Satellite data acquisition* – Remote sensing data acquisition based on the coordination of field sampling efforts with sensor overpass of the sample site builds on the spatial and spectral resolution considerations, adding repeat coverage cycles and image quality constraints into the procedure. The effort to coordinate field data collection is to control for scene/reflectance contrasts and external environmental influences that might contaminate the statistical model. Determining optimal time/date periods for reliable data collection and valid statistical analysis is complicated by satellite repeat cycles and local weather conditions. In the absence of coordinate data acquisition, temporal matching is needed to identify imagery that closely corresponds

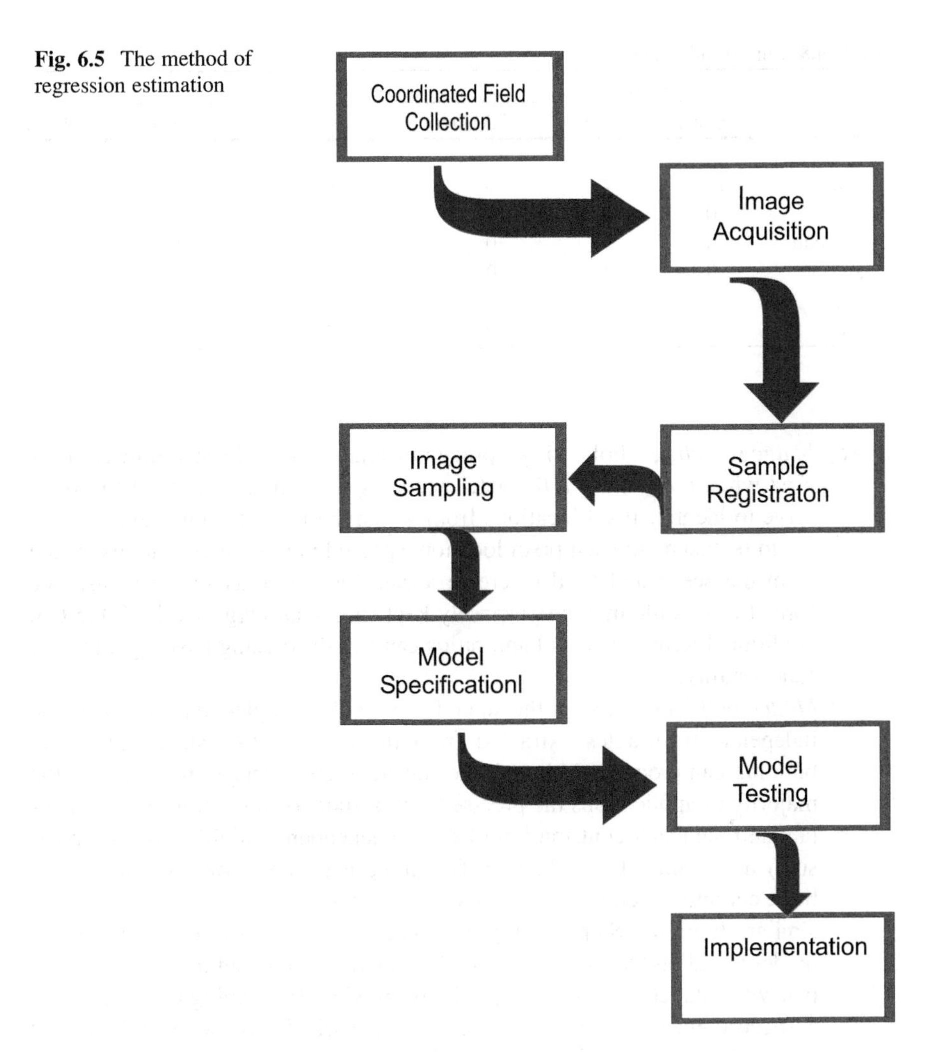

Fig. 6.5 The method of regression estimation

to the scene and environmental conditions that existed when field sampling took place.

(c) *Sample registration* – Registration requires establishing geographic reference between sample site locations taken in the field and their corresponding locations in the remotely sensed image. Through the use of GPS, the geographic position of filed sample sites can be noted, giving each sample site measurement value an (*X*, *Y*) coordinate in a standard geographic referencing system such as latitude/longitude. Once registered to a geographic reference system, the field samples can be projected onto a digital base map and used as an overlay coverage to sample the pixels in the geometrically corrected image.

Table 6.8 Simple tally matrix

	As classified					
	Urban	Forest	Agriculture	Grassland	Water	Total
As sampled						
Urban	12	4	0	0	1	17
Forest	0	0	0	1	0	1
Agriculture	0	0	0	0	0	0
Grassland	0	2	0	3	0	5
Water	0	2	0	1	4	7
Total	12	8	0	5	5	30

(d) *Image sampling* – Following a procedure analogous to the sampling strategy used when conducting a thematic accuracy assessment, the field locations serve to identify pixel locations from which the required reflectance values are to be taken. At each pixel location, spectral measurements are extracted from the scene and used to create a data file for statistical analysis. The format of this file may be purposely kept simple as suggested in Table 6.8. Additional terms, such as band ratios can be added using basic spreadsheet functionality.

(e) *Model building* – Using the data file of field sampled data (*Y*) and the independent variables extracted from the imagery (*X*), statistical model building can proceed. Although the statistical test selected may vary, in the majority of applications the precise form of functional relationship is uncertain and the exact contribution of each independent variable (spectral measure) in the model is unknown. Obtaining a parsimonious solution under these conditions relies on the application of stepwise multiple regression or a similar algorithm. Stepwise regression allows variables to enter or exit the model based on their degree of fit. This facility can be an important advantage when developing a sound predictive model. By stepping in or removing a spectral measure, the contribution of a variable to the overall fit and explanatory power of the model can be examined. The variables that remain in the relationship become those that best characterize the pattern of the condition under investigation.

(f) *Model testing* – A statistically significant model that explains sufficient variations of the dependent variable can be employed to identify and quantify surface conditions elsewhere over the image or applied more universally over time. With a robust model, periodic site evaluations and measurements that may be acquired from remotely sensed sources are used in a characterization study.

(g) *Implementation* – The final step in this solution is to put the model to work in an applied setting whether in place of traditional field sampling or as an early warning indicator that helps to prioritize more detailed field inspections.

2. *Image-based constructs* – Factor analysis and related data reduction methods are statistical techniques used to describe variability among the observed variables (Byrne et al. 1980; Bierman et al. 2011). In statistical modeling factor analysis is frequently used to analyze interrelationships among a large number of variables and to explain these variables in terms of their common underlying dimensions (factors). Factor analysis is also used to verify the analyst's conceptualization of a construct of interest. These constructs are hidden in the data and form from the unique linear combination of a set of explanatory variables selected for analysis. To illustrate its use in remote sensing, consider the example of "degraded" land. The term "degraded" is a construct that we might understand but find difficult to measure in absolute terms. In our imagery, we observe distinct variations in spectral response across the landscape. Is it possible to identify areas that are "degraded" based on the observed patterns of spectral response in our imagery? The underlying hypothesis guiding us is that variations in three or more observed spectral bands (or band combinations) explain the variations in a single unobserved variable; "degraded" land. Factor analysis searches for such joint variations in response to the unobserved latent variable. The observed variables are modeled as linear combinations of the potential factors, plus an error term. Mathematically, the factor analysis model is often expressed algebraically as:

$$
\begin{aligned}
Y_1 &= \alpha_{11}F_1 + \alpha_{12}F_2 + \cdots + \alpha_{1m}F_m,\\
Y_2 &= \alpha_{21}F_1 + \alpha_{22}F_2 + \cdots + \alpha_{2m}F_m,\\
Y_3 &= \alpha_{31}F_1 + \alpha_{32}F_2 + \cdots + \alpha_{3m}F_m,\\
&\ldots\\
&\ldots\\
&\ldots\\
Y_n &= \alpha_{n1}F_1 + \alpha_{n2}F_2 + \cdots + \alpha_{nm}F_m,
\end{aligned}
$$

where Y = a variable with known data, α = a constant, and F = a function, *f*() of some unknown variables.

According to this expression, factor analysis attempts to represent a set of observed variables $X_1, X_2, \ldots, X_n$ relative to a number of common factors plus a factor that is unique to each variable. These common factors define the latent structures hidden in the data which are hypothetical and explain how correlations among the observed variables organize to explain a new dimension. Relationships develop as linear functions where the goal is to solve for the coefficient (α) which best reproduce the observed variable from the factor (F). There are two basic methods to accomplish this goal:

(a) *Principal component analysis* – This method provides a *unique solution* such that the original data can be reconstructed from the results. It examines the total variance among the variables, so the solution generated will include as many factors as there are variables. There is only one method for completing a principal components analysis; this is not true of any of the other multidimensional methods.

(b) *Common factor analysis* – This family of techniques uses an estimate of common variance among the original variables to generate the factor solution. Based on this approach, the number of factors will always be less than the number of original variables.

Factor analysis has enjoyed a broad range of applications relevant to the problem of environmental characterization (Subbarao and Subbarao 1996; Kaplunovsky 2005; Esengun et al. 2006). As a means of statistical estimation, its use supports the desire to (1) explain a complex pattern through a smaller number of variables, (2) define the nature of the derived latent character of the data, and (3) express the degree to which simply explanations can summarize the variability observed. In a remote sensing context, we can review the example introduced by Doerffer and Murphy (1989). In this study, the objective was to characterize the near-shore environment in an area focused on the Wadden Sea. Using Landsat TM data, factor analysis was applied to spectral reflectance values in order to derive representations of three hypothetical variables (constructs): (1) topography, (2) water content, and (3) surface temperature. The rationale for adopting the factor analysis approach was based on the observation that variations in spectral reflectivity of the near-shore environment could be attributed to contrasts in sediment type, water content, and the presence of organisms such as algae, mussel shells, and benthic diatoms on the surface. The spatial distribution of these variables was assumed to correspond with reflectivity patterns with a high degree of covariance that would fall within a set of common factors. Factor analysis was applied to the Landsat TM data, and the factors were mapped using the factor scores calculated by the algorithm for each pixel in the data. By mapping the factor scores, the value of each pixel on the common factor could be interpreted and evaluated. The results of this analysis produced a three factor solution that fit the hypothetical constructs and explained 82% of the total variability exhibited across all seven Landsat TM bands. Through this approach, factor analysis demonstrated that a set of proximate factors could be derived that characterized surface environments and produced useful indicators that enables the assessment of environmental change.

6.5 Summary

Typically, environmental characterization involves the assembly of available physical and biological data and information in electronic form at a common geographic scale and converting this data into the input formats required by the analytical models. Characterization can be explained in two ways: (1) the baseline environmental characterization that captures the current condition of the system and (2) a template condition which is often regarded as a representation of the restoration potential of the environmental system. Characterization is used as a reference condition against which to compare against, to define change, and indicate a direction for needed improvement. In most cases, it is not a management alternative. It is generally an historic reconstruction adjusted for irreversible (anthropogenic or natural) to the normative condition. In this chapter, the application of remote sensing to the characterization problem was examined together with a review of landscape indicators that can be derived from remotely sensed data and fed into the characterization process.

References

Anderson, J., Hardy, E., Roach, J. and Richard, E. (1976) A Land Use and Land Cover Classification System for Use with Remote Sensor Data. Geological Survey Professional Paper 964. United States Government Printing Office.

Artiola, J., Pepper, I. and Brusseau, M. (2004) Environmental Monitoring and Characterization, Elsevier, Amsterdam, 410p.

Avery and Berlin (1992)

Babu, S. and Reidhead, W. (2000) Monitoring Natural Resources for Policy Interventions: A conceptual Framework, Issues and Challenges, Land Use Policy, 17, 1–11.

Berberoglu, S. and Akin, A. (2009) Assessing Different Remote Sensing Techniques to Detect Land use/Land Cover Changes in the Earthern Mediterranean, International Journal of Applied Earth Observation and Geoinformations, 11, 46–53.

Boteguilha, A. and Ahern, J (2002) Applying Landscape Ecological Concept and Metrics in Sustainable Landscape Planning, Landscape and urban Planning, 59, 5–93.

Bierman, P. Lewis, M., Ostendorf, B and Tanner, J. (2011) A Review of Methods for Analyzing Spatial and Temporal Patterns in Coastal Water Quality, Ecological Indicators, 11, 103–114.

Byrne, G., Crapper, P. and Mayo, K. (1980) Monitoring Land-cover Change by PCA of Multitemporal Landsat Data, Remote Sensing of Environment, 10, 175–184.

Cook, P. (2006) Site Characterization, Proceedings, CO2SC Symposium Lawrence Berkeley National Laboratory, Berkeley, California, 3–9.

Cushman, S., McGarigal, K. and Neel, M. (2008) Parsimony in Landscape Metrics: Strength, Universality an Consistency, Ecological Indicators, 8, 691–703.

Dale,V. and Beyeler, S. (2001)Challenges in the Development and Use of Ecological Indicators, Ecological Indicators, 1, 3–10.

Deering, D., Rouse, J., Haas, R., Schell, A. (1975) Measuring Forage Production of Grazing Units from Landsat MSS Data, Proceedings: Tenth International Symposium on Remote Sensing of Environment, ERIM, 2, 1169–178.

DiGregorio, A. (2005) UN Land Cover Classification System (LCCS) (http://www.glen-lccs.org)

Di Gregorio, A. and Jansen, L. (2000) Land Cover Classification System (LCCS): Classification Concepts and User Manual, Environment and Natural Resources Service, GCP/RAF/287/ITA FAO, Rome, 179p

Doerffer, R. and Murphy, D. (1989) Factor Analysis and Classification of Remotely Sensed Data for Monitoring Tidal Flats, Helgoländer Meeresuntersuchungen, 43, 275–293.

Esengun, K., Gunduz, O., Akay, M., Cicek, A. (2006) Assessment of the Local Government's Efforts in Challenging to Environmental Problems Using Factors Analysis, Journal of Applied Sciences, 6, 1052–1058.

Frohn, R. (1998) Remote Sensing for Landscape Ecology, Lewis Publishers, Boca Raton, FL, 99p.

Frohn, R and Hao, Y. (2006) Landscape Metric Performance in Analyzing Two Decades of Deforestation in the Amazon Basin of Rondonia, Remote Sensing of Environment, 100, 237–251.

Hammond A. (1995) Environmental Indicators, World Resources Institute, Washington, DC.
Helming, K., Pe´ rez-Soba, M., Tabbush, P. (Eds.), 2008. Sustainability Impact Assessment of Land Use Changes. Springer-Verlag, Berlin.
Herold, M., Latham, J. DiGregorio, A, and Schmullius, C. (2006) Evolving Standards in Land Cover Characterization, Journal of Land Use Science, 1, 157–168.
Herold, M., Scepan, J. and Clarke, K. (2002) The Use of Remote Sensing and Landscape Metrics to Describe Structures and Changes in Urban Areas, Environment and Planning A, 34, 1143–1458.
Homer, C., Huang, C., Yang, L., Wylie, B. and Coan, M. (2004) Development of a 2001 National landcover Database for the Unites States, Photogrammetric Engineering and Remote Sensing, 70, 829–840.
Huete, A. (1988) A Soil-Adjusted Vegetation Index (SAVI), Remote Sensing of Environment, 25, 295–309.
Jensen, R. and Binford, M. (2004) Measurement and Comparison of Leaf Area Index Estimators Derived from Satellite Remote Sensing Techniques, International Journal of remote Sensing, 25, 4251–4265.
Jensen, R. and Hardin, P. (2005) Estimating urban leaf area using field measurements and satellite remote sensing data. Journal of Arboriculture, 31, 21–27.
Kaplunovsky, A. (2005), Factor analysis in environmental studies, HAIT J. Sci. Eng. B, 2(1 – 2), 54–94.
Kaufman, Y. J., Tanre, D. (1992) "Atmospherically resistant vegetation index (ARVI) for EOS-MODIS," in _Proc. IEEE Int. Geosci. and Remote Sensing Symp. '92_, IEEE, New York, 261–270.
Kauth, R. J. and Thomas, G.S. (1976) "The tasseled cap–A graphic description of the spectral-temporal development of agricultural crops as seen by Landsat," _Proceedings of the Symposium on Machine Processing of Remotely Sensed Data_, Purdue University, West Lafayette, Indiana, pp. 41–51.
Kays, B. (2000) Environmental Site Assessment: Site Characterization Methodologies, Managing Soils in an Urban Environment, Agronomy Monograph no, 39 American Society of Agronomy, Madison, 69–92.
Kogan, F. (1995) Application of Vegetation Index and Brightness Temperature for Drought Detection, Advances in Space Research, 15, 91–100.
Kriegler, F., Malila,W., Nalepka, R. and Richardson, W.(1969) Preprocessing transformations and their effects on multispectral recognition. In Proceedings of the 6th International Symposium on Remote Sensing of the Environment, Ann Arbor, Michigan, USA. Ann Arbor, MI, University of Michigan. Institute of Science and Technology, 97–109.
Landres, P. (1992) Ecological Indicators: Panacea or Liability, in McKenzie, D. (ed) Ecological Indicators, Vol. 2, Elsevier, London.
Lein, J. (2006) Toward the Rapid Characterization of the Built Environment within the Wildland-Urban Interface: A Soft Classification Strategy, GIScience and Remote Sensing, 43, 179–196.
Lein, J. (2004) Assessing the Applicability of Satellite Imagery for Environmental Monitoring and Regulatory Compliance assessment, International Journal of Environmental Technology and Management, 4, 157–173.
Melesse A., Weng, Q., Thenkabail, P., Senay, G. (2007) Remote Sensing Sensors and Applications in Environmental Resources Mapping and Modeling, Sensors, 7, 3209–3241.
McGarigal, K. and Marks, B. (1995) FRAGSAS: Spatial Pattern Analysis Program for Quantifying Landscape Structure, General Technical Report PNW-GTR-351, U.S. Department of Agriculture, Forest Service, Pacific Northwest Research Station, Portland Oregon.
Nagler, P., Glenn, E. and Huete, A. (2001) Assessment of Spectral Vegetation Indices for Riparian Vegetation in the Colorado River Delta, Mexico, Journal of Arid Environments, 49, 91–110.
National Research Council (2000) Ecological Indicators for the Nation, National Academy Press, Washington DC.

Niemeijer, D and deGroot, R. (2008)A Conceptual Framework for Selecting Environmental Indicator Sets, Ecological Indicators, 8, 14–25.

Noss, R. and Cooprrider, A. (1994) Saving Nature's Legacy: Protecting and Restoring biodiversity, Island Press, Washington, DC.

Renetzeder C., Schindler, S. Peterseil, J., Prinz, M. Mucher, S., Wrbka, T. (2010) Can W Measure Ecological Sustainability? Landscape Pattern as an Indicator for Naturalness and Land Use Intensity at Regional, National and European Level, Ecological Indicators, 10, 39–48.

Riley, J. (2000.)Summary of the discussion session contributions to topic 1: what should a set of guidelines with regard to indicators contain? In: Riley, J., Harwood, K. (Eds.), UNIQUAIMS Newsletter no. 10. Harpenden,. IACR-Rothamsted, UK, pp. 5–6.

Riitters, K., O'Neill, R., Hunsaker, C., Wickham, J., Yankee, D., Timmins, S, Jones, K. and Jackson, B. (1995) A Factor Analysis of landscape Pattern and Structure metrics, Landscape Ecology, 10, 23–39.

Rouse, J. Haas, R. Shell, J. and Deering, D. (1974) Monitoring Vegetation Systems in the Great Plains with ERTS, Third Earth Resources Technology Satellite-1 Symposium, Vol. 1, 309–317.

Schindler, S. Poirazidia, K. and Wrbka, T. (2008) Toward a Core Set of Landscape Metrics for Biodiversity Assessments: A Case Study from Dadia National Park, Greece, Ecological Indicators, 8, 502–514.

Schultz, G. and Engman, E. (2000) Remote Sensing in Hydrology and Water Management. Springer, New York.

Smeets, E. and Weterings, R. (1999) Environmental Indicators: Typology and Overview, European Environmental Agency, Technical Report No. 25, Copenhagen, Denmark, 20p.

Subbarao,C. and Subbarao, N. (1996) Characterization of groundwater contamination using factor analysis, ENVIRONMENTAL GEOLOGY, 28, 175–180.

The Nature Conservancy (1994) Standardized National Vegetation Classification Systems, United States Department of the Interior (http://biology.usgs.gov/npsveg/nvcs.html)

Walker, P. and Peters, P. (2007) Making Sense in Time: Remote Sensing and the Challenges of Temporal Heterogeneity in Social Analysis of Environmental Change, Human Ecology, 35, 69–80.

Xie, Y., Sha, Z. and Yu, M. (2008) Remote Sensing Imagery in Vegetation Mapping: A Review, Journal of Plant Ecology, 1, 9–23.

Zheng, G. and Moskel, L. (2009) Retrieving Leaf Area Index (LAI) Using Remote Sensing: Theories, Methods and Sensors, Sensors, 9, 2719–2745.

Chapter 7
Environmental Monitoring and Change Detection

In the previous chapter, discussions focused on the application of satellite remote sensing as a means to document and describe the baseline status of the environmental system. Through the use of image-derived indices that characterized critical aspects of the landscape, important environmental relationships could be summarized and the initial condition of the surface could be established against which changes may be evaluated. In this chapter, we will expand on this preliminary treatment of change and broaden the scope of environmental characterization by introducing the application of remote sensing technology in the temporal domain. Sensing in the temporal domain recognizes the dynamic nature of the landscape, and with the addition of time in the equation, we can adopt a more process-oriented view of the environmental system. Understanding process is obviously central to the study of environmental change, and the capacity of satellite systems to provide multitemporal images of land surface objects greatly enhances our ability to observe the progression of human and natural forces as they act on the landscape. We will begin our examination of this subject with a brief review of the principles that guide environmental monitoring efforts, with this background the role of remote sensing is explored leading to a detailed survey of the methods developed to sense change. The chapter culminates with an overview of anomaly detection and the integration of temporal data in predictive analysis and environmental modeling.

7.1 Principles of Environmental Monitoring

The use of the term monitoring is widespread in both the remote sensing and environmental assessment literature (Rachon 2003; Duro et al. 2007). In both cases, the idea of monitoring implies the systematic collection of data over time. With this simple explanation, monitoring in the context of remote sensing is comparatively easy to conceptualize, given the regularity of satellite orbits and the constant recording of reflected or emitted electromagnetic from the surface by

J.K. Lein, *Environmental Sensing: Analytical Techniques for Earth Observation*,
DOI 10.1007/978-1-4614-0143-8_7,

earth observational sensors. Recording, mapping, and sampling radiance, however, only provides a basis for monitoring, and this non-trivial reality can cause confusion when we overlook the fact that monitoring is a methodology, not simply a satellite's revisit schedule. To appreciate monitoring as a methodology, it can be useful to clarify certain terms that are often used interchangeably but take on subtle differences when related to the detection and documentation of environmental change. One term frequently used in conjunction with remote sensing is the word surveillance. For the purpose of our discussion, surveillance can be defined as the systematic measurement of variables and processes over specific temporal period in order to establish a series of data in time (Spellerberg 2005). There is no implied goal to the act of surveillance, and although it is systematic, it involves simply collecting data over an interval of time. Earth orbiting satellites are instruments of surveillance, collecting measurements over their operational life spans. The concept of monitoring also requires the systematic measurement of variables and processes over time, but carries one important distinction. Monitoring is purposeful in that the variables subject to measurement have been selected for a well-defined reason, and the collection of data is undertaken to fulfill clearly stated goals and objectives (Spellerberg 2005). Monitoring is also conducted to collect data according to a set of standards or requirements that are problem-centric and designed to provide specific information on the characteristics of given situation and how that situation changes over time. For example, a physician will systematically monitor a patient's heart rate to determine whether it improves over time in response to the administration of a new blood thinning drug. Because there is an optimal heart rate for this patient, monitoring that variable can determine how successful this medication is in terms of improving the patient's health. In response to the monitored information, the physician may then increase or decrease the dosage until a satisfactory result is sustained. Monitoring, as suggested in the remote sensing literature, does not always meet the criteria of being operationally purposeful at the level of rigor as our medical example implies, which invites confusion. Therefore to carry our discussion forward, we will define environmental monitoring in more exacting terms as the systematic collection of environmental data in a standardized manner, at regular intervals over time, to provide information to direct decision making in a well-defined context. With this focus on the provision of decision-relevant information several styles of monitoring can be noted (Brydges 2004):

- *Simple monitoring* – which involves the recording of a single variable at one point over time
- *Survey monitoring* – that replaces the lack of a historical record with a survey of the current environmental conditions in both the affected and non-affected areas
- *Surrogate monitoring* – which involves the use of proxy information to infer changes and a way to compensate for the absence of previous monitoring
- *Integrated monitoring* – that employs detailed sets of environmental information

Although the level of detail and degree of sophistication associated with these general categories may vary, each develops out of a common methodology aimed at

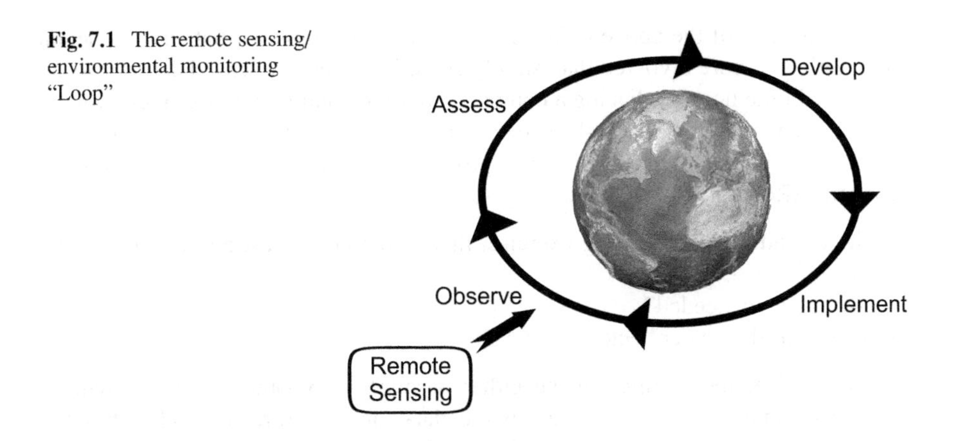

Fig. 7.1 The remote sensing/environmental monitoring "Loop"

deriving an objective appraisal of tangible environmental benchmarks. The benchmarking concept adds to further distinguish the process of monitoring from the environmental surveillance activities. Any environmental monitoring effort enables participants to engage in a learning process by offering a basis to that promotes adaptive decision making. Therefore to make efficient use of the learning opportunity, the results obtained via environmental monitoring need to be integrated into the way things are done, whether that describes crop planting and harvesting, forest management, or urban design and planning. As an engaged activity, monitoring becomes part of an active "loop" with remotely sensed data feeding the learning process with timely and relevant information (Fig. 7.1).

An environmental monitoring strategy that facilitates learning directs attention to seven practical issues that also help refine how remote sensing technology is applied (Usher 1991):

1. *Definition of objectives* – the basic question to resolve pertains to how one determines which aspects of the environmental system are to be assessed and how this process change is to be defined and expressed.
2. *Indicator selection* – based on the stated objectives and which characteristics provide the most concise and relevant answer to the monitoring question.
3. *Method* – what is the optimal means for measuring and observing the indicator.
4. *Measurement frequency* – what is the temporal interval needed to identify meaningful environmental trends, yet infrequent enough not to saturate the process with data overload.
5. *Program assessment* – as monitoring ensues, ongoing assessment of the objectives, indicator measurement frequency, and methods of analysis are required to insure consistency and validity of results.
6. *Data analysis* – what methods enable the assessment and analysis of change and which offer the greatest capacity to explore trends in the data.
7. *Evaluation* – as information is acquired from the monitoring program, how well does it support decision needs.

Giving each of the above points due consideration suggests that environmental monitoring is more involved than simply acquiring remotely sensed data across an analytical time horizon. Facing a complex environmental system together with the wide variety of factors that influence natural process, monitoring not only demands a strategy, but also a well-planned program framed around three organizing concerns (Roberts 1991):

- Why conduct a monitoring exercise; there must be a question that monitoring can answer – what is it?
- What data is needed?
- How will data be collected?

In practice, monitoring can be either descriptive-hypothetical where data is collected, analysis seeks patterns in the data, and interpretation identifies the possible causes, or monitoring may be post hoc correlative focused on collecting data and the formulation of a question followed by an analysis that will explore how well the data "fits" the answer (Roberts 1991). In either case, the monitoring activities are prone to inefficiencies unless guided by a well-articulated conceptual model that helps to organize information, elucidate critical component of the environmental system, and provide a scientific framework around which the monitoring program takes shape. Developing a clear conceptual model is a challenge for most monitoring programs. Because environmental relationships are often complex, communicating complexity to audiences with contrasting needs, levels of expertise, and expectations requires sensitivity and a flexibility of approach, particularly when remote sensing technology is involved (Lookingbill et al. 2007). Meeting the communication challenge has been greatly simplified by a conceptual modeling building schema introduced by Lookingbill et al. (2007). The model is designed to improve information gathering by combining scientific theory with pragmatic management considerations focused squarely on the identification of monitoring endpoints (Fig. 7.2). Modeling building according to this design begins by: (1) gathering relevant information pertaining to environmental functioning and the likely management concerns, then continues by (2) identifying specific resources of concern for the management area that influence long-term sustainability, (3) deriving lists of stressors into scenarios that define specific treats to sustainability, and culminates in (4) evaluating the usefulness of this conceptual model based on a set of assessment criteria (Table 7.1). Overall a useful conceptual model is the one that

- Articulates important processes and variability
- Highlights interactions between environmental processes
- Identifies critical links between drivers, stressors, and system responses
- Facilitates the selection and justification of monitoring variables
- Supports evaluation of data from the monitoring program
- Communicates process to technical and non-technical audiences.

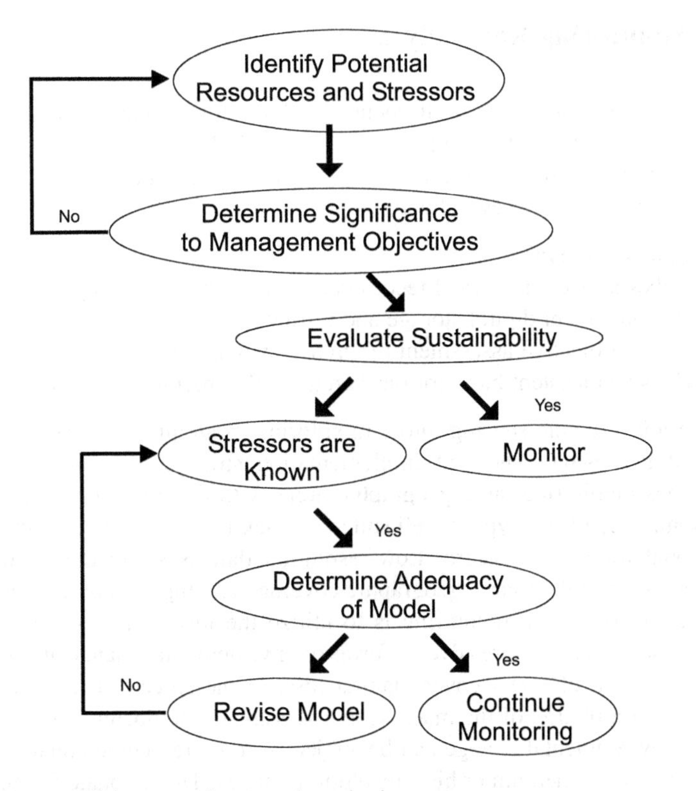

Fig. 7.2 Conceptualizing an environmental monitoring system

Table 7.1 Assessment criteria guiding environmental monitoring
Correspondence between model prediction and monitoring observation
Applicability to decision-making needs
Reliability

Although a monitoring program need not concentrate on developing quantitative environmental models or engage in policy formulation, producing a set of realistic and focused conceptual models is an essential ingredient for designing strategies that clearly identify critical environmental trends, directing environmental management operations while contributing to sound policy review mechanisms (Gross 2003).

7.2 Monitoring Remotely

Satellite remote sensing is a well-documented tool for monitoring the environmental system (Spitzer 1986; Leimgruber et al. 2005; Jat et al. 2008; LeMarie et al. 2006; Dymond et al. 2001). Satellite data enjoys a comparative advantage over other methodologies in several ways:

- Large areal coverage
- Describes a recent historical record dating from the 1970s to present
- Offers convenient digital storage and retrieval
- Facilitates objective assessment of environmental conditions
- Provides a consistent basis for measurement that permits the analysis of change

Integrating remote sensing into a monitoring program can be accomplished in either of two ways. The first implementation strategy, a top-down approach, encourages monitoring large geographic areas with low-resolution satellite imagery. Monitoring of this type is well suited to track programs aimed at international or national topics of concern. Low-resolution data is selected for this type of analysis owing to their large geographic coverage and high temporal repeat cycles. The goal of top-down monitoring is to utilize the low spatial resolution data to identify and indicate where adverse forms of environmental change are active, such as the reduction of forest cover, habitat loss, or the spread of land degradation. Following completion of the macroscale assessment, the documented locations of adverse environmental change can be subjected to more detailed analysis through the acquisition of medium or high-resolution images. The rationale for engaging in medium or high-resolution assessments is to provide additional information regarding the causal activities that are responsible for the observed trends and to investigate the modification in finer detail. A top-down monitoring strategy is well suited to situations where the drivers of environmental change have already been well documented and environmental policies or legal thresholds designed to remedy the offending actions have been established.

The bottom-up method for monitoring the environment using satellite imagery focuses on the regional environmental situation and relies on guidance or direct involvement of local stakeholders or governmental bodies that have identified a specific need or problem that requires monitoring. Examples of this meso- to microscale examination of local trends might include local programs aimed at watershed restoration activities, wetland preservation, strip mine reclamation, or urban growth management issues. In these examples, specific repair, restoration, or policy directives are at work, and the activities they describe demand regular assessment to determine their effectiveness. Additionally, the active processes modifying the landscape may be too difficult to detect with low-resolution imagery, and finer spatial detail is required. Typically environmental monitoring targets have been established as part of the larger management program and performance indicators are in place that may be used to track and report progress toward the desired policy goal. Alternatively, bottom-up monitoring may be called upon in

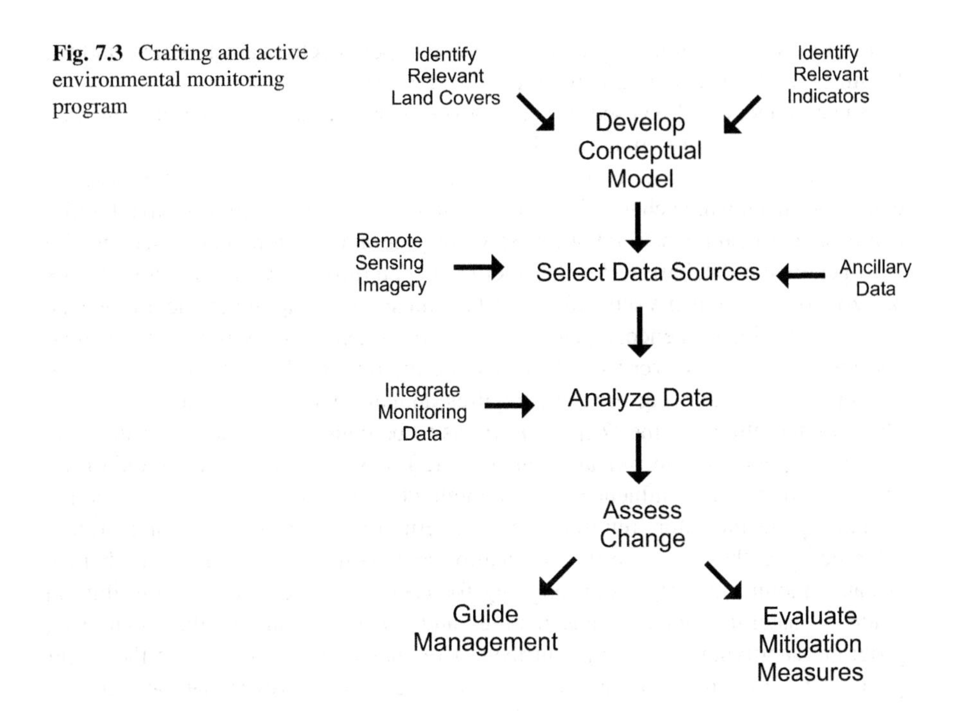

Fig. 7.3 Crafting and active environmental monitoring program

response to an anomalous event or an ongoing novel situation such as the aftermath of a natural hazard event, oil spill, or sudden infestation. In these examples, the active process often falls below the geographic scale of coarse or medium resolution sensors.

A comprehensive strategy for implementing a monitoring program built around a well-defined conceptual model that incorporates remotely sensed data has been introduced by Jones et al. (2009). While this strategy is directed toward the assessment of land use/land cover change in areas surrounding national parks in the United States, certain aspects of their contribution in environmental monitoring can be assembled into a more generic design (Fig. 7.3). As suggested in Fig. 7.3, there are four main phases to a monitoring program based on the use of remotely sensed data. Remote monitoring begins with the creation of a conceptual model that defines all the relevant elements that must be incorporated in the program in order to achieve meaningful results. Remotely sensed imagery is introduced in the second phase of this design where the sources of data for monitoring the environmental indicators are identified. The selection of image data focuses attention around a series of questions that underscore the practical considerations that affect the type of data acquired and the analytical procedures needed to produce a final decision product. The utility of satellite imagery hinges on the selection of the optimal sensor given the specifics of the monitoring problem. Here, guidance from the conceptual model is critical as issues pertaining to spatial resolution, spectral resolution, swath width, geographic coverage, repeat cycle characteristics, and the length of the historic data archive must be addressed.

Since imagery will be used to either develop or serve as an indicator, how well the sensor and the raw image product match the requirements for environmental monitoring is a nontrivial matter. Once the data product is selected, the baseline conditions from which change is to be assessed must be produced. Data availability can be a significant constraint at this stage of program development. Since the aim is to generate a database to fill the interval as determined by the monitoring time horizon, or to establish a schedule for data acquisition to capture the next time step in the monitoring sequence, repeat coverage cycles, external environmental factors, and modes of data delivery stand to frustrate the flow of data into the monitoring program. If a monitoring program requires annual data beginning at time T, the phased integration of time ($T_{+1} \ldots T_{+n}$) has to be considered when allocating the resources (personnel, technical, financial, etc.) to support long-term monitoring. Data selection is also influenced by the method(s) that will be employed to define change in the indicator, reintroducing concerns related to accuracy and ground referencing as they relate to the verification and documentation of actual change. Because monitoring implies an ongoing focused data collection, the contribution made by satellite imagery runs parallel and complimentary to the monitoring program; supplying an information feed at regular intervals to support the larger goals of the monitoring program. Over the long-term, however, data availability and the continuity of the image product can become a limiting factor as satellite systems fail and operational constraints conspire to create an uncertain future.

7.3 Sensing Change

Mapping and detecting environmental change using remotely sensed data has been approached using a variety of techniques and applied to an extensive list of topics. Recent examples include from Rogan et al. (2002), Kleinod et al. (2005), Koltunov and Ustin (2007), Wang and Xu (2010), and Verbesselt et al. (2010). Based on this selective sample, change detection has become both a major application of remote sensing technology and an active area of research and development (Ridd and Liu 1998). Our interest in environmental change detection recognizes that terrestrial ecosystems are in a state of permanent flux, and the status or description of change varies across spatial and temporal scales. Digital change detection describes those methodologies that support the quantification of temporal trends and alterations in surface phenomena from imagery acquired by satellite sensors employed in an environmental monitoring program. Detecting change in a digital image is a complex task. In the previous chapter, we engaged the concept of change and formed a working definition that provides a basis to its remote evaluation. In that discussion, environmental change was explained as (1) an alteration in the characteristics or components that describe the land surface and (2) as a spectral movement or contrast of a pixel entity over time. Both aspects of this definition are anchored by the pixel, the fundamental unit of sampling electromagnetic energy reflected or emitted by the surface. The disposition of a pixel from one time step to other evidences a trajectory within what we assume is a well articulate thematic context. The differences observed in a pixel, expressed as a function of time, loosely explain a rate of change which can be abrupt/dramatic or subtle with transitions occurring gradually as suggested by measured dissimilarities in radiance values. When the observation is made that the environment has "changed," the alteration in a pixels radiance value could explain a categorical shift (i.e., a pixel classified as forest becoming a pixel now classed as urban) or a continuum, such as a numerical trend expressed by an index (Coppin et al. 2004). Change may also be characterized at the surface using terminology such as a conversion or modification that attempt to express underlying process moving from an origin. When addressing the concept of change, conversions at the surface imply a purposeful redirection in use of cover where one land type has been replaced by another. Modifications, however, suggest subtle shifts that affect the character of the surface without directly replacing the type of surface involved. Determining whether change is dramatic or subtle, a modification or a conversion follows after the detected differences in radiance can be measured. Once measured, characterizing change requires careful interpretation of the patterns revealed in the data and the actors involved as the surface that express an active process taking place.

The list of adjectives that can be employed to describe change in the status of a pixel and indirectly the environment is numerous. When viewed within the spatial confines of our imagery, four expressions of change can be noted (Khorram et al. 1999). First is the conventional notion of change where a land type at time T becomes a different category at time T_{+1}. This follows a second expression of change where an observed land type expands, shrinks, or alters its shape by time T_{+1}. Alternately a land type may (3) shift position or it may (4) fragment or coalesce when observed at time T_{+1}. Unfortunately, given the numerical nature of our digital image, the basic assumption guiding change detection is that change at the surface is a direct function of differences in radiance values of a pixel viewed over time. This fundamental explication of change is further refined by the supposition that differences in radiance due to surface change are more pronounced (larger) than difference in radiance caused by other factors such as contrasts in atmospheric conditions, soil moisture content, or sun angle (Mas 1999). Consequently, not all detectable (measurable) changes are equally important. Therefore, our ability to monitor change in the environment depends on:

- How well baseline conditions have been established
- How well variability can be expressed in directional terms
- How well the spatial extent and context of change can be understood

The main challenges facing the satellite monitoring and detection of environmental change have been summarized by Coppin et al. (2004) as expectations that serve as specific requirements for a sensor system to:

1. Detect modifications in addition to conversions
2. Monitor rapid and abrupt change in addition to progressive and incremental changes

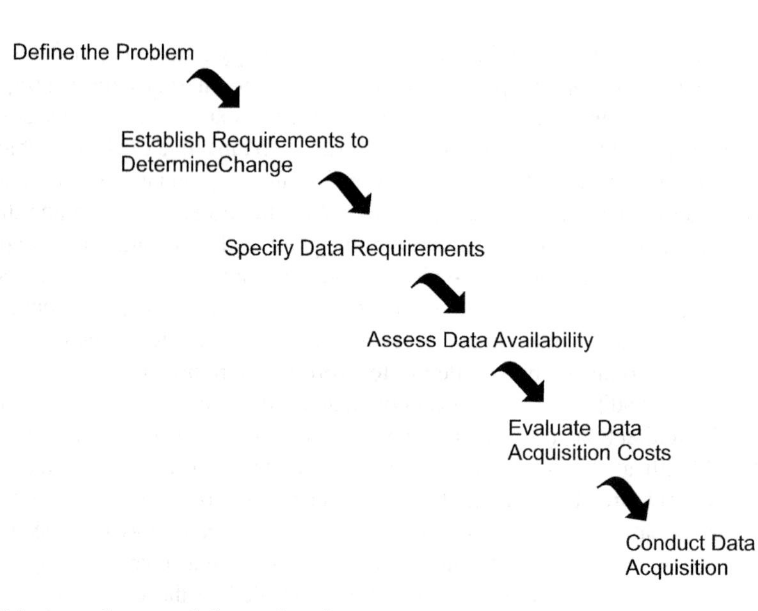

Fig. 7.4 An environmental change detection strategy

3. Separate interannual variability from longer-term trends
4. Correct for scale dependence influences that confuse statistical estimations
5. Match the temporal sampling rate of satellite observations to the temporal scale of the process of interest

Taken together these system requirements advance the idea that a useful exercise in change detection is the one that allows the analysis to clearly distinguish the real change (phenomenology-based differences in the features of the surface) from noise (differences resulting from influences external to the problem). Achieving this distinction develops out of a well-crafted change detection methodology (Lu et al. 2004).

The analytical approach for implementing a change detection methodology in conjunction with an environmental monitoring program has been well documented in the remote sensing literature (Lyon et al. 1988; Muchoney and Haack 1994; Lunetta et al. 2004). Change assessment is conducted according to two main phases. The first follows a series of steps that concentrate on project formulation (Fig. 7.4), which continues into an operational phase that focuses on image analysis (Fig. 7.5). The initial phase of change assessment takes shape during the design of the monitoring program. During this phase the objectives are specified that will determine the direction of the analysis of change and the question(s) that will be answered using the imagery. A precisely defined problem contributes to the next step in the implementation process, product specification. Product specification concerns the actual deliverables of analysis and how well they convey information

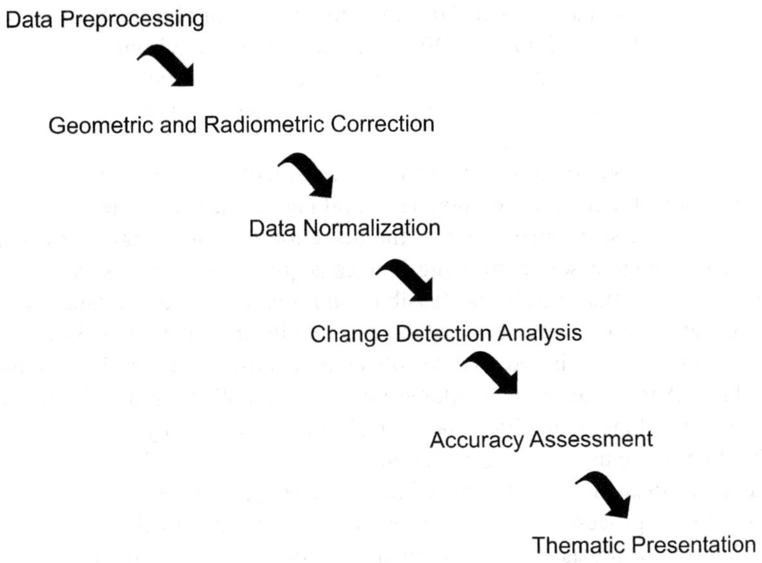

Fig. 7.5 An operational remote sensing monitoring methodology

to the end users. The deliverable, either as a hard copy cartographic display or a digital file to be incorporated into a spatial database, typically requires thoughtfully developed standards for thematic and geometric accuracy. Standards will insure that utility of the product as a reliable source of information and will enforce consistency over the monitoring time horizon. With a set of minimum standards established, the data requirements needed to support the project can be entertained. For environmental monitoring applications, this involves consideration of the spectral, spatial, and temporal resolution needs that will ultimately determine which sensor system or class of sensor is the most appropriate for the application. With respect to spectral resolution, thought must be given to the available wavelengths of the electromagnetic spectrum and which combinations are needed to monitor the phenomenon in question. Next, consideration should be given to the bandwidth and how that will provide sufficient spectral detail to detect the phenomena. Spatial resolution corresponds to geographic scale and how well the scale implied by the data will represent the phenomena as a measurable object. We can think of spatial detail in a manner similar to the concept of a minimum mapping unit and how well an object assumes geometric form. Finally, temporal resolution asks us to consider how the phenomena behave with respect to time. Time implies seasonality, progression, and how process becomes discernable over the time horizon. Because our interest is capturing the pattern of real change, sufficient time must elapse for a change at the surface to be detectable. The spacing in time varies for natural process and human activities. Deciding on the optimal time-step for effective analysis and detection demands sensitivity to the nature of the phenomena involved in the context of the monitoring problem as defined. The temporal dimension of a change study is also influenced by data availability. Often, due to external environmental effects or sensor perturbations gaps can appear in the digital archive that may render a particular date or sequence of dates, unusable. Where gaps exist or in situations where external factors, such as cloud cover or haze, conspire to reduce the available imagery, adjustments to the data acquisition cycle have to be made. Although adjustments are common, temporal lags in the time series can introduce variations in the scene that frustrate the detection of real change. Ideally, monitoring and change assessment using images acquired on anniversary dates is the goal; however, that rarely is feasible and reality forces reliance of close approximations that manage to satisfy the monitoring objectives. For example, a monitoring program based on data collected in mid-April every 3 years may find that (1) no April imagery is available forcing data collection to rely on an early May or early June acquisition date, or (2) the collection cycle for year 3 was unable to provide usable imagery for any of the suitable months and data for year 4 was used, instead. Adjustment, while a necessity, introduces error and uncertainty into the process and underscores the need to careful calibration. The final consideration in the assessment of change introduces the practical consideration of the costs associated with data acquisition. Environmental monitoring programs are designed to be ongoing and systematic data collecting operations. Acquiring data from reliable instruments that are consistent and accessible is critical to the success of the monitoring program. Cost factors can significantly retard aggressive environmental monitoring activities, unless funding sources are committed to data acquisition over the life-cycle of the monitoring program.

Moving from project formulation to the actual analysis of time-sequenced imagery introduces six additional steps into the change detection procedure. These additional steps describe the analytical treatment of the raw satellite imagery and the data processing activities required to produce a meaningful deliverable. Taken in turn, the processing tasks include:

1. *Data preparation* – This step concerns the activities followed in order to assemble imagery into a data set that facilitates time-sequenced analysis. The tasks required at this stage center around subsetting or mosaicking the imagery to the geographic specifications as dictated by the study area together with the operations that address image quality issues such as cloud cover, and haze removal and the treatment of radiometric artifacts that extend across the analytical time horizon.
2. *Feature selection* – A multitemporal data set assembled from multispectral imagery can produce a file containing numerous spectral bands. Depending on the application, not all spectral bands for a image/date may be required for analysis or useful based on their spectral contrast. Histogram analysis can be used to examine the individual bands in the data set and eliminate redundant or undesirable bands to reduce processing requirements and form a more parsimonious data set.

3. *Geometric registration and radiometric correction* – Accurate geometric registration is critical to any image-based change detection study. Because analysis is performed on a per-pixel basis, pixel misregistration will contribute to error and incorrect results. Registration is accomplished by selecting on date (image) to serve and the geometric anchor to which all other images (dates) will be fit. Using either ground control points or moving directly to image-to-image registration, the goal is to resample the data set to the lowest root mean square (RMS) error value as practical. Once geometric shifts between dates have been compensated for, radiometric corrections can be undertaken to enhance the fidelity of the data across the time horizon. Despite the fact that change detection demands the use of anniversary data to reduce atmospheric and environmental scene anomalies, variability in radiance can be expected; therefore, using radiometric corrections acts to normalize interscene and temporal variability reducing noise in the change signal. Several of the more common approaches to correct radiometric shifts are described in Table 7.2.
4. *Change analysis* – Once a suitable data set has been assembled and normalized to insure geometric and radiometric consistency over the analytical time horizon, assessment can proceed to the selection and execution of a change detection algorithm. There are several algorithms available to analyze change in digital imagery, and the topic of change detection has been extensively reviewed in the remote sensing literature (Lu et al. 2004; Mas 1999; Gao 2009). We will examine the more common methods in a separate section. In general, the methods of change analysis follow either of two strategies: (1) preclassification spectral change detection or (2) postclassification categorical change analysis (Lunetta and Elvidge 1999).
5. *Accuracy and quality assurance* – The challenge in change analysis is to identify real, substantive, and significant differences in the environment across the multitemporal data set. The successful execution of an algorithm does not imply that this challenge has been met. Indeed, all of the complicating factors and pitfalls associated with thematic analysis of remotely sensed data magnify during a change study which places paramount importance on the question of accuracy. Change detection accuracy should be assessed by examining the results obtained at selected sampling locations in a manner similar to the procedures followed when conducting a thematic accuracy assessment (Gao 2009). Assuming that the baseline characterization which established the initial start date of the monitoring exercise meets the established accuracy standard and that normalization error was kept at a suitable minimum, actual change in the scene can be quantified, and illogical differences (pixels that display anomalous values, spurious, or categorical designations that conflict with underlying theory) should become readily apparent. Calculating the ratio of illogical difference (change) to total detected change offers a useful statistic for quality assurance purposes in an environmental monitoring program. Additional methods have been reviewed by Gao (2009) and include the use of error propagation theory, correlational analysis, and qualitative evaluation approaches based on visualization techniques and computer animation (Radke et al. 2003).

Table 7.2 Common radiometric correction methods

Method	Description
Dark object subtraction	Examine brightness values in an area of shadow or for a very dark object (such as a large clear lake) and determine the minimum value. The correction is applied by subtracting the minimum observed value, determined for each specific band, from all pixel values in each respective band. This method is based on the assumption that the reflectance from these features, if the atmosphere is clear, should be very small, if not zero
Contrast enhancement	Contrast enhancement involves changing the original values so that more of the available range is used, increasing the contrast between targets and their backgrounds
Linear contrast stretch	This involves using the minimum and maximum brightness values in the image. A linear stretch uniformly expands a small range to cover the full range of values from 0 to 255. Enhances the contrast in the image with light-toned areas appearing lighter and dark areas appearing darker, making visual interpretation much easier
Histogram equalization	Assigns more display values to the frequently occurring portions of the histogram. In this way, the detail in these areas will be enhanced relative to those areas of the original histogram where values occur less frequently
Spatial filtering	Spatial filters are designed to highlight or suppress specific features in an image based on their spatial frequency
Convolution filtering	Involves moving a "window" (e.g., 3 × 3, 5 × 5, etc.) over each pixel in the image, applying a mathematical calculation using the pixel values under that window, and replacing the central pixel with the new value. Image has been filtered and a "new" image has been generated. By varying the calculation performed and the weightings of the individual pixels in the filter window, filters can be designed to enhance or suppress different types of features
Image transformations	Involve the manipulation of multiple bands of data to generate "new" images from two or more sources which highlight particular features or properties of interest, better than the original input images

7.4 Change Detection Algorithms

The core problem in change detection involves the separation of pixels that are "significantly" different from those that are not across preceding images in a multitemporal sequence. The pixels identified as being different comprise the "change mask" which can be the consequence of a combination of underlying factors, several having no relevance to the environmental process(es) involved. Ideally, the method of analysis should produce a change mask that does not contain unimportant or nuisance forms of change that have been induced by senor–scene–environmental irregularities (Radke et al. 2003). Because the distinction between "significantly different" and "unimportant" varies with application, selecting the appropriate change detection algorithm can be a difficult task. In a remote sensing application the method of detecting change is therefore genuinely distinct from the more complicated problem of understanding change. Understanding change places the temporal shifts displayed by a pixel over time into a semantic context that connects radiometric evidence to theory, process, and the dynamics of the environmental system.

The utility of a change detection algorithm is determined principally by its capacity to delineate apparent image change. Delineation can follow either an algebraic or classification-based scheme, and each broad approach has its advantages and limitations (Singh 1989; Lu et al. 2004).

7.4.1 Algebraic Techniques

This family of change detection algorithms employs mathematical operations to produce a change mask from remotely sensed imagery. Methods that fall into this category include the following:

- *Image differencing* – Intuitively appealing, the image differencing technique involves subtracting on date (T_1) of imagery from a second date (T_2) that have been subject to precise geometric registration (Fig. 7.6). The process, suggested by the relation:

$$\Delta_{ij} = (T_1) - (T_2),$$

  proceeds on a pixel-by-pixel basis creating a image expressing the results of subtracting the brightness value of pixel i and T_1 from the brightness value of pixel i and T_2. This widely applied algorithm performs well in a range of settings. The logic is also comparatively straightforward. A pixel that has undergone no change will record a value of zero. Differences between dates are defined when Δ moves in either a positive or negative direction from the initial date/value of pixel i. Although easy to comprehend, the value of Δ can be small, such as when T_1 and T_2 are not well separated spectrally (i.e., $40 - 43 = -3$). Therefore, a critical aspect of this form of change detection involves the selection of a change threshold; a value chosen by the analyst that is used to clearly separate real (interesting) change from no change (noise). A secondary and related issue with the image differencing approach, which is common to most algebraic change detection techniques, surrounds the problem of interpretation. Even with the selection of a change threshold, what determines "interesting" change? Since Δ can assume positive and negative values, ascribing meaning to Δ relies on the context of the problem. In general, certain directional trajectories can suggest unique change patterns or trends. For example, positive values

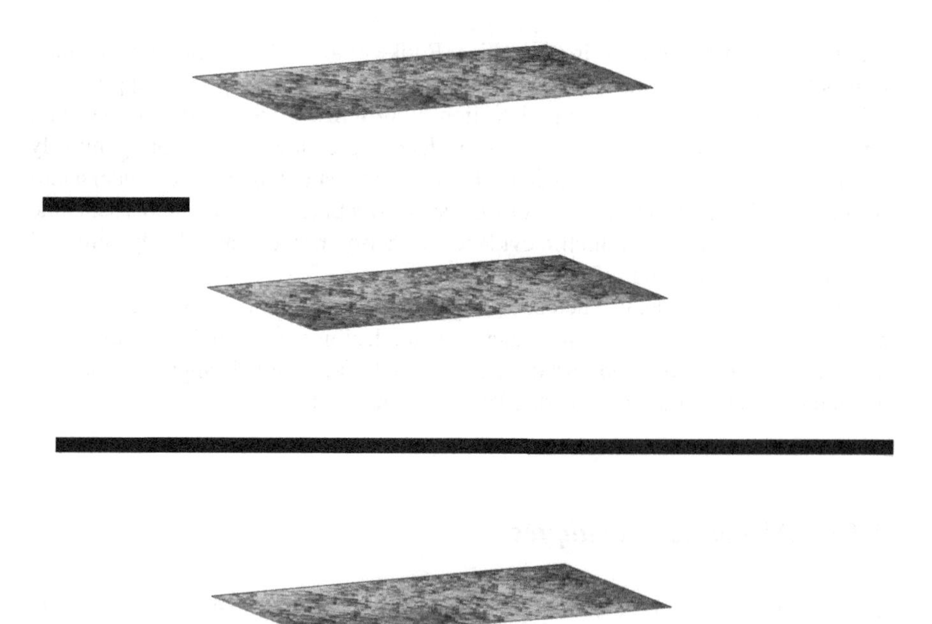

Fig. 7.6 Simple image subtraction

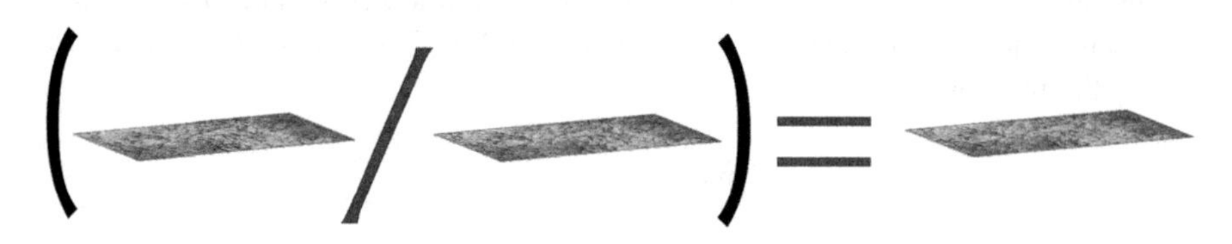

Fig. 7.7 Simple band ratio approach

imply movement from relatively low reflectance (radiance) surfaces to higher reflectance (radiance) surfaces. Such transitions may be indicative of vegetative cover changing to exposed soil or to a concretized state. Conversely, high reflectance to low reflectance shifts may suggest the maturation of vegetative cover over time or the aging of built surfaces such as roof tops or roads.

- *Image ratioing* – This method calculates the ratio of two registered images on a pixel-by-pixel basis. The division of

$$\Delta_{ij} = (T_1)/(T_2),$$

yields a quotient where the no-change condition is expressed as 1, and the values less than or greater than 1 can be interpreted as some indicator of change (Fig. 7.7). Because band ratioing is also an image enhancement operation, this method can also reduce the impacts associated with sun angle contrasts, shadow, and topography. Similar to the image differencing technique, ratioing is also dependant on the selection of a change versus no-change threshold and subject to the same interpretation issues.

- *Image regression* – Using a regression-based approach to the detection of change is based on the assumption that a pixel from time T_1 is a linear function of the time T_2 pixel. Regression analysis establishes the relationship between the two dates, then estimates pixel values of the second date image. Subtracting the regressed image from time T_1 forms an expression of difference (change) according to:

$$Diff x_{ij} = X_{ij}(T_2) - X_{ij}(T_1),$$

where X is the regressed image. Once again a threshold is selected to partition $Diffx_{ij}$ to identify area of real (interesting) change.

- *Change vector analysis* – According to this method, the brightness value of a pixel occupies a location, an (x, y), in measurement space defined by the brightness ranges expressed by two or more spectral bands. When that value changes over time, the location of that pixel (P) will also change. This difference in position is explained by a change vector (C_i) such that:

$$C_i = P_{(i,y)} - P_{(i,z)}.$$

For pixel i, its shift is the difference between year y and year z. The magnitude of this change is expressed as the Euclidean distance separating y and z. The direction of the change vector C_i is measured by the direction of the angle between year y and year z in measurement space. The direction of this angle suggests the nature of the change process. Typically, the no-change condition is described by a zero-length vector; however, the trajectory that defines interesting change requires the selection of a change threshold.

7.4.2 Classification-Based Techniques

Classification-based methods of change detection eliminate the reliance on radiance differences over time as the determinant of change. Instead, change over time is expressed as categorical shifts in the labels assigned to pixels. When compared to algebraic approaches, classification techniques exploit informational changes in the disposition of classes rather than the spectral differences explained by pixel measurements over time. Three classification-based approaches to change detection dominate:

- *Postclassification change detection* – This change detection procedure requires the independent classification of a multitemporal data set into a sequence of thematic maps. The success of this approach depends on the thematic comparison of the classified images on a pixel-by-pixel basis. Postclassification change detection demands the adoption of a standard system of classification as well as the use of well-defined accuracy guidelines to insure that map quality is consistent across the analytic time horizon. Lack of consistency will frustrate comparative analysis and produce unreliable estimates of change. Comparative analysis generally employs some form of cross-image tabulation to generate a detailed matrix expressing categorical differences between dates. When compared to algebraic approaches, postclassification avoids the need for strict image normalization and provides important flexibility since the assessment of change can be tailored to informational categories that are relevant to the monitoring program. The use of categorical distinctions as the basis of change versus no-change simplifies the problem of designating whether or not real change has occurred and eliminates the use of change thresholds. Categorical designations are support the extraction of essential "From–To" information from the data which can make significant contributions to understanding change in relation to theory and process.

- *Unsupervised change detection* – A variant of the postclassification method, this technique selects spectrally similar groups of pixels, applies a clustering algorithm to the time T_1 image to produce primary class, then proceeds to identify spectrally similar groups in the time T_2 image to form primary class at that time step in the data. Change, according to this approach, is based on differences in pixel membership to the primary classes (clusters) between time T_1 and time T_2. Although the method is relatively simple, the unsupervised method can make it difficult to identify and label change classes. This bottleneck often hampers creation of a complete matrix of change information.
- *Multidate composite analysis* – This technique combines images from two or more dates into a single multilayer data file (stack) (Fig. 7.8). The composite multilayer image can then be subjected to either cluster analysis to identify change and no-change categories or principle component analysis (PCA) to produce a linear combination of the scene over time to create a change composite (Healey et al. 2005; Lasaponara 2006; Cakir 2006).

7.5 The Thresholding Question

The methods of change detection reviewed in the previous section are by no means exhaustive. The techniques selected provide tractable approaches that fit well within the larger problem of environmental monitoring. The advantages and limitations of each are clearly apparent (Table 7.3), and without radiometric calibration, geometric registration, methodological consistency, and a sustained sense of purpose with respect to the enabling monitoring program, useful characterization of processes acting on the environment will remain elusive. Perhaps one of the more vexing

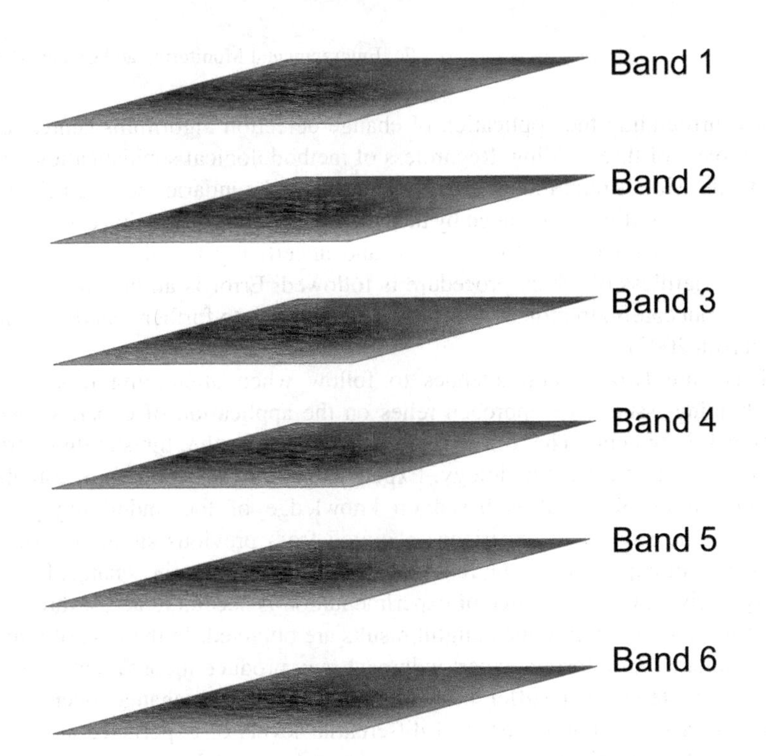

Fig. 7.8 A multilayer image stack

Table 7.3 Advantages and disadvantages associated with the selected change detection algorithms

Method	Advantage	Disadvantage
Multidate composite image	Requires only a single classification	Non-quantitative and no to–from information is generated
Image algebra	Efficient and does not require classification of individual scenes	Requires careful selection of thresholds and no to–from information is generated
Postclassification	Provides to–from information and produces thematic maps for each time period	Success depends on accuracy of the individual thematic classifications and requires two independent classifications
Change vector analysis	Uses all bands and does not require classification of individual scenes; vector information can provide to–from information	Computationally intensive and difficult to interpret

issues surrounding the application of change detection algorithms centers around the problem of thresholding. Regardless of methodological sophistication, at some point a decision is required in order to establish the boundaries between change and no-change conditions as dictated by the data. Although this decision can be deferred using postclassification methods, error and uncertainty remain embedded in the results regardless of which procedure is followed. Error is an unavoidable reality and without careful treatment its effects can migrate into further analytic operations (Rogerson 2002).

There are two general avenues to follow when attempting to resolve the thresholding issue. One approach relies on the application of expert subjective–technical judgment. The other means of resolving the thresholding problem employs a statistical methodology. Expert judgment runs the risk of establishing arbitrary thresholds, but is based on knowledge of the underlying processes involved and subjective experience acquired from previous situations. Based on the quality of expert judgment, reasonable thresholds separating change from noise can be derived. Often a degree of experimentation is needed to adjust thresholds to refine the solution before meaningful results are obtained. In the case of long-term environmental monitoring, expert judgment may produce inconsistent results over time. Consistency can suffer particularly when experts change over time and opinions vary as a consequence of differential levels of experience and training. Reducing the potential limitations imposed by reliance on human expertise as led to the adoption of auto-thresholding methods.

Auto-thresholding techniques draw on the statistical properties on the change data. One simple method of auto-thresholding uses standard deviation units. According to this approach "interesting" change identifies conditions that occur beyond one standard deviation from the mean value. Other more detailed methods include:

- *Otsu's Histogram Shape-Based Method* (Otsu 1979) – This strategy is based on discriminant analysis and uses the zero-and first-order cumulative moments of the histogram to calculate the value of the thresholding level.
- *Tsai's Moment-Based Method* (Tsai 1985) – This method determines the threshold that preserves the first three moments of the input image.
- *Kapur's Entropy-Based Method* (Kapur et al. 1985) – This method considers the thresholding problem as two classes of events, each characterized by a probability density function. Using probability, this technique maximizes the sum of the entropy of the two probability density functions to identify a point of convergence that become the threshold value.
- *Kittler's Histogram Shape Method* (Kittler and Illingworth 1986) – This method determines a threshold value by approximating the histogram as a Gaussian distribution and identifies a cut-off point using a cost function derived from Bayes classification rule.

7.6 Summary

Environmental change, defined as an alteration or disturbance of the environment by human or natural ecological processes, and the monitoring of Earth's environments to detect those changes was the focus of this chapter. Remote sensing technologies have played an important role in environmental monitoring. Identifying the principles that guide the implementation of a monitoring program and integrating remote sensing into that schema directed discussion to the question of sensing change and the methods developed to study change using remotely sensed data. A selection of change detection algorithms were examined to complete the analysis of change and the systematic monitoring of key indicators in the landscape system over time.

References

Brydges, T. (2004) Basic Concepts and Applications of Environmental Monitoring, in Wiersma, G. (ed) Environmental Monitoring, CRC Press, Boca Raton, FL, 83–109.

Cakir, H., Khorram, S. and Nelson, S.A. (2006) Correspondence Analysis for Detecting Land Cover Change, Remote Sensing of Environment, 102, 306–317.

Coppin, P., Jonckheere, I., Nackaerts, K., Muys, B. and Lambin, E. (2004) Digital change detection methods in ecosystem monitoring: A review. International Journal of Remote Sensing, 25(9), 1565–1596.

Gao, J. (2009) Digital Analysis of Remotely Sensed Imagery, McGraw-Hill, New York, 645p.

Gross J.E. (2003) Developing Conceptual Models for Monitoring Programs. National Park Service US Department of the Interior. [Cited 5 Jul 2011] Available from URL:http://science.nature.nps.gov/im/monitor/docs/Conceptual_Modelling.pdf.

Duro, D., Coops, N., Wulder, M. and Han, T. (2007) Development of a large area biodiversity monitoring system driven by remote sensing, Progress in Physical Geography, 31, 235–260.

Dymond, J. Begue, A. and Loseen, D. (2001) Monitoring land at regional and national scales and the role of remote sensing, JAG 3, 162–175.

Healey, S.P., Cohen, W.B., Yang, Z.Q. and Krankina, O.N. (2005) Comparison of Tasseled Cap-based Landsat data structures for use in forest disturbance detection. Remote Sensing of Environment, 97, 301–310.

Jat, M., Garg, P. and Khare, D. (2008) Monitoring and modeling of urban sprawl using remote sensing and GIS techniques, International Journal of Applied Earth Observation and Geoinformation, 10, 26–43.

Jones, D., Hansen, A., Bly, K., Doherty, K., Verschuyl, J., Paugh, J., Carle, R. and Sory, S. (2009) Monitoring land use and cover around parks: a conceptual approach, Remote Sensing of Environment, 113, 1346–1356.

Kapur, J., Sahoo, P. and Wong, A. (1985) A new method for gray level picture thresholding using entropy of the histogram, Computer Vision and Graphics Image Processing, 29, 273–285.

Khorram, S., Biging, G., Chrisman, N., Colby, D., Congalton, R., Dobson, J., Ferguson, R., Goodchild, M., Jensen, J. and Mace, T. (1999) Accuarcy assessment of Remote Sensing

Devised Change Detection, ASPRS Monograph, American Society for Photogrammetry & Remote Sensing, Bethesda, Maryland, 64 p.

Kittler, J. and Illingworth, J. (1986) Minimum error thresholding, Pattern Recognition, 19, 41–47.

Kleinod, K., Wissen, M. and Bock, M. (2005) Detecting vegetation changes in a wetland area in Northern Germany using earth observation and geodata Journal for Nature Conservation, 13, 115–125.

Koltunov, A., Ustin, S. (2007) Early fire detection using non-linear multitemporal prediction of thermal imagery Remote Sensing of Environment, Vol. 110, Issue: 1, September 14, 2007. pp. 18–28.

Lasaponara, R. (2006) On the use of principal component analysis (PCA) for evaluating interannual vegetation anomalies from SPOT/VEGETATION NDVI temporal series Ecological Modelling, 194, 429–434.

Leimgruber, P., Christen, C. and Laborderie, A. (2005) The impact of Landsat Satellite Monitoring on Conservatoin Biology, Environmental Monitoring and Assessment, 106, 81–101.

LeMarie, M., van der Zaag, P., Menting, G., Baquete, E. and Schotanou, D. (2006) The use of remote sensing for monitoring environmental indicators: the Case of the Incomati estuary, Mozambique, Physics and Chemistry of the Earth, 31, 857–863.

Lookingbill, T., Gardner, R., Townsend, P. and Carter, S. (2007) Conceptual Models as Hypotheses in Monitoring Urban ladscapes, Environmental Management, 40, 171–182.

Lu, D., Mausel, P., Brondizio, E. and Moran, E. (2004) Change detection techniques International Journal of Remote Sensing, 25, (37):2365–2401.

Lunetta, R. and Elvidge, C. (eds) (1999) Remote Sensing Change Detection, CRC Press, Boca Raton, FL, 350p.

Lunetta, R. Johnson, D., Lyon, J. and Crotwell, J. (2004) Impacts of Imagery Temporal Frequency on Land Cover Change Detection Monitoring, Remote Sensing of Environment, 89, 4444–454.

Lyon, J., Yuan, D., Lunetta, R. and Elvidge, C. (1988) A Change Detection Experiment Using Vegetation Indices, Photogrammetric Engineering and Remote Sensing, 64, 143–150.

Mas, J. (1999) Monitoring land-cover changes: a comparison of change detection techniques, International Journal of Remote Sensing, 20, 139–152.

Muchoney, D. and Haack, B. (1994) Change Detection for Monitoring Forest Defoliation, Photogrammetric Engineering and Remote Sensing, 60, 1243–1251.

Otsu, N. (1979) A threshold selection method from ray-level histograms, IEEE Transactions on Systems, Man and Cybernetics, 9, 2–66.

Radke, R., Andra, S. Al-Kofahi, O. and Roysam, B. (2003) Image Change Detection Algorithms: A Systematic Survey, IEEE Transactions on Image Processing, 14, 294–301.

Ridd, M. and Liu, J. (1998) A Comparison of Four Algorithms for Change Detection in an Urban Environment Remote Sensing of Environment, 63, 95–100.

Roberts, K.A. (1991) Field monitoring: Confessions of an addict, in Goldsmith, B. (ed) Monitoring for conservation and ecology, Chapman & Hall, London, 179–211.

Rogerson, P. (2002) Change Detection Thresholds for Remotely Sensed Images, Journal of Geographical systems, 4, 85–97.

Rachon, G., Johannsen, C., Landgrebe, D., Engel, B., Harbor, J., Majumder, S. and Biehl, L. (2003) Remote sensing as a tool for achieving and monitoring progress toward sustainability, Clean Technologies and Environmental Policy, 5, 310–316.

Rogan, J., Frankln, J. and Roberts, D. (2002) A comparison of methods for monitoring multitemporal vegetation change using Thematic Mapper imagery Remote Sensing of Environment, 80, 143–156.

Singh, A. (1989). Digital change detection techniques using remotely sensed data. International Journal of Remote Sensing, 10(6), 989–1003.

Spellerberg, I. (2005) Monitoring Ecological Change, Cambridge University Press, 410p.

Spitzer, D. (1986) On the Application of Remote Sensing for Environmental Monitoring, Environmental Monitoring, 7, 263–271.

Spitzer, D. (1986) On the Application of Remote Sensing for Environmental Monitoring, Environmental Monitoring, 7, 263–271.

Tsai, W. (1985) Moment-preserving thresholding, Computer Vision and Graphics Image Processing, 29, 377–393.

Usher, M.B. (1991) Scientific requirements of a monitoring programme, in Goldsmith, B. (ed) Monitoring for conservation and ecology, Chapman & Hall, London, 15–32.

Verbesselt, J., Hyndman, R., Newnham, G. and Culvenor, D. (2010) Detecting trend and seasonal changes in satellite image time series Remote Sensing of Environment, 114, 106–115.

Wang, F. and Xu, Y. (2010) Comparison of remote sensing change detection techniques for assessing hurricane damage to forests, Environmental Monitoring and Assessment, 162, 311–326.

Chapter 8
Thermal Sensing and Anomaly Detection

Thermal remote sensing directs our attention to the role of temperature and an indicator of land surface conditions and contrasts (Diak 1995; Xian and Crane 2006; Quattrochi and Luvall 1999). Recognizing that all surface objects with temperatures above absolute zero (0 K) emit electromagnetic radiation, thermal remote sensing concentrates on the techniques involved in measuring the electromagnetic energy emitted by objects at the surface and the physical mechanisms that influence how object behave within the thermal portion of the electromagnetic spectrum. Our interest in temperature as an environmental indicator and medium for land surface characterization is somewhat analogous to the role temperature plays in the diagnosis of human health. Features and surface objects that comprise the environmental system emit radiant energy as a function of their temperature, should the thermal characteristics of surface objects deviate from an ambient state, numerous questions can be entertained to account for the observed thermal differences and the spatial patterns they create (Kustas and Anderson 2009; Weng 2009). Temperature in this regard suggests the presence of active processes that modify how objects absorb heat energy, store heat energy, and radiate heat energy. Sensing the thermal disposition of the environment can therefore reveal certain clues and patterns indicative of behaviors or changes that cannot be sensed using the reflective portions of the electromagnetic spectrum (Luvall et al. 2000). In this chapter, we will examine the principles of thermal remote sensing, describe how thermal energy is measured by satellite sensors, and explore the role of thermal data in environmental analysis and anomaly detection.

8.1 Principles of Thermal Remote Sensing

Thermal energy has been widely used in geologic and military remote sensing applications (Jensen 2007). The first declassified thermal data obtained from satellite sensors were used primarily for meteorological observation and weather forecasting. The civilian use of thermal imagery for land surface remote sensing

J.K. Lein, *Environmental Sensing: Analytical Techniques for Earth Observation*,
DOI 10.1007/978-1-4614-0143-8_8, © Springer Science+Business Media, LLC 2012

Table 8.1 Characteristics of the TIMS scanner

Swath width	31.3 km
Spatial resolution	50 m
Spectral resolution	Wavelength (μm)
Band 1	8.2–8.6
Band 2	8.6–9.0
Band 3	9.0–9.4
Band 4	9.4–10.2
Band 5	10.2–11.2
Band 6	11.2–12.2

dates roughly to the early 1980s with the development of the thermal infrared multispectral scanner (TIMS). This thermal scanner acquired infrared energy in thermal six bands (Table 8.1). The success of this program contributed to the deployment of thermal sensors on Landsat Thematic Mapper series (Landsat TM 4 and Landsat TM 5) beginning in 1982. Presently, thermal imagery is acquired by several Earth observational satellites and sensors including ASTER, MODIC, and the AVHRR. The thermal products generated by these systems, however, are not interpreted in the same manner as the measurements obtained from the reflected bands (Jensen 2007). Rather, treatment of thermal data requires an understanding of the governing processes that control how thermal energy interacts with features of the land surface and how these properties of the surface direct the rate and intensity of these interactions.

Thermal sensors measure the radiant energy of Earth's surface features. This form of electromagnetic flux is an external manifestation of an objects energy state based on it temperature. The energy emitted by the object is used to quantify its radiant temperature. Radiant temperature is strongly influenced by the ability of an object to absorb heat energy either by conduction, which explains the transfer of heat through a material by molecular contact, convection, describing the transfer of heat energy through the physical movement of heated material, or radiation, which explains heat transfer in the form of electromagnetic waves. Objects at the Earth's surface receive thermal energy primarily through radiative transfer from the sun. However, conductive processes also transfer heat from below the surface. Although the magnitude of this flux is appreciably small, there are important geographic variations in this energy flow and can be detected remotely. In many of these regional examples, the energy transferred by conduction are indicative of thermal springs, volcanoes, and patterns of subsurface heat flow that are modified by local geologic features and activities (Sabins 2007).

The relative power of an object or material to emit heat by radiation is determined by its emissivity. Emissivity is defined as the ratio of energy radiated by a black body at the same temperature as the object of interest. A black body is an idealized object that absorbs all of the electromagnetic radiation that it receives. This theoretical object would have an emissivity (ε) of one while any environmental surface would display an emissivity less than one. Although emissivity is a dimensionless quantity (has no physical unit related to it), objects that are composed of

Table 8.2 Emissivity from selected surfaces and objects

Water	0.960
Fresh snow	0.986
Coniferous needles	0.971
Leaves	
Corn, beans	0.940
Cotton, tobacco	0.980
Sugar cane	0.940
Dry peat	0.970
Wet peat	0.983
Dry fine sand	0.949
Wet fine sand	0.962
Thick green grass	0.986
Thin green grass on wet clay soil	0.975
Forest, deciduous	0.950
Forest, coniferous	0.970
Fur, hair	
Mouse	0.940
Squirrel	0.980
Hare, wolf	0.990
Human skin	0.980
Glass	0.940

Emissivity expressed as a fraction of the radiant energy of a blackbody as the same temperature from Geiger, R., Aron, R., and Todhunter, P. (2003) The Climate Near the Ground, Rowman and Littlefield

dark or dull materials tend to define values of emissivity approaching one while more reflective surfaces tend to describe lower emissivity values (Table 8.2). In general, emissivity depends on several factors including temperature, the angle of emission and wavelength. In the majority of applications of thermal imagery, it is assumed that an objects spectral emissivity and absorptivity do not depend on wavelength. This assumption, referred to and the gray body assumption, allows values of emissivity to be held constant. However, there are important differences in the fraction of radiation absorbed by an object relative to wavelength that does produce subtle contrasts in emissivity. These deviations from the ideal behavior of a black body are resolved by taking into consideration both the structural geometry and chemical composition of an object as expressed according to Kirchoff's Law of Thermal Radiation ($1 - \varepsilon =$ reflectivity). For this reason, the radiant temperature of an object at the surface will always be less than its kinetic temperature (the temperature we measure and make common reference to using a thermometer).

The radiant temperature of an object describes its radiant flux. This definition of temperature can be measured remotely using sensors designed to detect electromagnetic radiation in the thermal infrared wavelengths of the spectrum (Fig. 8.1). The thermal portion of the spectrum occupies wavelengths occupying the range from 3.0 to 14.0 μm. However, not all of the wavelengths in this interval are

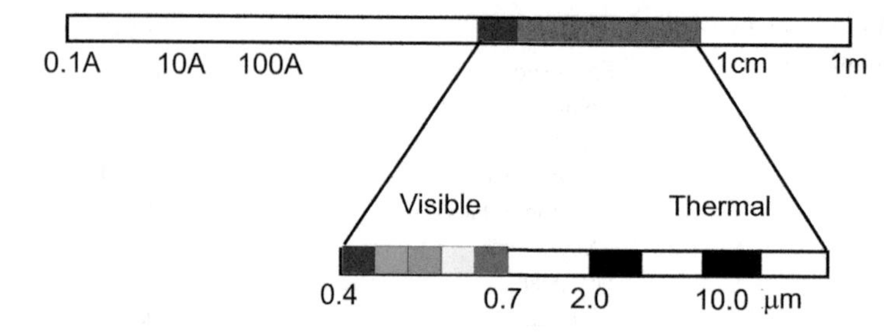

Fig. 8.1 The thermal portion of the electromagnetic spectrum

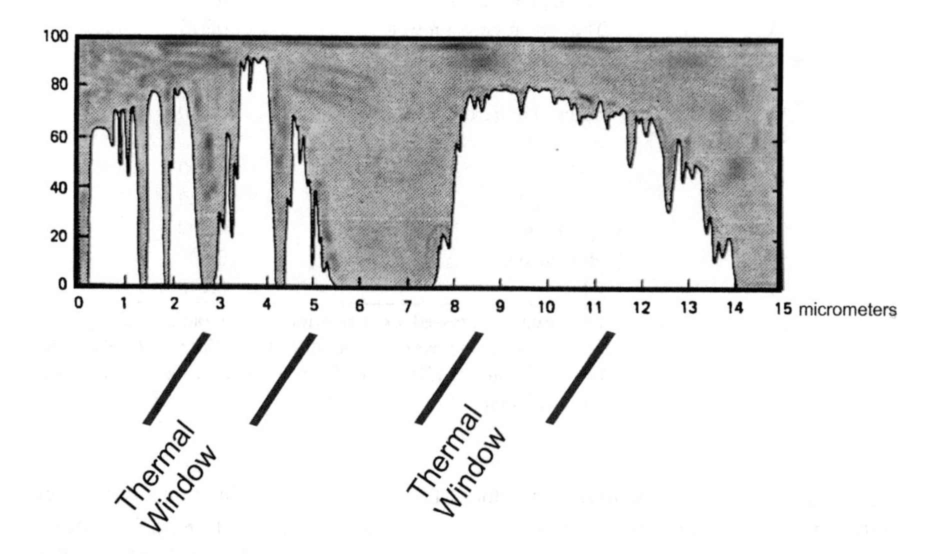

Fig. 8.2 Thermal windows useful for remote sensing applications

transmitted uniformly through the atmosphere. As with reflected energy, the atmosphere acts to degrade the intensity and effect the spectral composition of energy recoded by a thermal sensor. Atmospheric constituents such as carbon dioxide, ozone, and water vapor absorb energy at particular wavelengths in the thermal region, while gases and suspended particles in the atmosphere can emit radiation based on their temperatures, which add to the quantity of the radiant energy that is being sensed. There are two atmospheric windows where radiant energy is transmitted with little or no absorption. The first thermal window falls with the spectral range of 3.0–5.0 μm. The second is located along the wavelengths between 8.0 and 14.0 μm (Fig. 8.2). At the top of Earth's atmosphere a narrow absorption band can be identified between 9.0 and 10.0 μm, corresponding with the Earth's ozone layer. To effectively record thermal energy satellite systems avoid this interval and tend to operate primarily within the wavelength interval from 10.5 to 12.5 μm.

Sensing the radiant flux of an object at constant temperature varies as a function of wavelength. Locating where in the thermal spectrum an object displays its radiant energy peak (λ_{max}), which is defined by Wien's displacement law, expressed as:

$$\lambda_{max} = \frac{2{,}897}{T_{rad}},$$

where T_{rad} is radiant temperature in degrees Kelvin (Absolute temperature) and the value 2,897 is a physical constant. Using this relationship, the wavelength where an object's radiant energy is at its highest can be found by substituting a value of temperature for T_{rad}. For example, the radiant energy peak for a roof top surface at 62°C (145°F) can be estimated by

$$\lambda_{max} = \frac{2{,}897}{336\,\mathrm{K}},$$

which solves to a wavelength of 8.62 μm. A thermal infrared sensor operating in the 8.0–14.0 μm region would detect this roof top surface. The thermal property of an object greatly influences the calculation of λ_{max} and frustrates the simple characterization of objects using radiant temperature. Because all surface materials have the capability to conduct, store, and exchange heat energy, the ability of a material to absorb heat energy becomes a direct function of its thermal (heat) capacity. The heat capacity of a object explains the quantity of energy needed to raise the temperature of 1 g of the material in question by 1°C. Differences in heat capacity influence how well energy is stored by a material and that ability to store energy has a direct effect on its radiant profile. Water, for example, has a high thermal capacity. Consequently, water bodies heat and cool more slowly than a sandy surface, which due to their heat capacity, tend to heat rapidly and cool rapidly. The ability of heat energy to pass through a material is a function of its thermal conductivity (Φ). This quantity is expressed as the number of calories that will pass through 1 cm^3 of a material in 1 s when two opposite faces of the material are maintained at a 1°C difference in temperature. The final thermal property of an object with relevance to thermal remote sensing is its thermal inertia ($\mathcal{F}$). Thermal inertia measures the thermal response of a material to changes in temperature and is expressed in calories per cm^2 per second square root per 1°C according to the expression:

$$\mathcal{F} = \sqrt{(\Phi * D * \xi)},$$

where Φ is the object's thermal conductivity, D represents its density in g/cm^{-3} and ξ is the object's thermal capacity.

8.2 Moving to Measurement

The potential of Thermal Infrared remote sensing in environmental analysis concentrates interest on the radiation emitted from terrain features as a means of mapping and characterization. Understanding how thermal energy is partitioned

Table 8.3 Factors influencing emissivity

Tone of objects	Darker objects are better absorbers and better emitters
Surface roughness	The rougher the surface relative to the wavelength the greater the surface area and greater the potential for absorption and re-emission
Moisture content	The more moisture content, the greater the ability to emit energy
Field of view	
Viewing angle	

across a landscape and the magnitude and contrasts displayed by the thermal characteristics of landscape features helps to explain the fluxes and redistributions of materials among these features (Quattrochi and Luvall 1999). As we noted earlier in Chap. 4, characterization implies measurement, and the temperature extremes as well as the heating and cooling rates of landscape features can supply information pertaining to the type and condition of an object. In general, when the land surface is observed using thermal infrared data, darker image tones represent cooler radiant temperatures while lighter tones in the image represent warmer radiant temperatures. This pattern is analogous to the concept of brightness as applied to the interpretation of reflected infrared bands. This common thread permits us to associate brightness to temperature at the surface and relate its status to a measurable quantity. Translating brightness into unit of measure that convey more meaningful information, however, requires detailed calibration of the thermal infrared data (TIR).

Radiometric calibration of TIR data first requires compensating for atmospheric absorption and emission in the radiance arriving at the sensor, then involves corrections that take into account surface emissivity effects. Several of the factors that influence the emissive power of surface objects are listed in Table 8.3. The process begins with the assumption that good energy absorbers are also good energy emitters and seeks to remove noise effects that hamper the retrieval of surface temperature estimates. A thermal sensor measures the emitted spectral radiance it receives (L_λ). This quantity is called the brightness temperature, symbolized as T_{B} and is expressed according to the relation:

$$T_{\mathrm{B}} = \frac{\lambda_4}{2kc} L_\lambda,$$

where λ = wavelength, k = Planck's constant, c = speed of light, and L = spectral radiance.

With knowledge of the emissivity (Θ) of an object or material, brightness temperature can be related directly to kinetic temperature by the formula:

$$T_{\mathrm{B}} = \Theta^{1/4} T_{\mathrm{kin}}.$$

This association suggests that brightness temperature can be used to monitor temperature as well as the properties of objects or materials based their emissivity.

Fig. 8.3 Typical thermal image (Landsat ETM+)

Compensating for atmospheric contamination of the spectral radiance received by the sensor requires some form of external referencing. One common approach uses empirical measurements taken in the field with either:

1. A thermometer measuring true kinetic temperature or
2. A hand-held radiometer measuring radiant temperature

Using the data sampled from the field, regression analysis is used to establish the degree of fit between field-sampled locations and the corresponding brightness temperature obtained via the thermal imagery. Although this approach is efficient, the empirical method does not take the atmosphere directly into account, nor does it adjust for emissivity contrasts. Incorporating emissivity into the calculation of brightness temperature is estimated using

$$T = a \times \Theta \times {T_{\mathrm{kin}}}^{4} + b,$$

where (*a*) is the slope of the relationship established using the regression fit and (*b*) is the *y*-intercept (Jensen 2007). Rewritten, this equation permits the calculation of "true" kinetic temperature for every pixel in the image such that

$$T_{\mathrm{kin}} = \left(\frac{\mathrm{BV}_{ij} - b}{a \times \Theta} \right)^{1/4}.$$

Solving for T_{kin} not only supplies an analysis with estimates on temperature, but facilitates the production of maps depicting the spatial distribution of surface temperature (Fig. 8.3). For the majority of environmental applications, particularly those designed to contribute to environmental monitoring, change detection or special purpose investigations such as those conducted in association with environmental compliance enforcement operations, collecting reference temperatures in the field may not be practical or feasible. In these situations alternative approaches to land surface temperature estimation are available (Sobrino et al. 2004; Dash et al. 2002; Schmugge et al. 1998). The alternative solutions to temperature estimation apply theoretically derived algorithms based on the linearization of the Planck function. The approaches available include

- Split-window algorithms – Using these techniques, brightness temperatures recorded on one thermal channel are corrected for atmospheric contamination using the linear difference between the brightness temperature in a second thermal channel. The general form of the split-window method can be expressed as

$$T_{\mathrm{LS}} = a_1 T_{\mathrm{i}} + b_1 T_{\mathrm{i}} + c_1,$$

  where the coefficients a_1, b_1, and c_1 are estimated from computational models (Prata 1994).
- Single-channel algorithms – This method utilizes the radiance in one thermal channel and corrects for atmospheric effects using a radiative transfer model (examples include the MODTRAN and LOWTRAN algorithms). The radiative transfer model solves for atmospheric transmission based on vertical and horizontal profiles of temperature and humidity supplied from vertical sounding instruments such as radiosonde, using data obtained from local weather forecasting centers. A correction algorithm that avoids dependence on radiosounding was introduced by Qin et al. (2001) for use with Landsat TM thermal data and a somewhat more tractable approach without specific sensor requirements has been offered by Jimenez-Munoz and Sobrino (2003). This generalized technique estimates land surface temperature from one thermal channel based on the equation:

$$T_{\mathrm{s}} = \gamma\,[\varepsilon - 1\,(\varphi L_{\mathrm{sensor}} + \varphi_2) + \varphi_3] + \sigma$$

  with

$$\gamma = \left\{ \frac{c_2 L_{\mathrm{sensor}}}{T_{2\mathrm{sensor}}} \left[\left(\frac{\lambda}{c_1} \right) L_{\mathrm{sensor}} + \lambda^{-1} \right] \right\}^{-1}$$

  and

$$\sigma = -\gamma L_{\mathrm{sensor}} + T_{\mathrm{sensor}},$$

where L_{sensor} = at satellite sensor radiance, T_{sensor} = at satellite sensor brightness temperature in K, and λ = effective wavelength (11.76 μm for Landsat TM Band 6).

$$C_1 = 1.19104 \times 10^8 \text{ W } \mu\text{m}^4 \text{ m}^{-2} \text{ sr}^{-1}$$

$$C_2 = 14387.7 \, \mu\text{m K}.$$

The atmosphere functions (symbolized as φ) are derived as a function of total atmospheric water vapor content (w). For Landsat TM Band 6 those equations are

$$\varphi_1 = 0.147114w_2 - 0.15583w + 1.1234,$$

$$\varphi_2 = -1.1836w_2 - 0.37607w - 0.52894,$$

$$\varphi_3 = -0.04554w_2 + 1.8719w - 0.39071.$$

Depending on the nature of the problem rigorous estimation of land surface temperature may not be required. In these examples, it may be appropriate to avoid atmospheric correction if thermal contrast rather than precise temperature measurement is sufficient for the purposes of analysis. Furthermore, it has been observed that the coefficients derived from the various correction algorithms are valid only for the data set used to derive them. A transferability issue acts to constrain the universal application of these functional relationships and introduces estimation error in the results when they are applied beyond the limitations imposed by the defining data set. Therefore, a set of thermal responses for a specific landscape phenomenon measured using a particular TIR sensor cannot reliably be extrapolated to describe the same TIR measurements either from other sensors or from images recorded at different times using the same sensor (Dash et al. 2002; Quattrochi and Goel 1995; Weng et al. 2004). A simplified temperature estimation technique can be offered that yields suitable results. This generic approach involves three main steps to retrieve brightness temperature values from TIR data:

1. Conversion of digital brightness values as recorded by the TIR sensor into units of spectral radiance.
2. Conversion of the spectral radiance values into values of "at-satellite" brightness temperature under the assumption of uniform emissivity at the surface.
3. Correction of the "at-satellite" brightness temperature using spectral emissivity based on land cover units.

A demonstration of this procedure using ASTER and Landsat TM thermal imagery follows. For the Landsat TM example, spectral radiance is calculated as

$$L_{\text{rad}} = 0.0370588 \times \text{ BV}_{ij} + 32,$$

where BV_{ij} is the digital TIR brightness value recoded by the sensor for pixel (i, j). Converting spectral radiance (l_{rad}) to "at-satellite" brightness temperature is accomplished from

$$T_{\mathrm{B}} = \frac{K_2}{\ln(K_1/L_\lambda + 1)},$$

where T_{B} is given in degrees Kelvin and K_1 and K_2 are pre-launch calibration coefficients. For Landsat ETM + the coefficient for $K_1 = 666.09$ and for $K_2 = 1{,}282.71$.

For the ASTER sensor conversion to spectral radiance is determined by

$$L_{\mathrm{rad}} = \mathrm{Gain} \times (\mathrm{BV}_{ij} - 1),$$

where the gain for each of the ASTER TIR bands are

$$\begin{aligned}
\mathrm{Gain}(10) &= 0.006882\\
\mathrm{Gain}(11) &= 0.006780\\
\mathrm{Gain}(12) &= 0.006590\\
\mathrm{Gain}(13) &= 0.005693\\
\mathrm{Gain}(14) &= 0.005225
\end{aligned}$$

Conversion to "at-satellite" brightness temperature is accomplished by the formula:

$$T_{\mathrm{B}} = \frac{K_2}{\lambda_{\mathrm{c}} \ln\left[K_1/\lambda_{\mathrm{c}}^5 (L_{\mathrm{rad}} + 1)\right]},$$

where, $K_1 = 1.91 \times 108$, $K_2 = 1.439 \times 104$, and $\lambda_{\mathrm{c}} =$ wavelength.

The final step involves correction for emissivity which will produce an estimate of land surface temperature according to

$$T_{\mathrm{s}} = \left(\frac{1}{\varepsilon^{1/4}}\right) T_{\mathrm{B}}.$$

Estimates of emissivity (ε) can be acquired from look-up tables based on land cover type or derived from NDVI measures based on empirical relationships (Van de Griend and Owe 1993). One generalized approximation for emissivity is given as

$$\mathrm{Emissivity} = \varepsilon_{ij} = 1.0094 + 0.047 \times \ln(\mathrm{NDVI}_{ij}).$$

Although TIR data has tremendous potential for a range of environmental applications, its utility must be balanced against the current methods by which TIR data is collected from Earth-orbiting satellites. First, it is important to recognize that thermal sensors detect radiation from the surface of ground objects, typically over the first 50 μm. Consequently, radiant temperature as recorded may not always be indicative of an object's internal temperature, despite careful corrections and calibrations. To understand the implications of this measurement disparity, we need to only consider the surface of a lake where surface evaporative cooling may produce temperatures cooler than those only a few centimeters below. The sensor can only measure surface radiant energy and not the warmer subsurface conditions. In addition, most operational satellite sensors that acquire easily accessible imagery in the TIR region of the spectrum make their geographic overpass between 10:00 and 11:00 AM local time during the daylight orbit and between 10:00 and 11:00 PM local time during their night-time orbits. Neither of these time periods are ideal for collecting TIR data remotely which is a significant operational constraint when developing a thermal application. Lastly, there are technical limitations that constrain TIR data collection and its overall usefulness. Because there is relatively little radiation emitted by the Earth's surface in the TIR portion of the electromagnetic spectrum, and because satellite sensors that detect TIR radiation have high orbital velocities, the spatial resolution that must be used to sample a scene using emitted radiation is larger than that used to acquire data in the reflective bands. Therefore, the relative coarse spatial resolution of TIR data can be a severe limitation for applications that demand fine spatial detail. Also, TIR detectors must be kept extremely cold during use and require regular calibration to ensure consistent data quality. Despite these potential limiting factors, TIR data offers the capacity to recover measurements about objects that cannot be acquired using reflective infrared bands; measurements that may signal uncharacteristic deviations in the status of the environmental system that move analysis into the realm of anomaly detection.

8.3 Anomaly Detection and Mapping

When conducting studies of the environmental system we conveniently assume the surface in an ambient state where objects that form the landscape of interest organize into what we define as an expected condition. The synoptic view provided by our imagery gives us a spatial perspective where our expectations of "normal" conditions support certain beliefs regarding how the processes constituting the environmental system perform. Thermal measurements along with the image-derived indicators introduced in Chap. 7, serve not only to document the ambient state of the environment but also to communicate curious, conspicuous, and unanticipated patterns that emerge from an otherwise homogeneous background. Thermal data can also exploit the differential heating of objects and surface due to heat capacity contrasts to reveal heat shadows and other aberrations in emittance that contain useful information that is not discernable using reflected infrared imagery. This form of detection highlights the presence of anomalies at the surface and enables their evaluation and spatial representation.

Thermal anomaly detection has received considerable attention in the remote sensing community due in part to the wider availability of thermal data at improved spatial resolutions (Koltunov and Ustin 2007; Kant and Badarinath 2002). With specific reference to environmental analysis and assessment, a thermal anomaly can be defined as a deviation or departure from the normal or expected pattern of

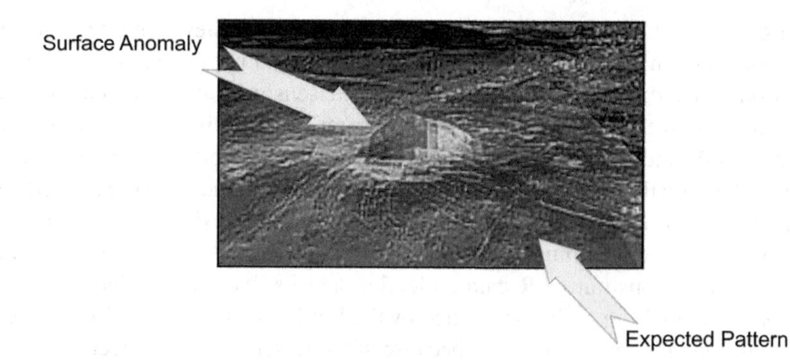

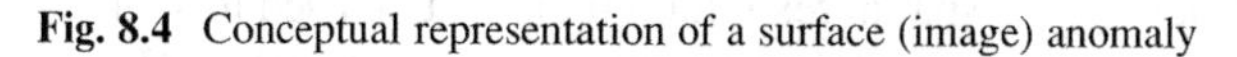

Fig. 8.4 Conceptual representation of a surface (image) anomaly

surface emittance. Such deviations, when detected, emerge from the background as peculiar, irregular, abnormal, and therefore difficult to classify (Ashton and Schaum 1998). Anomaly detection attempts to locate and identify irregular or abnormal patterns at the surface based on the image. In the measurement space defined by thermal imagery, an anomaly can be explained as any pixel that is spectrally different from the background. Conceptually, we can visualize anomalous pixels as spikes or troughs in brightness values whose unique characteristics cannot be attributed to noise or error (Fig. 8.4). Distinctiveness, however, relies on statistical measures that separate difference by applying anomaly thresholds to the imagery. Using the convention of a threshold, pixels can be extracted from the background pattern thereby reducing the likelihood of returning false positives. Conducted in this manner, anomaly detection can be an important component of environmental characterization and monitoring programs, since it is the irregularities and departures from expected conditions that signal reason for concern. We can liken this to the familiar dental X-ray where the shadow on a tooth communicates the presence of decay. At the surface the observed anomaly may be the simple consequence of a random event, or the anomaly may identify

1. The consequences of an action taken (or not taken)
2. A failure in a human construction (such as a pipeline or containment barrier)
3. The footprint of a malicious action or event (such as an illegal dumping of chemical solvents or the clearing of vegetation)
4. The initiation or progression of an unexpected or unanticipated process (such as a chemical spill, wildfire, or infestation)
5. The occurrence of environmental outliers or extremes
6. The presence of a unique environmental feature (such as a ground water seep or rock outcrop)

In each of these examples the detection of the anomaly serves to document environmental conditions and assist early warning or emergency response activities. Identification also presents the anomalous pattern as a spatial object that can be subject to further analytical operations.

While intuitively appealing, the challenge with anomaly detection relates to identification of "extreme" or "out of place" pixels in the digital imagery. Anomalous pixels do not conform to expected values. Depending on the application, nonconforming pixels may explain outliers, discordant observations, exceptions, aberrations, surprises, peculiarities, or contaminants (Chandola et al. 2009). Distinguishing among these various manifestations of an anomalous pixel is somewhat complicated by the presence of human error, natural deviations in populations, changing behaviors of the system under study, and systematic faults (Hodge and Austin 2004). Moving from an abstract concept to the tangible detection of nonconformity with an expected condition begins by first, defining what is "normal" given the nature of the problem and the range of states (radiances) than encompass "normality" and follows next by filtering out noise in the data which may parrot the actual anomaly. Formulation of the detection problem is influenced by several factors. Central among them are the actual nature and quality of the imagery, the constraints and requirements introduced by the problem, and the availability of labels (classes) that can be associated with a pixel that can determine if that pixel is normal or anomalous (Chandola et al. 2009). When spectral patterns are observed the nature of the pixel anomaly is likely to fall into one of three descriptive categories:

1. *Point anomalies* – where an individual pixel or region of pixels are spectrally anomalous with respect to the rest of the image/scene
2. *Contextual anomalies* – where the context of the pixel (its classification label) is anomalous with respect to those constituting a bounding neighborhood
3. *Collective anomalies* – where a collection of measures is anomalous with respect to the entire scene

Depending on how well the outlier investigation problem can be structured there are three methodological strategies that can be used to guide detection of one of all of these anomaly classes (Hodge and Austin 2004; Chandola et al. 2009):

- *Type 1 detection* – This methodological approach determined outliers with no prior knowledge of the data. This technique follows an unsupervised clustering logic and is based on the assumption that normal patterns of spectral radiance are more frequent in the images than are anomalies.
- *Type 2 detection* – This direction to the outlier problem is based on developing a model of normality and abnormality with the context of the problem. The goal of this approach is to apply labels using supervised classification methods to tag pixels as either normal or abnormal based on training data or information gathered prior to analysis that assists in the characterization of outlier patterns.
- *Type 3 detection* – This technique operates in a semisupervised classification mode that supplies training data (prior information) for only the normal class. According to this method, a pattern is normal if it falls within the boundary of the known "normal" class/state. Any pattern beyond this boundary is assumed to be anomalous.

Selecting the appropriate outlier detection strategy depends on two conditions:

1. The availability of an algorithm that can reliably model the data distribution and highlight outliers for clustering, classification, and recognition.
2. The identification of the spectral or informational class boundary that delineates normality together with thresholds for establishing when patterns are anomalous.

8.4 Anomaly Detection Algorithms

The algorithms developed to assist with outlier and anomaly detection have been reviewed extensively by Chandola et al. (2009) and Hodge and Austin (2004). In this section, we will examine a selection of the available algorithms that are germane to the objectives of environmental remote sensing. Here, in the image domain, outlier detection is concerned with changes in an image over time or delineating regions within a static scene that appear abnormal (Coolbaugh et al. 2007). The anomaly can therefore be spatial or temporal taking the form as outlying points in the data distribution, or contextually different patterns that suggest novel features or features that have been misclassified (although not necessarily erroneous). In either instance the outlier is inconsistent and translates to actionable information about the environmental system. The algorithms developed for these applications are primarily statistical techniques that fit a numerical model to the data and employ methods of statistical inference testing to determine if a pixel in the scene belongs to the model or not.

One of the simplest statistical methods used to detect univariate and multivariate anomalies employs the box plot rule. A box plot portrays image data using descriptive statistics to summarize the minimum, maximum, median, and range of pixel values. These statistics are used to establish limits beyond which n observation (pixel) will be treated as an anomaly. A similar logic is applied in the Grubbs Test, which calculates a Z-value as the difference between the mean value for the pixel based on the formula

$$Z = \frac{|x - \bar{x}|}{S},$$

where x and S are the mean and standard deviation of the data sample. The Z-value is compared to a 1 or 5% significance level and the pixel value (p) is considered anomalous if

$$Z > N - \frac{1}{\sqrt{N}} \times \sqrt{T^2\alpha/(2N), n - 2/N - 2 + T^2\alpha/(2N), n - 2},$$

where N is the sample size and $T^2\alpha/(2N)$, $N - 2$ is a threshold taken from the t-distribution at a significance level of $\alpha/2N$. A variation of this test based on the Mahalanobis distance compares a pixel vale (p) to the sample mean to define the outlier anomaly such that

$$Y^2 = (x - \bar{x})' \,\mathrm{Cov}^{-1}(x - \bar{x}),$$

where Cov is the sample covariance matrix. Similar proximity-based methods have been introduced as well that use either the K-nearest neighbor algorithm or simple Euclidean distance.

When observations are compared over time anomaly detection based on regression analysis have been explored. The regression-based approach consists of two steps. Initially, a regression model is fit to the image data which is followed by an evaluation of the residuals. The residuals are used to determine the presence of outliers in the relationship. Anomalies are those observations (pixels) that are not explained well by the regression model as defined by the magnitude of the residual. The value of the residual therefore serves as a score separating static condition from an anomalous response pattern. For example, the pattern of illegal clear cutting between two dates may be detected where the "normal" pattern of forest is interrupted by residual values well beyond what might be considered the typical spectral response of forest cover for a specific time of the year. The main imitation with this strategy is the underlying assumption that the observations (pixels) are normally distributed. More flexible and autonomous approaches using nonparametric techniques are available so that all the data determine model structure.

The most straightforward nonparametric technique uses the histogram to form a profile of the data distribution. Using this approach, anomaly detection begins by calculating the histogram for the image followed next by a check to examine if an instance of data falls into one of the bins of the histogram. According to this strategy, brightness values that fail to organize with a histogram bin are labeled anomalous. The size of the bin use to perform the containment check determines how well anomalies are detected in the image. If the histogram bins are set to too small a width, pixels will fall in an empty or rare bin and thus return a false positive. Therefore, selecting an optimal bin size that reduces false positives is critical to the success of this method. An additional nonparametric technique for anomaly detection employs Kernel functions to approximate a probability density function (PDF) based on Parzen-window estimation. Parzen-window density estimation is essentially a data-interpolation technique (Duda et al. 2001; Simonoff 1996). Given an instance of the random sample, x, Parzen-windowing estimates the PDF of $P(x)$ from which the sample was derived. The PDF value $P(x)$ is the sum total of the contributions from the observations to this window. The Parzen-window estimate is defined as:

$$P(x) = \frac{1}{n}\sum_{i=1}^{n}\frac{1}{h_n^d}K\left(\frac{x - x_i}{h_n}\right),$$

where $K(x)$ is the window function or kernel in the d-dimensional space.

Kernel-based methods estimate the density distribution of the input data and identify outliers as observations (pixels) lying in regions of low density. A common detection algorithm using kernel density estimation employs the Gaussian mixture models of the form:

$$P_{(t/x)} = \sum_{j=1}^{m} \alpha_j(x)\, \phi_j\, (t/x).$$

An alternative technique adapted from extreme value theory uses the Gaussian mixture model to represent the data distribution defined by the image. Anomalies, based on this method, are those pixels evidencing extreme values located at the tail of the distribution. Extreme value theory examines the tails of the distribution to estimate the probability that a given observation (pixel) is an extreme value according to

$$P_{(\text{extremex})} = \exp\left[-\exp\left(\frac{X_{\text{m}} - \omega_{\text{m}}}{s_{\text{m}}}\right)\right].$$

Anomaly detection can be dramatically improved when multiple spectral bands locate anomalies (outliers) that are essentially very bright or very dark according to their relative location in multispectral measurement space (Chang and Chiang 2001). One of the more widely applied detectors of this variety is the $R(x)$ algorithm developed by Reed and Yu (1990). The basic $R(x)$ algorithm is defined according to the equation:

$$\Omega_{\text{RDX}}(r) = (r - \Gamma)^T K_{L\times L}(r - u),$$

where r is the vector of pixel spectral values, $\Gamma\iota\sigma$ the mean spectral vector for the area of interest (the mean of each spectral band), L is the number of spectral bands and K is the spectral covariance matrix. The algorithm performs in a manner similar to how a human analyst would visually search for outliers in a single band image by identifying bright or dark pixels. A human analyst would, of course, be challenged to identify outliers simultaneously across several spectral bands, the $R(x)$ algorithm, however, by implementing the Mahalanobis distance formula, establishes a multivariate search space that facilitates detection using either multi- or hyperspectral data. Anomaly detection using the $R(x)$ algorithm begins by calculating the mean of each band using a sliding dual concentric window (a small interior window centered within a larger outer window) (Fig. 8.5). The dimensions of the interior window are assumed to be the size of the feature of interest in the image. The algorithm proceeds to calculate the spectral covariance matrix before solving for the value of Ω as shown by equation (see above). Alternately, the $R(x)$ formula can be normalized using either the squared distance (magnitude) of the pixel vector to the mean

$$\Omega_{\text{nrxd}}(r) = (r - \Gamma)^T K_{L\times L}(r - u)/(r - \Gamma)^T(r - u)$$

or to the simple distance determined by,

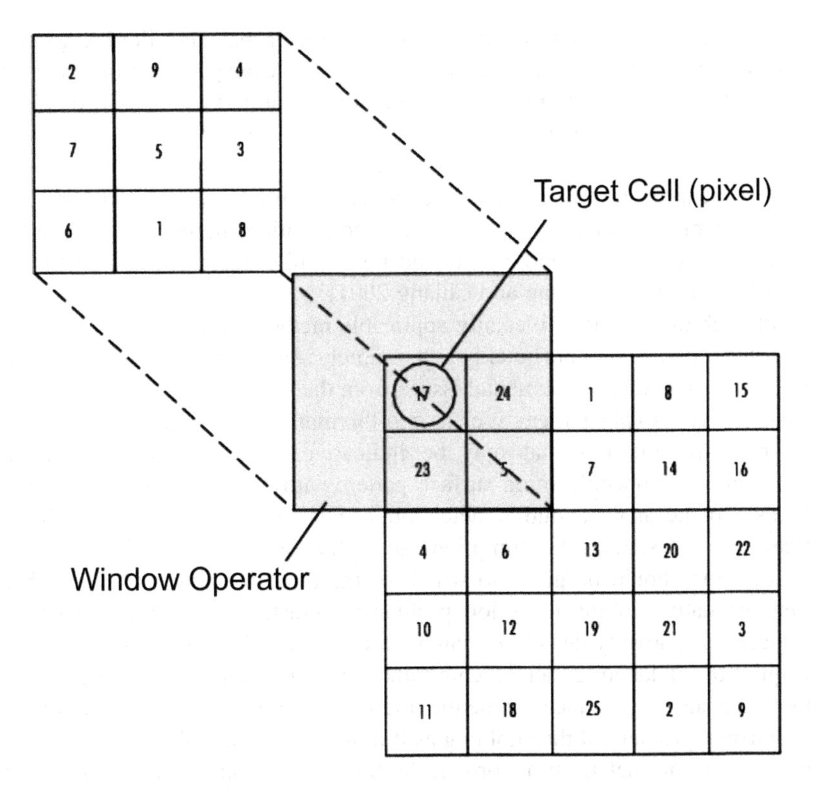

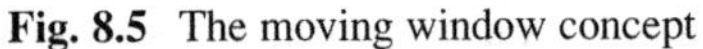

Fig. 8.5 The moving window concept

$$\Omega_{\mathrm{mrxd}}(r) = (r - \Gamma)^T K_{L\times L}(r - u)/\sqrt{(r - \Gamma)^T(r - u)}.$$

The calculated value of $\Omega(r)$ derived from either of the three methods produces an array of outlier pixels. Because the $R(x)$ algorithm involves computation of the mean and covariance matrix, real-time processing of the image data is infeasible. In order to enable real-time implementation of anomaly detection, the Causal $R(x)$ algorithm was introduced. The Causal $R(x)$ algorithm employs the sample correlation matrix and solves for $\Omega(r)$ such that

$$\Omega_{\mathrm{crx}}(r) - r_k T R^{-1}(r_k)\, r_k.$$

Unlike the $R(x)$ model, which requires knowledge of all of the data samples to populate the covariance matrix prior to processing, CR(x) processes and updates either in line-by-line or pixel-by-pixel fashion across the image. Also, since the sample correlation matrix accounts for both first-order and second-order statistics, CR(x) can capture spectral variability more efficiently that the $R(x)$ model which is limited to only second-order statistics. This quality of the CR(x) approach can be advantageous in remote sensing applications where imagery is generally nonstationary.

The $R(x)$ algorithm and its variants are a form of matched filtering where the signal to be matched is unknown. Through the use of each pixel vector as the signal to be matched, outlier search space is greatly simplified by the use of distance formula to partition spectral space into decision regions. Extensions to the $R(x)$ model include the addition of a Uniform Target Detector algorithm that provides a measure to express how close each pixel matches a uniform spectral pattern, and the Low Probability Detection algorithm. Incorporating these into the solution compensates for background effects and noise while improving the performance of the $R(x)$ algorithm (Chang and Chiang 2001).

Although there is no universally applicable method for anomaly detection, the ability to recognize outliers quickly in a remotely sensed image can offer important clues for a range on environmental issues from the simple identification of thermal hot-spots elated to phenomena such as geo-thermal activity or subsurface coal fires to surface peculiarities that may be indicative of contaminant flows, insect infestations, or nonconforming surface patterns attributable to discordant human actions. To the analyst, outlier detection begins by selecting an algorithm that agrees with the statistical assumptions about the data. From this initial decision, consideration should be given to which of the three fundamental approaches fit given the nature of the detection problem: a clustering strategy, classification approach, or a novelty detection search. Once a general strategy has been chosen, the question of labeling can be entertained. Labeling address the large issue of naming the anomaly, a facet of the methodology that is closely related to the more fundamental problem of thresholding as it influences what outlier conditions in the data separate normal from abnormal. As the analyst approaches the problem of anomaly detection, prior specification of the factors which determine a discordant observation allows the results of a processing algorithm to be translated into meaningful information.

8.5 Summary

Remotely sensed data has become important for the analysis, characterization and modeling of a wide array of land surface processes, whether biophysical or human, in the Earth system sciences. These processes range from changes in vegetation over time, the impact of human activities as well as more complex changes in ecosystem dynamics such as deviation in nutrient cycling, biomass, or water levels. Although remote sensing based on reflected electromagnetic energy is recognized as a powerful tool in the collection, analysis, and modeling of environmental data, less attention has been given to the use of thermal, and especially thermal infrared, remote sensing. This chapter provided a review of the use of emitted electromagnetic energy with a focus on the use of thermal data in anomaly detection and mapping. In this capacity thermal remote sensing is especially useful for understanding the fluxes and redistribution of materials as a key aspect of land surface processes, where deviations in those processes may signal abnormal or problematic events. Directing attention to the concept of an anomaly, the extreme or "out of place" relationships identifiable in the thermal patterns expand the application of remote sensing as a method of environmental assessment.

References

Ashton, E. and Schaum, A. (1998) Algorithms for the Detection of Sub-pixel Targets in Multispectral Imagery, Photogrammetric Engineering and Remote Sensing, 64, 723–731.

Chandola, V., Banerjee, A. and Kumar, V. (2009) Anomaly Detection: A Survey, ACM Computing Surveys, 41 1–72.

Chang, C. and Chiang, S. (2001) Anomaly Detection and Classification for Hyperspectral Imagery, IEEE Transactions on Geoscience and Remote Sensing, 40, 1–12.

Coolbaugh, M., Kratt, C. Fallacaro, A., Calvin, W. and Taranik, J. (2007) Detection of geothermal anomalies using Advanced Spaceborne Thermal Emission and Reflection Radiometer (ASTER) thermal infrared images at Bradys Hot Springs, Nevada, USA Remote Sensing of Environment, 106, 350–359.

Dash, P., Gottsche, F. Olesen, F. and Fischer, H. (2002) Land Surface temperature and Emissivity Estimation From Passive Sensor Data: Theory and Practice: International Journal of Remote Sensing, 23, 2563–2594.

Diak, G. (1995) Algorithms for Extracting Information from Remote Thermal-IR Observations of the Earth's Surface Remote Sensing of Environment, 54, 168–181.

Duda, R., Hart, P. and Stork, D. (2001) *Pattern Classification*, Wiley.

Geiger, R., Aron, R. and Todhunter, P. (2003) The Climate Near the Ground, Rowman and Littlefield, 642p.

Hodge, V. and Austin, J. (2004) A Survey of Outlier Detection Methodologies, Artificial Intelligence Review, 22, 85–126.

Jimenez-Munoz, J. and Sobrino, J. (2003) A generalized Single-channel method for Retrieving Land Surface temperatures From Remote Sensing Data, Journal of Geophysical Research D: Atmospheres 108, ACL2-1 – ACL 2–9.

Jensen, J. (2007) Remote Sensing of the Environment: An Earth Resource Perspective, Prentice Hall, 608p.

Kant, Y. and Badarinath, K. (2002) Sub-pixel fire detection using Landsat-TM thermal data Infrared Physics and Technology, 43, 383–387.

Koltunov, A. and Ustin, S. (2007) Early fire detection using non-linear multitemporal prediction of thermal imagery Remote Sensing of Environment, 110, 18–28.

Kustas, W. and Anderson, M. (2009) Advances in thermal infrared remote sensing for land surface modeling, Agricultural and Forest Meteorology, 149, 2071–2081.

Luvall, J., Rickman, D., Estes, M., Laymon, C. and Howell, B. (2000) A Decision Support System for Urban Landscape Management Using thermal Infrared Data, Photogrammetric Engineering and Remote Sensing, 66, 1195–1207.

Prata, A. (1994) Land surface determination from satellites, advances in Space Research, 14, 15–26.

Qin, Z., Karnieli, A. and Berliner, P. (2001) A Mono-window Algorithm for Retrieving Land Surface Temperature from Landsat Data and its Application to the Israel-Egypt Border Region, International Journal of Remote Sensing, 22, 3719–3746.

Quattrochi, D. and Goel, N. (1995) Spatial and Temporal Scaling of Thermal Infrared Remote Sensing Data, Remote Sensing Reviews, 12, 255–286.

Quattrochi, D. and Luvall, J. (1999) Thermal infrared remote sensing for analysis of landscape ecological processes: methods and applications Landscape Ecology, 14, 577–598.

Reed, I. and Yu, X. (1990) Adaptive Multi-band CFAR Detection of an Optical Pattern with Unknown Spectral Distribution, IEEE Transactions on Acoustics, Speech and Signal Processing, 38, 1760–1770.

Simonoff, J. (1996) Smoothing Methods in Statistics, Springer.

Sobrino, J., Jimenez-Munoz, J. and Paolini, L. (2004) Land surface Temperature Retrieval From Landsat 5 TM, Remote Sensing of Environment, 90, 434–440.

Schmugge, T. Hook, S. and Coll, C. (1998) Recovering Surface Temperature and Emissivity from Thermal Infrared Multispectral Data, Remote Sensing of Environment, 65, 121–131.

Sabins, F. (2007) Remote Sensing: Principles and Interpretation, Waveland Press Inc, 512p.

Van de Griend, A. and Owe, M. (1993) On the relationship between thermal emissivity and the Normalized Difference Vegetation Index for natural surfaces, International Journal of Remote Sensing, 14, 1119–1131.

Weng, Q. (2009) Thermal Infrared Remote Sensing for Urban Climate and Environmental Studies: Methods, Applications and Trends, ISPRS Journal of Photogrammetry and Remote Sensing, 64, 335–344.

Weng, Q., Lu, D. and Schubring, J. (2004) Estimation of Land Surface Temperature-Vegetation Abundance Relationships for Urban Heat Island Studies, Remote Sensing of Environment, 89, 467–483.

Xian, G. and Crane, M. (2006) An analysis of urban thermal characteristics and associated land cover in Tampa Bay and Las Vegas using Landsat satellite data Remote Sensing of Environment, 104, 147–156.

Chapter 9
Hyperspectral Sensing

Multispectral remote sensing has enjoyed widespread use for well over 40 years and in the previous chapters we have explored the sensors and techniques that have been developed to exploit the remote measurement of environmental attributes using this technology. While multispectral analysis provides information to guide our assessment of environmental processes, the inherent limitations imposed by the comparatively broad spectral resolution of multispectral sensors restricts the level of detail that can be extracted from the data. As interest in environmental remote sensing continues to develop, imaging capabilities that extend measurement options beyond the wide wavelength bands common to multispectral sensors can expand environmental analysis and characterization efforts. In recent years, hyperspectral remote sensing has emerged as an important complement to multispectral image analysis which demonstrated potential in application areas ranging from geologic mapping, agricultural analysis, ecological science, forestry, and urban planning (Treitz and Howarth 1999; Plaza et al. 2009; Madden 2004; Schaepman et al. 2009). In this chapter, we will introduce the principles that support hyperspectral remote sensing and the foundation knowledge required to employ this approach in land surface mapping. In our discussion emphasis will be given to those aspects of hyperspectral imagining that distinguish it from the more familiar multispectral approaches and the methodologies that have been developed to guide the application of this technology in environmental analysis.

9.1 Moving to the Hyperspectral

Although the concept of hyperspectral remote sensing was introduced in the mid-1980s, with the deployment of satellite sensors and accessible data archives, hyperspectral image analysis has become an exciting advancement in remote sensing over the last decade (Goetz and Alexander 2009). Hyperspectral data are measurements of reflected infrared radiation made across hundreds of wavelengths. When compared to multispectral imagery, the distinguishing characteristic of hyperspectral data is that

J.K. Lein, *Environmental Sensing: Analytical Techniques for Earth Observation*,
DOI 10.1007/978-1-4614-0143-8_9,

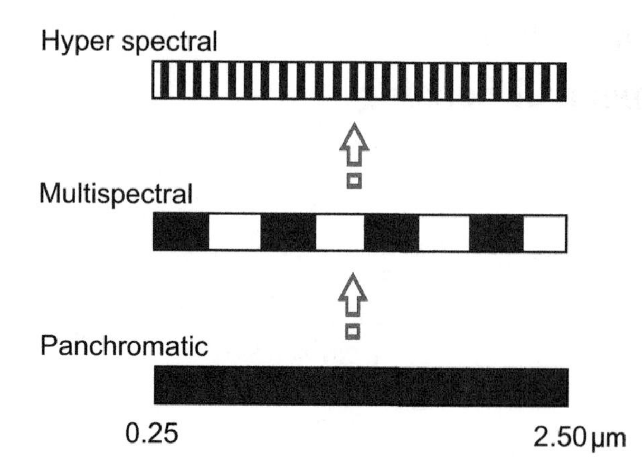

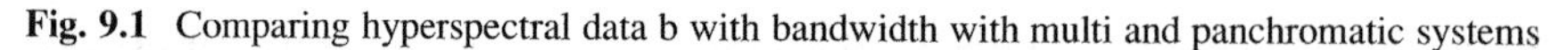

Fig. 9.1 Comparing hyperspectral data b with bandwidth with multi and panchromatic systems

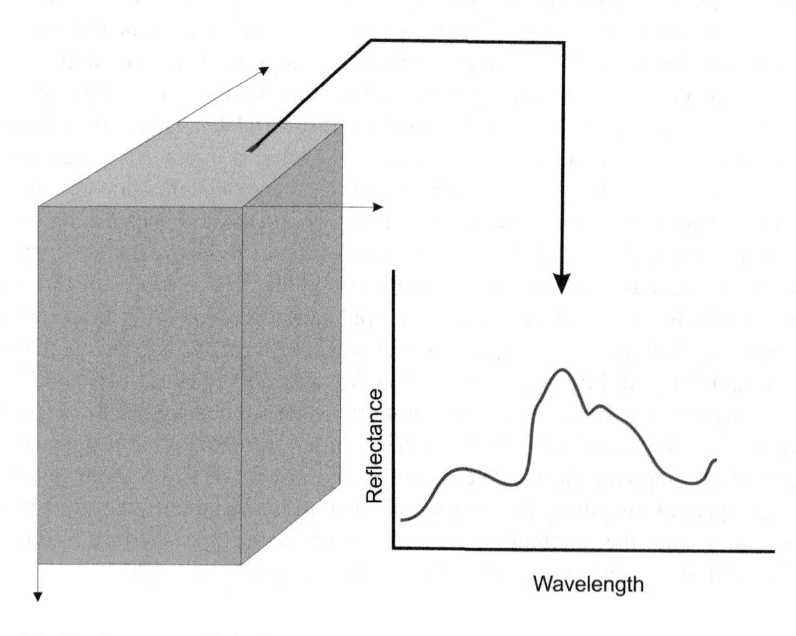

Fig. 9.2 The hyperspectral pixel

reflectance measurements are acquired over very narrow and continuous wavelength intervals (Fig. 9.1). Therefore, the reflectance curve produced by hyperspectral measurements assumes a more complex geometric shape that preserves the subtle details in reflection and absorption that are far more generalized when multispectral reflectance patterns are observed. By collecting data simultaneously across hundreds of narrow, adjacent spectral bands, hyperspectral measurements produce a continuous spectrum for each pixel comprising the image (Fig. 9.2). In terms of raw

Fig. 9.3 The measurement "Depth" of a hyperspectral scene

information content, a pixel in a hyperspectral image defined by continuous spectra, evidences considerably more information than is available for that same pixel in a multispectral image.

With greater spectral information available to explain patterns of energy reflectance, the capacity to learn more about the nature of the reflecting surface is enhanced. The spectra capture by the hyperspectral image defines a spectral resolution that is far more intelligible than the coarse signature common to the multispectral image. This feature of hyperspectral imagery is significant and suggests that for each pixel in the image a continuous spectrum of radiance is sample with a level of spectral discreteness that enables the identification of spectral signatures that correspond to the material composition of the pixel (Aspinall 2002). Typically, hyperspectral images are composed of spectral bands that are 0.005–0.01 μm wide, with sensors systems designed to detect reflected electromagnetic energy in the range from 0.40 to 2.50 μm (Fig. 9.3). Therefore, by capitalizing on the derivation of reflectance spectra, image classification using hyperspectral imagery is less focused on indentifying categories of land cover or broad descriptions of surface types or conditions. Instead, hyperspectral analysis concentrates on the detection of surface characteristics related to biogeochemical composition.

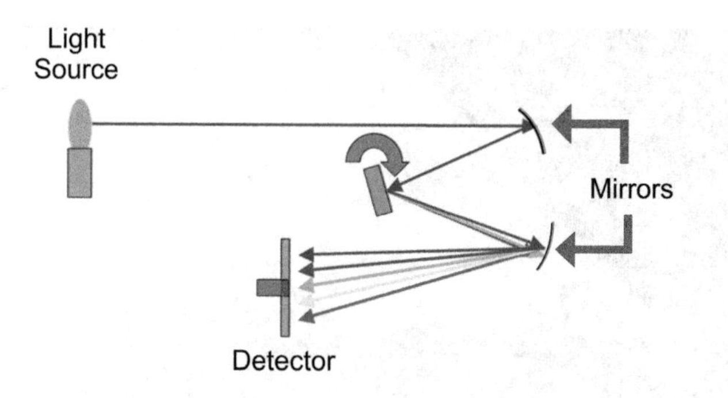

Fig. 9.4 General schematic of a spectrometer

The contrast between classification and detection suggests that the objectives of hyperspectral analysis differ, which is important to understand since it influences the types of application problems well suited for this form of remote sensing. With hyperspectral data, applications center on the problem of discovery as opposed to the task of labeling. Although this distinction is subtle, if we consider a pixel in multispectral space, the information we have to process allows us to name the pixel at a very general level, such as rock or vegetation. The detail added to our analysis by the continuous spectra of our hyperspectral image allows us to not only name pixels that are rock, but to discover which of those rock pixels are dolomite or which of those vegetated pixels are chickweed. In this context, detection or discovery is more concerned with the capacity to identify the existence or occurrence of a condition (Shaw and Burke 2003). Composition-specific analysis extends the capabilities of environmental remote sensing to engage problems pertaining to the presence or absence of a phenomena, the spatio-temporal status of an object or feature, the constituent elements that comprise a pixel, or the proportion of materials that characterize a given landscape arrangement. The spectra observed for a pixel imaging a concrete surface in a hyperspectral image displays characteristics identical to the spectra for concrete that would be measured in a laboratory. The intensity of radiation separated in this manner is measured by means of an instrument called a spectrometer (Fig. 9.4). This relationship is no accident (Govender et al. 2007; Ben-Dor et al. 2009; Fava et al. 2009). Rather, it establishes the theoretical basis behind hyperspectral remote sensing and promotes the use of hyperspectral imagery as an information source for:

- Object/feature detection.
- Material mapping.
- Material identification.
- Mapping details of land surface properties.

9.2 Imaging Spectroscopy

Spectroscopy is a science that uses the absorption, emission, or scattering of electromagnetic radiation by atoms or molecules to qualitatively or quantitatively understand materials and physical processes and has been a powerful tool for investigating environmental and biological agents for many years (Schmidtlein 2005; Swayze 2000). The science of Spectroscopy involves the study of spectra and how matter interacts with electromagnetic radiation to form spectra that enables the identification of substances by observing how energy is emitted or absorbed. In principle, the science behind spectroscopy is comparatively straightforward. When matter is excited by the application of thermal, electrical, or radiant energy, electromagnetic radiation is emitted as the object of matter relaxes back to its original state. The spectrum of radiation emitted by a substance that has absorbed energy is referred to as its emission spectra. This is the emission feature of matter that is studied by spectroscopy. Using emission spectroscopy, the interaction between matter and electromagnetic radiation can be explored by concentrating a continuous range of radiation on a substance or material and examining which frequencies or wavelengths of energy it absorbs. The resulting spectrum derived from the material displays the original range of radiation that was focused upon it with dark spaces corresponding to those particular frequencies (or wavelengths) that are missing or absorbed. The spectrum produced by this process is termed an absorption spectrum (or spectra) and is used to detect individual absorption features of materials created when energy interacts with specific chemical bonds in a solid, liquid, or gas. Generally, either the emitted or absorbed radiation is analyzed by separating radiation into various frequency or wavelength components (Plaza et al. 2009).

A spectrometer measures the properties of radiation and typically operates over a selective portion of the electromagnetic spectrum. As it receives the energy radiated by a material, the spectrometer produces a graph (spectra) that describes the intensity of emitted or absorbed radiation over the portion of the spectrum it is designed to operate within. Spectra, quantified by means of a spectrometer, can fall into one of three broad categories: (1) continuous, (2) line, or (3) band. Heated solids and the Sun produce continuous spectra where the emitted radiation contains all the wavelengths/frequencies within a particular region of the electromagnetic spectrum. Excited atoms in the gas phase form line spectra and only certain wavelengths/frequencies are produced.

Band spectra are produced by excited molecules emitting radiation in groups of closely spaced lines that merge to form bands. Taken together these categories of emission and absorption spectra contain useful information regarding the chemical structure and composition of materials. Because every material object is formed by chemical bonds, the potential to detect substances using spectroscopy is very real and promising. The actual detection of materials, however, somewhat complicated and dependent on factors such as:

- Spectral coverage.
- Spectral resolution.

- The signal to noise ratio of the spectrometer.
- The abundance of the material in a sample.
- The strength of absorption features for the material in the wavelength region subject to measurement.

Taking these factors into consideration, it is both the absorption features displayed by the material along with the overall shape of the spectra that helps distinguish one substance from another. Several spectra for common materials are illustrated in Fig. 9.5. As we examine these spectra variations in the observed peaks and troughs produce signature-like patterns that clearly differentiate materials. In many respects the shape and characteristics of these spectra suggest to the eye that when materials are examined over a continuous range of wavelengths, distinctive pattern-response sequencing become sufficiently evident to facilitate the identification of that material.

As a method of material identification, spectroscopy can be used in laboratories, in the field and on platforms ranging from aircraft to satellite sensors. In the laboratory spectra are produced when a source of continuous radiation is applied to a material. The intensity of radiation is measured by the spectrometer and a spectra is recorded. In remote sensing, imaging spectroscopy is employed to acquire reflectance data in the form of spatially gridded spectra. The gridded spectra produce a two-dimensional image, one for each spectral channel across a series of continuous bands (Fig. 9.5). The spectrum obtained from one pixel, therefore, closely resembles the spectra of a material obtained through laboratory spectroscopy. The close correspondence between the pixel and the material spectra obtained in the laboratory is significant for the purposes of mapping and environmental analysis. Given two sources of information that display a signature pattern produced by the interaction between matter and energy, a comparison can be made, in this case between the image spectra and a laboratory spectra, that enables the identification and verification of the substance involved.

Hyperspectral remote sensing is a form of imaging reflectance spectroscopy. Reflectance spectroscopy differs from other categories of spectroscopy in two fundamental ways (Schaepman et al. 2009). First, energy is reflected or scattered by object materials to the sensor as opposed to other methods that require energy to pass through a substance suspended in a gaseous or liquid medium. This quality of reflectance spectroscopy emphasizes differences in the physical processes of energy reflection vs. energy transmission critical to effective measurement. Second, reflectance spectroscopy for remote sensing applications is limited to operating within the 0.4–2.5 μm portion of the electromagnetic spectrum where atmospheric windows provide optimal transmission of the energy source. Despite this limitation numerous materials and substances display unique spectral reflectance signatures in the 0.4–2.5 μm wavelength interval, owing largely to the absorption of photons by specific electronic, vibrational, or scattering processes that are produced by common earth materials in this portion of the spectrum. (Curran 1994; Vane and Goetz 1988) For example, reflectance characteristics in the 0.4–1.0 μm spectral range define wavelengths that are influenced by the presence of transition metals

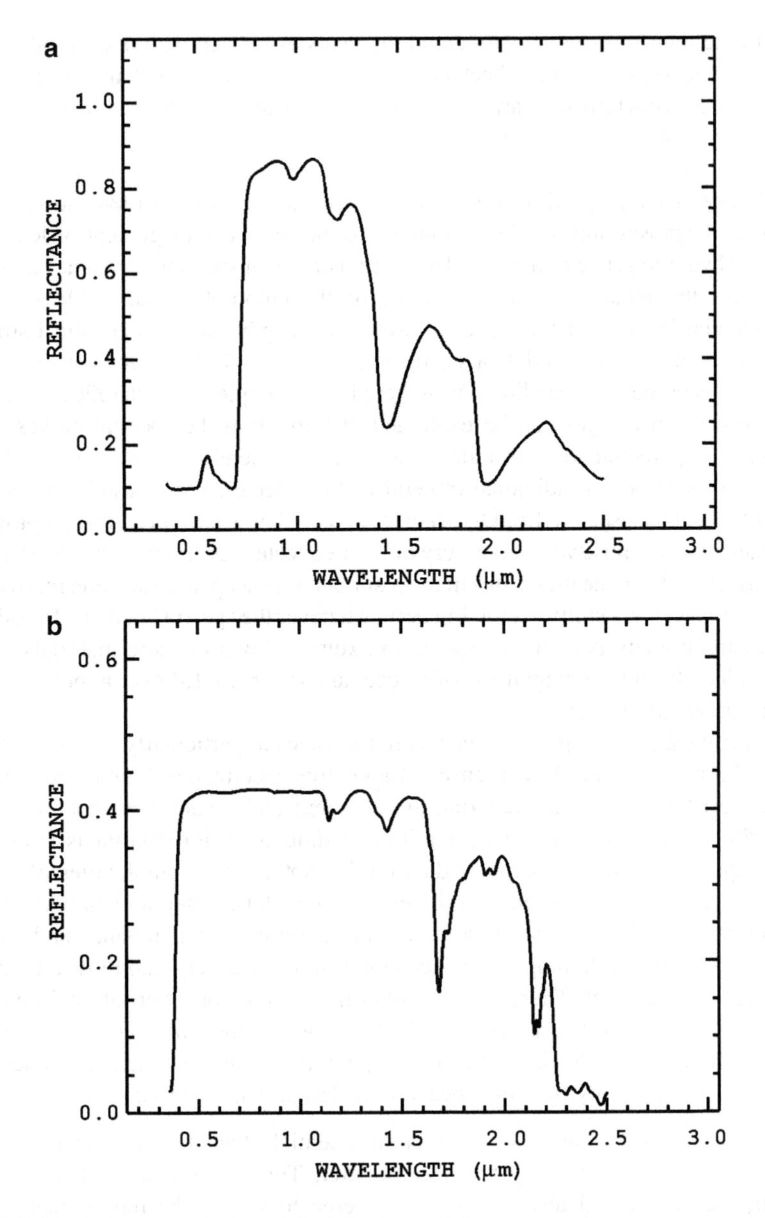

Fig. 9.5 What a spectra looks like: (**a**) oak leaf and (**b**) fiberglass roof

such as iron. This property makes this interval of diagnostic value for mineral detection and identification. Unique absorption patterns around 0.9 μm are indicative of electronic transitions resulting in absorption in mineral crystal fields. This characteristic facilitates their identification in an image. Vegetation, defined chiefly by chlorophyll absorption features at 0.48–0.68 μm that are the product of electronic transitions in the carotenoid pigments associated with photosynthesis can also be understood in the reflectance spectra. Vegetation reflectance also displays diagnostic characteristics part at 1.0 μm. Beyond this wavelength two reflectance peaks associated with vegetation can be noted. The first occupies the 0.8-μm region and the second is found at 1.3 μm; areas referred to as the infrared plateau. High reflectance properties in this area are associated with leaf tissue structure and provide diagnostic information pertaining to the cellular arrangement of leaf tissue and the hydration state of the leaf. The steep increase in spectral reflectance at 0.8 μm identifies the location of the red edge of the chlorophyll band. Movement of the spectral location of this edge to lower wavelengths suggests chemical stress in the leaf which can be used as a surrogate for a range of environmental stressors to which vegetation is responding. From these brief examples specific reflectance peaks and absorption troughs can be examined directly from the spectral curves which enable the potential identification of a material. Therefore, for relatively "pure" materials, such as an individual mineral or tree species, it is possible to construct useful reflectance spectra from hyperspectral data. These spectra can be employed in a "matching" exercise with laboratory or field-collected spectra in order to extract the material type from the image in a manner similar in concept to a thematic map of land cover produced via multispectral imagery, wherever the spectra "match." In addition, with careful analysis of image spectra, mixtures of two or more materials can be determined from the components of a spectral curve recorded over a more complex land surface arrangement.

Imaging spectroscopy, however, is not a panacea, particularly when conducted via airborne or satellite-based sensors. Laboratory spectra, developed to identify the characteristics of a given material, are acquired under controlled conditions and describe substances in the purest condition. Although this information is essential to hyperspectral sensing, laboratory data rarely captures the complexities of a landscape where mixtures of soil, wood, vegetation, metals, water, and numerous other materials may be present in a pixel. Furthermore, contaminants and natural variations within a homogeneous scene can introduce very subtle differences in the spectra of a pixel. Therefore to exploit the remote collection of environmental spectra, those factors that influence reflectance, absorption, and scattering need to be taken into account. The data collection capabilities of an imaging spectrometer are determined principally by four aspects of its design (Clark 1999):

1. *Spectral range* – spectral range is an essential characteristic that defines the wavelengths captured by the spectrometer. This range insures that sufficient diagnostic spectral absorptions are covered to satisfy the requirements for a given application. Certain spectral ranges are commonly used:

Ultraviolet	0.001–0.4 μm
Visible	0.4–0.7 μm
Near-infrared	0.7–3.0 μm
Mid-infrared	0.3–30.0 μm (thermal)
Far-infrared	30.0 μm to 1.0 mm

2. *Spectral bandwidth* – Bandwidth defines the width of an individual channel of the spectrometer. Bandwidth is an important parameter since the narrower the spectral bandwidth, the narrower is the absorption feature that can be measured by the instrument. Bandwidth is analogous to spectral resolution such that at wider bandwidths subtle spectral detail cannot be detected. Typically, bandwidth sample greater than 0.025 μm lose the ability to resolve absorption features of some materials and thus produce less useful spectra when compared to their laboratory counterparts.
3. *Spectral sampling* – spectral sampling explains the distance separating the spectral bandpass profiles for each channel of the spectrometer expressed in units of wavelength. Bandpass describes the wavelength region each channel has been designed to operate within. A channel, under optimal conditions, only accepts reflectance within its wavelength-sensitive range. However, energy can leak in from beyond this window. Most spectrometers employ a Gaussian curve filter to control for bandpass aberrations. Sampling, to reduce error and bias must be close enough spectrally to measure the peak and valley locations along the curve without overlap. Spectrometers usually sample at one-half a full width at a half-maximum of the Gaussian profile (termed the Nyquist frequency) or at half Nyquist.
4. *Signal to noise ratio* – the signal to noise ratio affects the ability of a spectrometer to record detail with precision. The signal to noise ratio needed to satisfy a given application is influenced by the strength of the spectral features under investigation and is dependent on detector sensitivity, spectral bandwidth, and the intensity of reflected or emitted energy from the surface.

The data collection capabilities of an imaging spectrometer can also be impacted by a suite of external environmental factors. Many of these factors are common to other forms of remote sensing, but when hyperspectral data are involved they exert greater influence on data quality and the level of detail defined by the image spectra.

Energy recorded by an imaging spectrometer must transmit through the atmosphere. Consequently, any attempt to measure the spectral properties of the material environment must consider the absorption patterns of the atmosphere that conspire to modify the spectra of a pixel. Critical absorption regions occur across the spectrum. Those with the greatest impact on hyperspectral data occur at wavelengths below 0.35 μm where ozone absorption is active. Ozone absorption is also a factor at 9.6 μm, while at 0.76 μm oxygen is the primary absorption factor. Absorption by carbon dioxide occurs at 2.01 and 2.06 μm. Water vapor is responsible for most of the absorption features throughout the remainder of the spectrum. When photons enter an absorbing medium they too are absorbed according to Beer's Law:

$$I = I_o e^{-kx},$$

where I is the observed intensity, and I_o is the original light intensity, k is an absorption coefficient, and x is the distance traveled through the medium.

When spectra are observed, the absorption bands we see are the product of either electronic or vibrational processes. Electronic absorption is attributed to the excitation of outer electrons which create changes in energy states. As an atom absorbs energy, electrons are promoted from their ground state to a state that is termed "excited." One of the more common electronic processes revealed in the spectra of minerals is the result of unfilled electron shells of transition elements such as iron, nickel, chromium and cobalt. Absorption bands can also be caused by charge transfers or interelement transitions where the absorption of a photon causes an electron to move between ions. Charge transfer absorptions are the main cause of the red color evidenced by iron oxides (Clark 1999). Vibrational processes involve the bonds in a molecular compound. At certain energy levels the atomic units held by the bonds are set into motion in either a back and forth or rotational manner. The frequency where a molecule absorbs energy depends on (1) the strengths of the bonds and (2) the masses of atom participating in the movement. Vibrational absorption is common in nonmetal materials and tends to produce complex spectra. It is the analysis of electronic or vibrational patterns of absorption and reflectance as captured by our spectroscopic data that provides essential clues that help to identify a material. The position of peaks and troughs as well as the overall shape of the spectra is not only instructive for the purpose of identification, but also facilitates the machine processing of hyperspectral imagery. In the following section a sample of spectra are examined to illustrate the variability inherent to our material environment and to demonstrate how the knowledge contained in a reflectance spectra can be assembled into a database that can assist with the processing and interpretation of hyperspectral imagery.

9.3 Endmember Spectra

An endmember is a spectrum chosen to represent a surface material in its pristine state (Dehaan and Taylor 2003; Rogge et al. 2007). In theory, an endmember is a hyperspectral signature defining a reflectance vector obtained from the measurement of a pure sample of a material obtained under conditions assumed to be comparable to those describing the hyperspectral image (Fig. 9.6). Two conditions underscore the endmember concept. First is the absence of "mixing" in producing the spectra. Second, it is common to consider the endmember representative of fundamental materials much the same way a dictionary provides reference definitions for words. Appreciating the role of an endmember as a reference connects us to the manner by which electromagnetic energy interacts with a surface. When energy interacts with materials and substances, energy at certain wavelengths are preferentially absorbed while at other wavelengths energy is transmitted in the material (van der Meer 2004). The position, shape, depth, and width of the absorption features evidenced by a spectrum are controlled by the chemical and crystal structure of the material. Therefore, the descriptive variables characterizing absorption features can

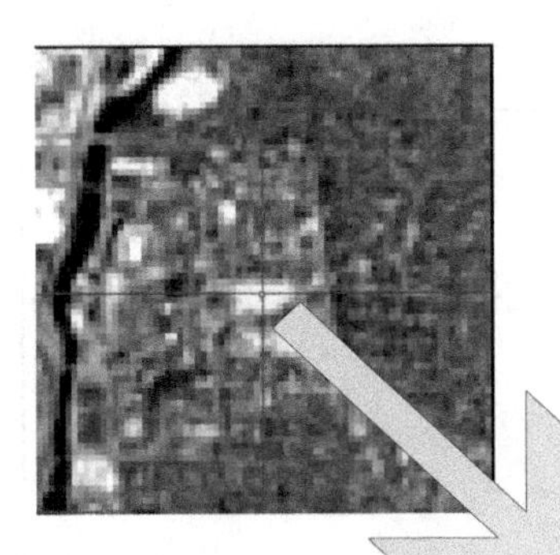

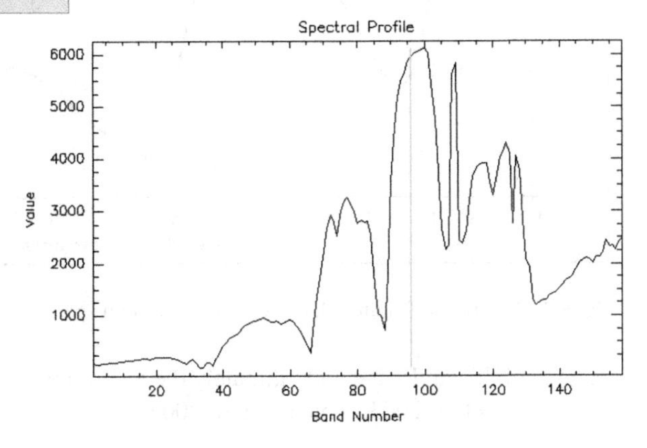

Fig. 9.6 An image endmember spectra

be directly related back to the physical structure and chemistry of the sample material. More importantly, absorption depth and asymmetry serve as useful indicator of the amount of the material causing the pattern or absorption described by the spectra.

To illustrate the diagnostic qualities of endmember reference spectra we can examine a series of material samples and compare the position, shape, depth, and width of the absorption features they evidence. This illustrative review begins with common types of vegetation and continues with an overview of human-made materials. Each of these examples was obtained from the United States Geologic Survey Digital Spectral Library (Clark et al. 1999). Through the careful analysis of these spectra information can be gathered pertaining to the nature of objects at the surface that helps the remote sensing analysts answer a series of fundamental questions, such as:

- What is it?
- What is it like?
- What is it made up of?

Placing these simple diagnostic questions into an environmental context gives us clues regarding the identification, status, and potential changes that characterize

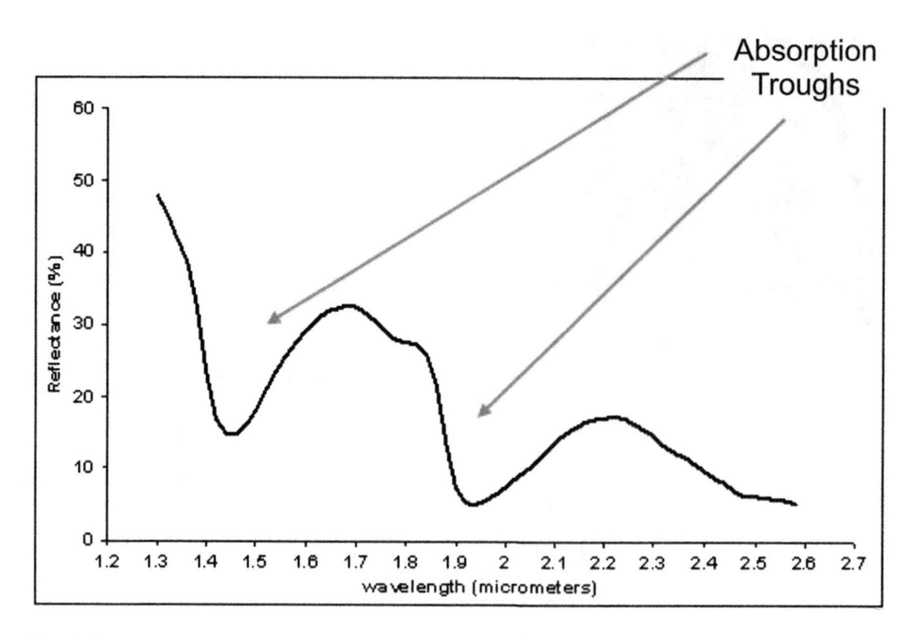

Fig. 9.7 Absorption features characterizing vegetation spectra

features of the landscape and ultimately the environmental system (Craig et al. 2006; Underwood et al. 2003; Younan et al. 2004).

- *Vegetation spectra* – typical green vegetation displays characteristic absorption features related to the presence of chlorophyll-a, chlorophyll-b, and carotenoids in leaves (Fig. 9.7). Peak absorption patterns related to chlorophyll-a occurs approximately at 0.42 μm while the chlorophyll-b absorption peak I found at 0.47 μm. Secondary absorption peaks are found at 0.65–0.67 μm, respectively. Carotenoids display a bi-modal absorption peak occurring at 0.48 μm and again at 0.51 μm. Consequently, when the reflectance spectra for typical green vegetation is observed, we find reflectance elevated between 0.75 and 1.3 μm with secondary reflectance peaking at 1.6–1.8 μm with a lesser reflectance increase located between 2.2 and 2.5 μm. This characteristic pattern extends across a range of vegetation samples producing nearly identical absorption bands over a variety of species including Fir tree (Fig. 9.8a), Blue Spruce (Fig. 9.8b), Blue Oak (Fig. 9.8c) and the familiar grass that comprise our lawns (Fig. 9.8d). Shifts or modifications in these spectral patterns would be diagnostic of important variations in the photosynthetic properties of vegetation, indicative of changes in water content, biomass, soil nutrients, and the overall concentration of chlorophyll in leaves (Im and Jensen 2008).
- *Material spectra* – the reflectance spectra for materials common to the human landscape are more diverse and varied when compared to vegetation endmembers. Diagnostic absorption features can be far more subtle and more difficult to generalize since reflectance patterns are controlled by the chemical and crystalline

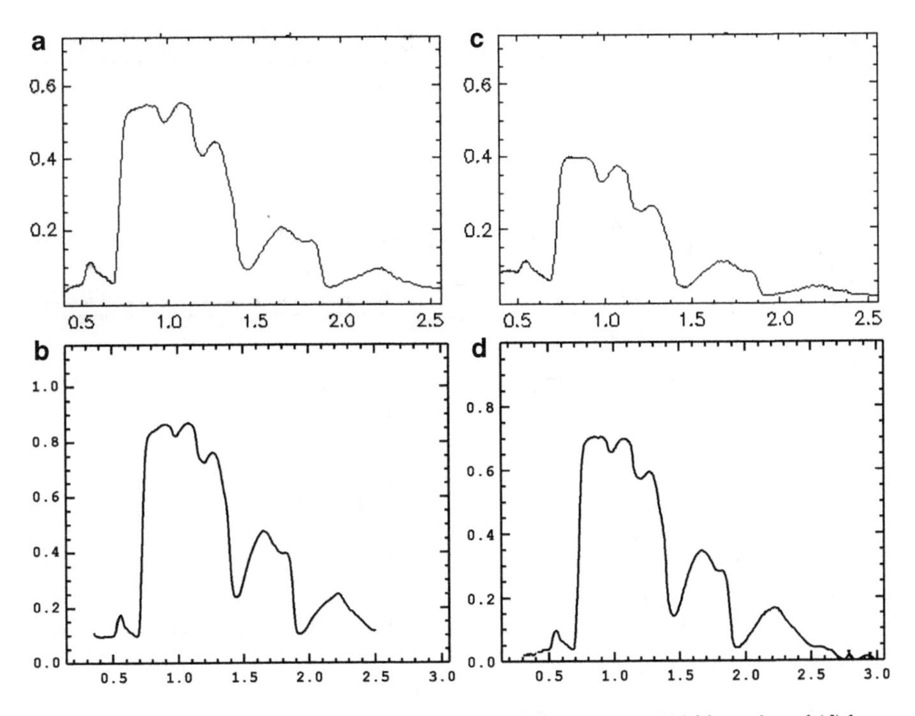

Fig. 9.8 Representative vegetation spectral: (**a**) fir tree, (**b**) blue spruce, (**c**) blue oak, and (**d**) lawn

structures of the materials. For example, asphalt tar, a material common to the roof tops of industrial and commercial land uses is noticeably absent of definable absorption bands (Fig. 9.9a). Rather, the reflectance spectra approximates a flat line with a very slight dip in reflectance at 0.90 μm. The endmember spectra for a terra-cotta roofing shingle, however, displays a more discernable patter with a small absorption trough located at 1.9 μm (Fig. 9.9b). Other materials frequently encountered in the built-environment include concrete road surfaces (Fig. 9.9c) with a diagnostic pattern evidencing high reflectance beginning at 0.60 μm through 2.1 μm and an absorption band falling at 1.9 μm. Galvanized sheet metal (Fig. 9.9d) shows a diagnostic absorption trough at 1.0 μm while materials, such as red raving brick evidence a spectral pattern similar in shape to concrete road surface (Fig. 9.9e). Several materials, such as fresh pine plywood (Fig. 9.9f) share characteristic patterns similar to those of vegetation; however, more exotic materials such as plastic (visqueen) advertise strong absorption bands between 2.3 and 2.5 μm (Fig. 9.9g).

Despite the limited breadth of this review several general principle regarding absorption features and diagnostic value of endmember spectra can be noted. For example, when mineral spectra are observed the shape of the spectral curve and the presence and location of specific absorption bands is greatly influenced by minerals' chemical composition and crystalline structure. The wavelength-dependent

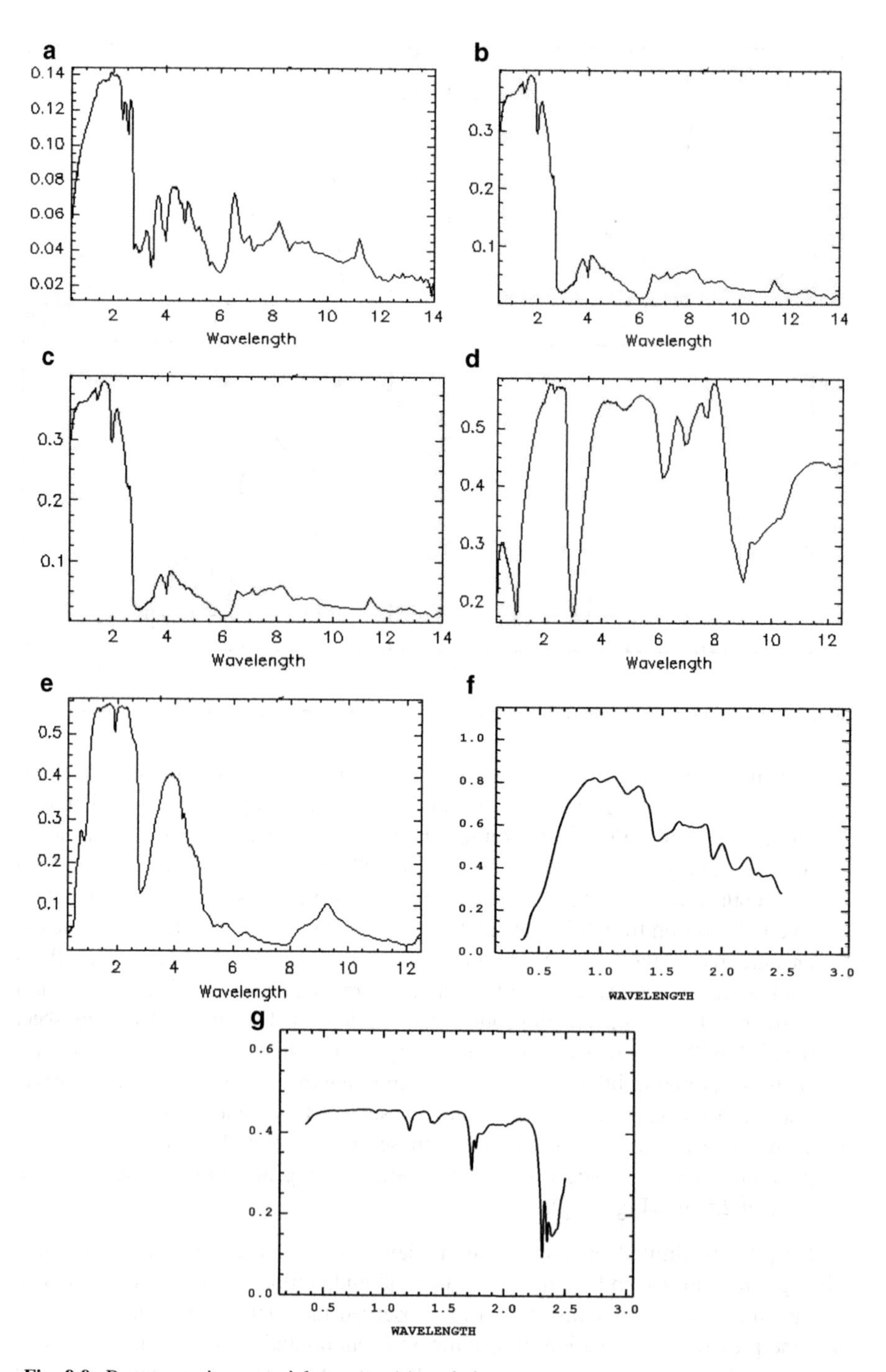

Fig. 9.9 Representative material spectra: (**a**) asphalt tar, (**b**) terra-cotta tile, (**c**) concrete road, (**d**) sheet metal, (**e**) brick, (**f**) plywood, and (**g**) plastic sheeting

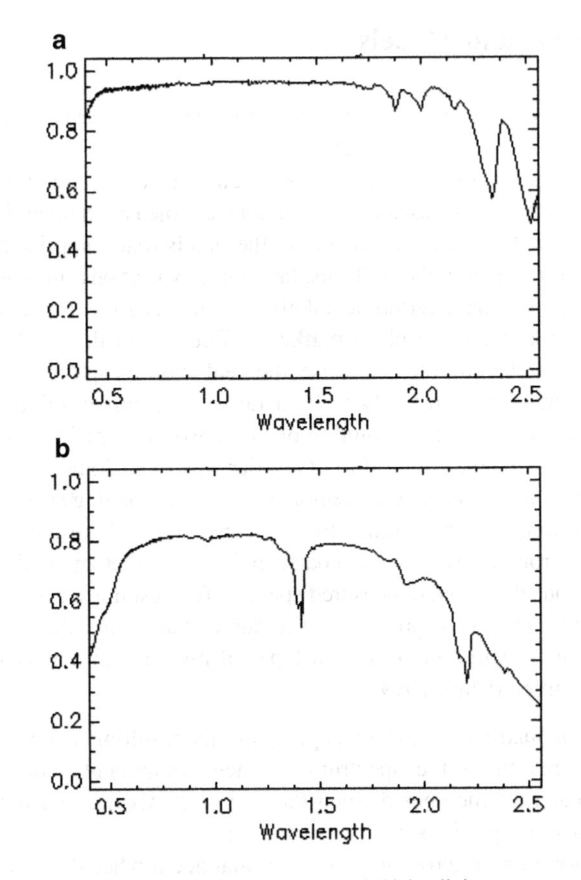

Fig. 9.10 Characteristic mineral spectra: (**a**) calcite and (**b**) kaolinite

absorption features typically result from the presence of chemical elements or ions, the ionic charge of the elements involved and the geometry exhibited by the chemical bonds between elements. Examples of these influences can be seen in minerals such as calcite, a major constituent of limestone, where the carbonate ion (CO_3^{2-}) produces a series of absorption bands between 1.8 and 2.4 μm (Fig. 9.10a), and Kaolinite, a clay mineral found in soils, describe absorption bands at 1.4 and 1.9 μm due to the presence of hydroxide ions (Fig. 9.10b). Vegetation spectra exhibit the controlling influence of various plant attributes where in the visible bands the shape of the endmember spectra is governed largely by the absorption effects of chlorophyll and other leaf pigments. Reflectance increased sharply across the boundary between the red and near infrared wavelengths, defining the location of the Red Edge. Higher reflectance in the near infrared wavelengths is directed primarily by the internal cellular structure of leaves. Beyond 1.3 μm reflectance declines as wavelength increases except of two pronounced water absorption bands at 1.4 and 1.9 μm.

9.4 Mixtures and Models

Although the spectra examined in the previous section provide a convenient way to visualize how differences in the chemical and crystalline structure of materials contribute to significant contrasts in their reflectance spectra, the materials we reviewed represent typical cases whose spectra were measured under ideal laboratory conditions. In reality, and certainly within the pixels that comprise a hyperspectral image, all natural materials will display some variations in composition and structures due to external environmental effects related to age, exposure, weathering, and contamination, which result in marked differences in their reflectance spectra. This fact is important if only because the real world captured by a pixel in a hyperspectral image is also likely to be a far more complex mixture of materials both as a result of the spatial resolution of the sensor and the inherent nature of the landscape and its diverse composition (van der Meer and Jong 2002). As we recall from our treatment of multispectral remote sensing, an imaging spectrometer records spectral measurements according to a defined spatial resolution. When the spectrometers ground resolution is coarse it is likely that more than one surface material will contribute to the measured spectra. The result is a mixed spectra where the endmembers explain a portion of the curve that define the pure spectra that contribute to the mixture. Four general types of mixtures in hyperspectral imagery have been identified (Clark 1999):

- *Linear/areal mixtures* – macroscopic patterns resulting from the sum of the fractional area times the spectrum of each component endmember. In this example, each endmember defined the end of a mixture and forms a "mixing space" whose proportions can be calculated.
- *Intimate mixtures* – microscopic mixtures that occur when different materials are in close contact and a single photon of energy interacts with more than one material.
- *Coatings* – mixtures produced when one material coats another and each coating acts as a scattering/transmitting layer whose optical properties vary with wavelength and the properties of the materials involved.
- *Molecular mixtures* – form at the molecular level when two liquids or a liquid and a solid are mixed together (i.e., mud).

Mixtures remind us that the spectra acquired from a hyperspectral sensor are a property of the ground features subject to measurement. Ideally we would like to measure those features accurately and precisely; however, the acquired spectra does not always resemble the reflectance characteristics of our reference spectra taken from the laboratory. Additionally, image spectra are also effected by:

- The spectra of the input solar energy.
- Interactions with the atmosphere.
- The geometry of ground illumination.
- The design characteristics of the sensor system.

These factors not only impact the receipt of accurate spectral reflectance, but also introduce within-scene variability that hampers comparisons between individual pixels in the same image.

In order to effectively compare image spectra with reference spectra the recorded radiance values on the image require conversion to reflectance values (unless they have already been encoded in units of reflectance). Typically, conversion must account for the solar source spectrum, illumination of the scene as a function of sun angle and topography, transmission through the atmosphere and sensor gain. Converting from radiance to reflectance becomes a mathematical procedure where the reflectance spectrum of the surface is multiplied on a wavelength by wavelength basis by the factors listed above. This form of reflectance modeling produces the measured radiance spectrum with the effect of sensor offset and path radiance due to atmospheric scattering included in additive fashion (Gao 2009). Several of the more commonly applied methods of reflectance conversion include the following:

1. *Flat field conversion* – An image-based method that uses an area on the image of uniform area that displays a comparatively flat spectral reflectance curve. From this area the image is converted to "relative" radiance by dividing each image spectrum by the mean spectrum of the flat field. Ideally, the flat field should be bright to reduce image noise which would allow the mean spectrum to approximate the combined effects of solar irradiance and atmosphere absorption and scattering. The drawback with this methods rest with the difficulty in locating areas that display a completely flat reflectance spectrum.
2. *Average relative reflectance conversion* – This method normalizes the image spectra by dividing by a mean spectrum, but calculates the mean spectrum from the entire scene. Prior to this operation radiance values in each image spectrum are scales to compensate for topographic shading and other conditions that produce variations in brightness.
3. *Empirical line method* – This technique uses image radiance and ground reflectance values to perform a linear fit to relate radiance to reflectance. The derived function quantifies the effects of sensor gain (slope) while the intercept of the function represents offset. The derived function is then employed to recalculate each image band to a value of apparent reflectance.
4. *Modeling methods* – This approach employs radiative transfer models of the atmosphere to simulate solar irradiance and computer scene radiance for the time/date of the image. This approach also corrects to atmospheric scattering and absorption. Several atmospheric correction algorithms have been developed to calculate concentrations of atmospheric gases from the spectral information contained in a hyperspectral image. Examples of commonly used models include *ACORN* (Atmosphere Correction Now), *FLAASH* (Fast Line of Sight Atmospheric Analysis of Spectral Hypercubes), *ATREM* (Atmosphere Removal) and *HATCH* (High Accuracy Atmosphere Correction for Hyperspectral Data (Kruse 2004).

9.5 Managing Endmembers

From the discussion presented in the previous sections of this chapter, we see that imaging spectroscopy centers around the measurement and analysis of spectra acquired as images. The use of these images to study environmental systems and processes is based on the capabilities of hyperspectral sensors to provide hundreds of spectral bands, creating one spectrum per pixel that can resolve constituent surface signatures (Martinez et al. 2006). However, hyperspectral data when applied to environmental problem-solving may enjoy high spectral detail, but are often frustrated by low spatial resolution. Consequently, numerous disparate materials can contribute to the spectrum measured for a single pixel, adding complexity to the task of surface mapping and environmental characterization. Since accuracy is paramount to the objectives of environmental remote sensing, the processing of hyperspectral imagery narrows to the problem of decomposing the mixtures inherent to the pixel into an estimation of the materials that contribute to the measured spectra. On face value resolving pixel mixtures appears to be the relatively straightforward task of identifying unknowns from the image spectra. However, the question of unknowns can be elusive to answer since contributions to a pixels spectrum can include all the absorbing and scattering components of the atmosphere as well as those of the surface that affect radiance. If the application problem has more unknowns than useful measurements, effective analysis can be undermined. Conversely, if there are more measurements than unknowns, the application is overdetermined and a meaningful solution should be attainable. Regardless of application area, we can generally assume that the objects or features we are most interested in understanding are not homogeneous or easily captured spatially by a pixel. Rather, our image spectra are likely to define the integration of signatures associated with the purest portions of the pixel. Therefore, the spectrum evidenced by a pixel is some fraction of the corresponding materials that it samples; a proportion that will vary in abundance based on the heterogeneity of the scene. Thematic extraction in this context concentrates on defining those fractions. The less complex the surface, the greater the likelihood a pixel explains "pure" spectra. As complexity at the surface increases, a pixel becomes a signature of macroscopically pure patterns that describe image endmembers.

Endmember extraction and the derivation of reference spectra is one of the more fundamental and critical tasks associated with hyperspectral remote sensing. As we noted previously in this chapter, a simple definition of the endmember concept remains elusive. Endmembers are often described as an idealized pure signature for a material or substance. In terms of thematic extraction, a more involved definition explains endmembers as the features recognizable in a scene. From this perspective, endmembers are meaningful to an observer and constitute abstractions of real objects that can be regarded as possessing uniform properties (Garcia-Aro et al. 1999). For the purposes of our discussion, it is perhaps more useful to consider an endmember as simply a clean, clear unambiguous spectra of a material that occurs somewhere in the image. These are the spectra we wish to identify and ultimately map.

A pixel that falls on a uniform surface, such as a large body of water, unbroken forest canopy, or open field, will display a spectrum that reveals an endmember spectra characteristic of that surface. Pixels falling on less uniform areas will have a spectrum composed of a mix of endmembers. As noted previously, most surfaces contain mixes of materials or conditions that produce spectral contrasts. These contrasts correspond to more than one endmember and the challenge is to determine the portion of each endmember responsible for the mixture. The complicating factor in this identification problem is the fact that one cannot tell what type of material a pixel is a sample of by simply looking at its spectra; some reference is needed. In this case a spectral reference.

Developing a spectral reference to guide the separation of pixel spectra into meaningful informational categories directs attention to the process of estimating endmembers. Several methods have been developed to help deduce endmembers. Two of the more common techniques involve (1) extracting endmembers from the image (image endmembers) or (2) deriving endmembers from field-collected or laboratory-measured spectra of known materials (ground-based endmembers).

- *Image endmembers* – Selecting reference spectra from a hyperspectral image is similar in concept to the signature collection procedures followed when conducting a supervised image classification operation (Dehaan and Taylor 2003). The task begins by selecting representative homogeneous pixels from the image that accurately and comprehensively describes surface materials or ground conditions germane to the application problem. For a given image spectra to be useful the pixel(s) sampled must be relatively pure. Pixels can be screened for purity either through careful visualization of the spectral scatter plots of candidate sample locations or by applying the Pixel Purity Index method. The Pixel Purity method utilizes an algorithm to allocate a score to each pixel in the image based on the number of times a pixel occupies a near-vertex position in repeated projections of the image data onto a random vector oriented through the mean of the data cloud (Chaudhry et al. 2006). Although the procedure sounds daunting, pixels that are relatively pure will have a high score on the Pixel Purity Index. This high score communicates that these pixels repeatedly occupy locations at the extremes of the data distribution and therefore represent spectra that is distinct. Having identified pure pixels based on their purity score, a sample of these can be acquired by referencing the data layer produced by the algorithm as a guide. Typically in a PPI image spectrally extreme pixels are displayed in bright tones while darker pixels are less spectrally pure. Using this image as reference locations containing pure (bright) pixels can be sampled. Sampling relies on gathering spectral profiles of the material of interest which are then written to files and complied to form a spectral library. From this sample mean spectra can be computed for each material or condition of interest. The resulting spectrum, if collected from an image that has been preprocessed to remove the contaminating influences of atmospheric scattering and absorption based on a sample that has been extracted from regions within the centers of known features, can serve as a reliable reference for hyperspectral classification operations. Extending image-based reference spectra beyond the original scene, however, may be problematic without additional calibration. One useful calibration procedure that facilitates comparison of reference spectra across images is the continuum removal method. Continuum removal normalizes reflectance spectra to permit comparison of individual absorption features from a common baseline. Following the continuum removal operation reference spectra can be scaled to provide better agreement ranges exhibited by image spectra.

- *Ground-based endmembers* – Ground-based collection of reference spectra can be undertaken either by means of field spectroscopy or through the utilization of spectroradiometer measurements made in a laboratory setting. Field spectroscopy requires measuring surface material in situ using "hand-held" spectroradiometers usually mounted on a pole or yoke to reduce contact with the operators body (Milton et al. 2009). A variety of field collection techniques have been introduced, some designed to measure the spectral properties of individual elements of the scene such as leaves or minerals, others designed to measure the reflectance characteristics of spatial assemblages such as soil surfaces or vegetation canopies, and lastly are those methods used to calibrate remote-sensed imagery. Most field collected data are acquired with the sensor element pointed vertically over the surface of interest and measurements are recorded by the device and stored as flat ACSII files to support computer processing. However, to maximize the utility of field measurements appropriate metadata should be collected and stored for each field-measured spectra (Table 9.1). Field-collected measurements are a passive optical method of acquiring spectra which requires optimal illumination conditions in the field. This requirement can often be problematic, but for the spectra to serve as useful reference, field-collected data demands conditions of low atmospheric water vapor content, low aerosol content, and solar zenith conditions that provide good illumination of the surface. In addition, when sampling materials in the field issues related to natural variability must be account for since the field environment can display subtle and complex within-class(material) variability. Laboratory spectra, as an alternative, are collected under more sterile conditions where external factors that could potentially contaminate a spectrum are subject to greater control. Ideally, in the laboratory, the direction of the light source, its incidence angle and other factors that are involved in data collection can be held constant. This advantage insures data consistency across all samples. Specimens of the materials or substances of interest are prepared and mounted on a stage, illuminated and then undergo measurement. Scans of the material are collected often from more than one sample and average to produce a single spectrum. The only limitation associated with collecting spectra in a laboratory setting is the problem of agreement. Since laboratory conditions are kept at an optimal during data collection, spectrum run the risk of being "too perfect" and not representative of what would be encountered in a real world setting. Therefore when laboratory spectra are compared to data collected in the field they are often overspecified which can pose problems during hyperspectral classification operations.

Table 9.1 Sample spectral library metadata file

TITLE: Russian_Olive DW92-4 DESCRIPT
DOCUMENTATION_FORMAT: PLANT
SAMPLE_ID: DW92-4
PLANT_TYPE: Tree
PLANT: Russian Olive
LATIN_NAME: *Elaeagnus angustifolia*
COLLECTION_LOCALITY: Denver West Office Complex, Golden, Colorado, USA
ORIGINAL_DONOR:
SAMPLE_DESCRIPTION:
Fresh leaves, stacked two on bottom, one on top, over a deep black sample cup
END_SAMPLE_DESCRIPTION
COMPOSITIONAL_ANALYSIS_TYPE: None
COMPOSITION_DISCUSSION:
END_COMPOSITION_DISCUSSION.
TRACE_ELEMENT_ANALYSIS:
TRACE_ELEMENT_DISCUSSION:
END_TRACE_ELEMENT_DISCUSSION.
SPECTROSCOPIC_DISCUSSION:
END_SPECTROSCOPIC_DISCUSSION.
SPECTRAL_PURITY: 1a2_3_4_ # 1= 0.2-3, 2= 1.5-6, 3= 6–25, 4= 20–150 microns

LIB_SPECTRA_HED:	where	Wave Range	Av_Rs_Pwr	Comment
LIB_SPECTRA:	splib04a r 5322	0.2-3.0μm	200	g.s.=
LIB_SPECTRA:	splib05a r 11813	0.2-3.0μm	200	g.s.=
LIB_SPECTRA:	splib06a r 30728			g.s.=

Source: Clark and others (2007) USGS Data Series 231 Spectral Library splib06a Sample Description, http://speclab.cr.usgs.gov/spectral.lib06/ds231/DESCRIPT/V/russianolive.dw92-4.html

9.6 Compiling Spectral Libraries

Reference spectra acquired by either an image-based or ground-based methodology are most useful when assembled into a spectral library. High quality spectral libraries are essential to the successful exploitation of the wealth of data contained in a hyperspectral image. As with any library, a spectral library serves as a repository of information that can be stored, searched, and accessed to support hyperspectral analysis. The library also represents a reference resource base that facilitates continued research on the collection of material spectra (Nidamanuri et al. 2010). The detail and accuracy of an environmental application using imaging spectroscopy is, in many respects, dependent upon the quality and extent of the spectral library. Not surprisingly, compilation of a spectral library reference is often the initial step in a hyperspectral investigation. Library compilation requires the selection of spectra for the materials of interest that are required given the nature and purpose of the investigation. These are the spectra that will function as endmember as analysis proceeds to the identification and mapping phases of the study.

To be useful the spectral library, acting as a database, should contain spectra of the major surface component materials that are likely to be encountered or whose presence or disposition in the image is of interest. The function of the library is to provide spectra as a common reference that is sufficiently representative to permit the accurate characterization of these materials within the study area. Spectral libraries can therefore form as a generic collection of spectra assembled for nonspecific application. In this arrangement the spectral library is simply a collection of assorted reflectance spectra that were produced without a specific connection to a mapping or detection problem. Alternatively, spectral libraries can be assembled from problem-specific examples collected for a well-defined purpose or application. Examples of generic spectral libraries that can be accessed for general applications include the following:

- The United States Geological Survey (USGS) Spectroscopy Lab Spectral Library (http://speclab.cr.usgs.gov/spectral_lib.html).
- The ASTER Spectral Library (http://speclib.jpl.nasa.gov).

Application-specific spectral libraries are assembled from reference spectra collected during field campaigns or laboratory investigations. In both of these instances "high-value" target materials have been selected from which spectra are measured. Examples may include phenomena such as a particular plant species, chemical agent, or land surface property that is unique or specialized and therefore not commonly found in a generic library. The creation of application-specific libraries and the issues guiding their collection have been well documented by Price (1995), Price (1998), and Herold et al. (2004).

Whether generic or "self-built," once assembled the spectral library becomes an invaluable tool for environmental analysis provided it addresses the user's primary needs. As hyperspectral sensing becomes more common, questions regarding the standardization of spectral libraries, the nature and content of library metadata, together with the practical issues introduced by the need for documentation, standard taxonomy, measures of accuracy, and user accessibility will demand critical evaluation. One promising solution to the general problem of library standardization and access has been introduced by Ferwerda et al. (2006) in the form of a prototype Web-based open source database for the distribution of hypespectral signatures.

9.7 Developing Hyperspectral Applications

Environmental analysis conducted using multispectral remote sensing provides useful information that supports broad types of questions concerning the identification and inventory of land surface objects (Craig et al. 2006; Govender et al. 2007; Ben-Dor et al. 2009). With the introduction of hyperspectral analysis a third possibility is to introduce analysis that directs our attention to the question regarding the condition of the landscape objects we have identified. Hyperspectral applications are typically developed around this third question, supporting investigative problems that are deductive in nature (Curran 2001; Aspinall et al. 2002). Deductive analysis facilitated by hyperspectral data directs interest to those types of problems where interest centers on understanding the composition of objects, their status and the identification of unique features. Examples can be as basic as studies designed to identify an invasive plant species within the scene, focused analysis to determine the status of concrete paving on highways, or investigations developed to investigate the composition of industrial gray or brown fields. In these examples the reflectance spectra obtained from imaging spectrometers aptly characterize the material environment and the library reference assembled to drive the application facilitate both the thematic identification of materials that are known, but also enable the assessment of unknowns, outliers, and anomalous patterns at the surface and the environmental complexes that envelop them.

Crafting a hyperspectral mapping exercise follows four fundamental processing steps:

- *Initial calibration* – this step in analysis requires the conversion of image data to radiance values.
- *Atmospheric correction* – operations here are aimed at removing image noise attributed to the atmosphere and converting radiance to apparent reflectance.
- *Endmember preparation* – a step involving the extraction of spectra from the image, removing the spectral continuum and building the spectral library of continuum-removed absorption features or acquiring library reference spectral through direct collection or by accessing a prepared spectral library.
- *Spectral classification* – classifying image pixels using specific processing strategies such as those preformed by spectral unmixing or spectral matching algorithms with reference to the spectral library.

Each of the processing step outlined above introduce detailed methodological considerations that will be explored and expanded upon in the next chapter. In general, the goal of hyperspectral analysis is no different than the principles that guide multispectral remote sensing: to derive the maximum information from the imagery, thereby minimizing reliance on other sources of data in order to move closer to the solution of a problem (Kalacska et al. 2009). Through careful project planning, based on clear objectives, we can recover more detail about the composition of the environment using hyperspectral data and undertake complex mapping applications that extend the utility of remote sensing technologies in environmental analysis. To the environmental community the promise of hyperspectral remote sensing as a means to identify the presence and abundances of specific diagnostic materials can better direct land management, environmental monitoring, and environmental assessment operations (Madden 2004; Pontius et al. 2008).

9.8 Summary

This chapter introduces the world of remote sensing beyond the multispectral sensors commonly employed in landscape analysis and mapping. Hyperspectral images are often referred to as spectrally over determined. Such imagery captures spectral information over an expanded sample of wavelengths that can be used to identify and distinguish between spectrally similar (but unique) materials. Hyperspectral imagery provides the potential for more accurate and detailed information extraction than is possible with other types of remotely sensed data; therefore, hyperspectral imagery provides opportunities to extract more detailed information than is possible using traditional multispectral data. In this chapter, the principles of hyperspectral analysis were introduced beginning with a treatment of spectroscopy to understand how materials absorb light energy in discernable ways. From this background the concept of an endmember was introduced than demonstrated how the chemical and material composition of surface objects cold be identified based on their spectra. With an understanding of these fundamental principles, the compilation of material spectra into libraries to guide analysis was examined.

References

Aspinall, Richard (2002) A geographic information science perspective on hyperspectral remote sensing, Journal of Geographical Systems, 4, 127–140.

Aspinall, R., Marcus, W., Boardman, J. (2002) Consideratons in Collecting, Processing and Analysing High Spatial Resolution Hyperspectral Data for Environmental Applications, Journal of Geographic Systems, 4, 15–29.

Ben-Dor, E., Chabrillat, S., Dematte, J., Taylor, G., Hill, J., Whiting, M., and Sommer, S. (2009) Using Imaging Spectroscopy to Study Soil Properties, Remote Sensing of Environment, 113, S38-S55.

Chaudhry, F., Wu, C., Liu, W., Chang, C. and Plaza, A. (2006) Pixel purity index-based algorithms for endmember extraction from hyperspectral imagery, in Chang, C. (ed) Recent Advances in Hyperspectral Signal and Image Processing, Transworld Research Network, 29–62.

Clark, R. (1999) Spectroscopy of Rocks and Minerals, and Principles of Spectroscopy, in Rencz, A. (ed) Manual of Remote Sensing, Volume 3, Remote Sensing for Earth Sciences, John Wiley and Sons, New York, 3–58.

Clark, R., Swayze, G., Gallaher, A., King, T., Calvin, W. (1999) The U.S. Geological Survery, Digital Spectral Library, Version 1, U.S. Geological Survey Open File Report 93–592.

Clark, R., Swayze, G., Wise, R., Livo, K., Hoefen, T., Kokaly, R. and Sutley, S. (2007) USGS Digital Spectral Library splib06a, U.S. Geological Survey, Data Series 231.

Craig, S., Lohrenz, S., Lee, Z., Mahoney, K., Kirkpatrick, G., Schofield, O., Steward, G. (2006) Use of Hyperspectral Remote Sensing Reflectance for Detection and Assessment of the Harmful Alga, Karenia brevis, Applied Optics, 45, 5414–5425.

Curran, P. (1994) Imaging Spectrometry, Progress in Physical Geography, 18, 247–266.

Curran, P. (2001) Imaging Spectrometry for Ecological Applications, International Journal of Applied Earth Observation and Geoinformation, 3, 305–312.

Dehaan, R. and Taylor, G. (2003) Image-derived spectral endmember as indicators of salinisation, International Journal of Remote Sensing, 24, 775–794.

Fava, F.; Colombo, R.; Bocchi, S.; Meroni, M.; Sitzia, M.; Fois, N.; Zucca, C. (2009) Identification of hyperspectral vegetation indices for Mediterranean pasture characterization, International Journal of Applied Earth Observations and Geoinformation, 11, 233–243.

Ferwerda, J. G.; Jones, S. D.; Du, Pei-Jun (2006) A Web-based open-source database for the distribution of hyperspectral signatures, Geoinformatics 2006: Geospatial Information Technology. Proceedings of the SPIE, 6421, 64210G-64210G-7.

Gao, J. (2009) Digital Analysis of Remotely Sensed Imagery, McGraw-Hill, 645p.

Garcia-Aro, F., Gilabert, M., Melia, J. (1999) Extraction of Endmembers from Spectral Mixtures, Remote Sensing of Environment, 68, 237–253.

Govender, M., Chetty, K., Bulcock, H. (2007) A Review of Hyperspectral Remote Sensing and its Application in Vegetation and Water Resource Studies, Water SA, 33, 145–151.

Goetz, Alexander F.H. (2009) Three decades of hyperspectral remote sensing of the Earth: A personal view, Remote Sensing of Environment, 113, S5–S16.

Herold, M., Roberts, D., Gardner, M., Dennison, P. (2004) Spectrometry for Urban Area Remote Sensing- Development and Analysis of a Spectral Library from 350 to 2400 nm, Remote Sensing of Environment, 91, 304–319.

Im, J. and Jensen, J. (2008) Hyperspectral Remote Sensing of Vegetation, Geography Compass 2/6 1943–1961.

Kalacska, Margaret E.; Bell, Lynne S.; Arturo Sanchez-Azofeifa, G.; Caelli, Terry (2009) The Application of Remote Sensing for Detecting Mass Graves: An Experimental Animal Case Study from Costa Rica*, Journal of Forensic Sciences, 54, I159–166.

Kruse, F. (2004) Comparison of ATREM, ACORN, and FLAASH Atmospheric Corrections sing Low Altitude AVIRIS Data of Boulder, Colorado, Proceedings of the 204 AVIRIS Earth Science and Applications Workshop, Jet Propulsion Laboratory, Pasadena, CA.

Madden, Marguerite (2004) Remote Sensing and Geographic Information System Operations for Vegetation Mapping of Invasive Exotics1, Weed Technology, 18, 1457–1463.

Martinez, P., Perez, R., Plaza, A., Aguilar, P., Cantero, M., Plaza, J. (2006) Endmember Extraction Algorithms from Hyperspectral Images, Annals of Geophysics, 49, 93–101.

Milton, E., Schaepman, M., Anderson. K., Kneubühler, M., and Fox, N. (2009) Progress in field spectroscopy Remote Sensing of Environment, 113, Supplement 1, s92–s109.

Nidamanuri, Rama Rao; Zbell, Bernd (2010) A method for selecting optimal spectral resolution and comparison metric for material mapping by spectral library search, Progress in Physical Geography, 34, 47–58.

Plaza, Antonio; Benediktsson, Jon Atli; Boardman, Joseph W.; Brazile, Jason; Bruzzone, Lorenzo; Camps-Valls, Gustavo; Chanussot, Jocelyn; Fauvel, Mathieu (2009) Recent advances in techniques for hyperspectral image processing, Remote Sensing of Environment, 113, S110–S122.

Pontius, Jennifer; Martin, Mary; Plourde, Lucie; Hallett, Richard (2008) Ash decline assessment in emerald ash borer-infested regions: A test of tree-level, hyperspectral technologies, Remote Sensing of Environment, 112, 2665–2676.

Price, J. (1995) Examples of high resolution visible to near-infrared reflectance spectra and a standardized collection for remote sensing studies, International Journal of Remote Sensing, 16, 993–1000.

Price, J. (1998) An Approach FOR Analysis of Reflectance Spectra, Remote Sensing of Environment, 4, 316–335.

Rogge, D.M.; Rivard, B.; Zhang, J.; Sanchez, A.; Harris, J.; Feng, J. (2007) Integration of spatial-spectral information for the improved extraction of endmembers, Remote Sensing of Environment, 110, 287–303.

Schaepman, Michael E.; Ustin, Susan L.; Plaza, Antonio J.; Painter, Thomas H.; Verrelst, Jochem; Liang, Shunlin (2009) Earth system science related imaging spectroscopy-An assessment, Remote Sensing of Environment, 113, S123–S137.

Schmidtlein, S. (2005) Imaging Spectroscopy as a Tool for Mapping Ellenberg Indicator Values, Journal of Applied Ecology, 42, 966–974.

Shaw, G. and Burke, H. (2003) Spectral Imaging for Remote Sensing, Lincoln Laboratory Journal, 14, 3–28.

Swayze, G. (2000) Using Imaging Spectroscopy to Map Acid Mine Waste, Environmental Science and Technology, 34, 47–57.

Treitz, P.; Howarth, P. (1999) Hyperspectral remote sensing for estimating biophysical parameters of forest ecosystems Progress in Physical Geography, 23, 359–390.

Underwood, Emma; Ustin, Susan; DiPietro, Deanne (2003) Mapping nonnative plants using hyperspectral imagery, Remote Sensing of Environment, 86, 150–161.

Van der Meer, F. (2004) Analysis of Spectral Absorption Features in Hyperspectral Imagery, International Journal of Applied Earth Observation and Geoinformation, 5, 55–68.

Van der Meer, F. and Jong, S. (2002) Imaging Spectroscopy: Basic Principles and Prospective Applications, Springer, 425p.

Vane, G and Goetz, A. (1988) Proceedings, Airborne Visible/Infrared Imaging Spectrometer (AVIRIS) Performance Evaluation Workshop, National Aeronautics and Space Administration, Jet Propulsion Laboratory Publication, Pasadena, CA.

Younan, N. H.; King, R. L.; Bennett Jr, H. H. (2004) Classification of Hyperspectral Data: A Comparative Study Precision Agriculture, 5, 41–53.

Chapter 10
Hyperspectral Classification

The detailed spectra defined in a hyperspectral images posses new image processing challenges and exciting opportunities. Unlike its multispectral counterpart, hyperspectral imagery captures a level of spectral resolution that contains unique compositional and structural information about the landscape not available in other forms of remotely sensed imagery. To exploit this source of information, thematic extraction based on hyperspectral data involves isolating spectral features in the image according to their reflectance properties followed by a comparison of these properties to those on known materials. In this chapter, we will review the methods employed to extract thematic information from hyperspectral imagery and examine the algorithms called upon to process image spectra.

10.1 Processing Fundamentals

The overarching goal in hyperspectral remote sensing is to achieve a successful identification of surface materials and conditions. With this goal in mind the classification problem using hyperspectral data reduces to the twofold task of (1) decomposing the spectral composition of a pixel and (2) matching pixel spectra to those of known materials of substances. Thematic extraction based on this comparatively simple logic produced a level of classification discreteness that also contrasts markedly with the thematic categorizations developed from multispectral data sets. Through the application of hyperspectral data, image classification focuses on the discrimination of surface types and their conditions as opposed to the naming of broad categories of land cover. At first glance this distinction may seem trivial, but it carries important and useful implications for environmental remote sensing applications. For example, multispectral classification is capable of producing information detailing the spatial extent of impervious surface over portions of a watershed. While this information is extremely useful in hydrologic and engineering studies, this same exercise performed using hyperspectral imagery can not only produce a map explaining the geographic pattern of impervious

J.K. Lein, *Environmental Sensing: Analytical Techniques for Earth Observation*,
DOI 10.1007/978-1-4614-0143-8_10,

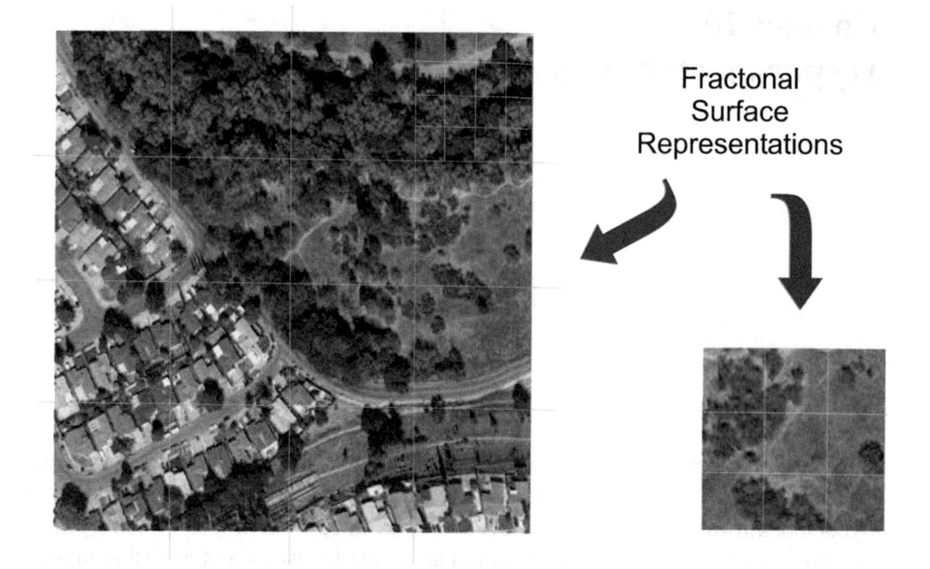

Fig. 10.1 Representation of surface fractions

surface, but also detail what surface types can be attributed to the pattern; whether impervious surface is composed of concrete, impervious asphalt, brick, or shake roof tops (Van der Meer 1998/1999, 2004, 2006). In this example, the addition of "type" in the categorization of the surface contributes significantly more information to the analyst and facilitates estimation of the abundance of these materials in the watershed.

The concept of abundance is a recurrent theme in hyperspectral classification. The concept explains the degree to which surface types that comprise elements of the landscape are evidenced in a pixel. The pixel, as a sample unit of fixed spatial dimension can be pure, describing a homogeneous surface, or a mixture of types commonly referred to as a composite. Abundance is an attribute of the pixel that is used to quantify the percentage or fraction of the pixel covered by a given material. We recall this idea from Chap. 9 during our discussion of spectral endmembers. Values of abundance are expressed along the numerical range from 0.0 to 1.0 and are displayed cartographically as a continuous surface communicating the same logic as a fuzzy or belief surface (Fig. 10.1). Using this numerical expression, the spatial distribution and relative amount of a given material at pixel locations over the scene can be understood. More importantly, abundance estimates can be transformed into categorical representations detailing the presences of a type or class of material, and through the selection of fractional thresholds, type or class dominance can be established and visualized (Fig. 10.2).

Hyperspectral classification may also proceed based on the logic of matching. Matching is a form of supervised pattern recognition logic where an unknown spectra is compared to that of a known material. Spectral matching, according to

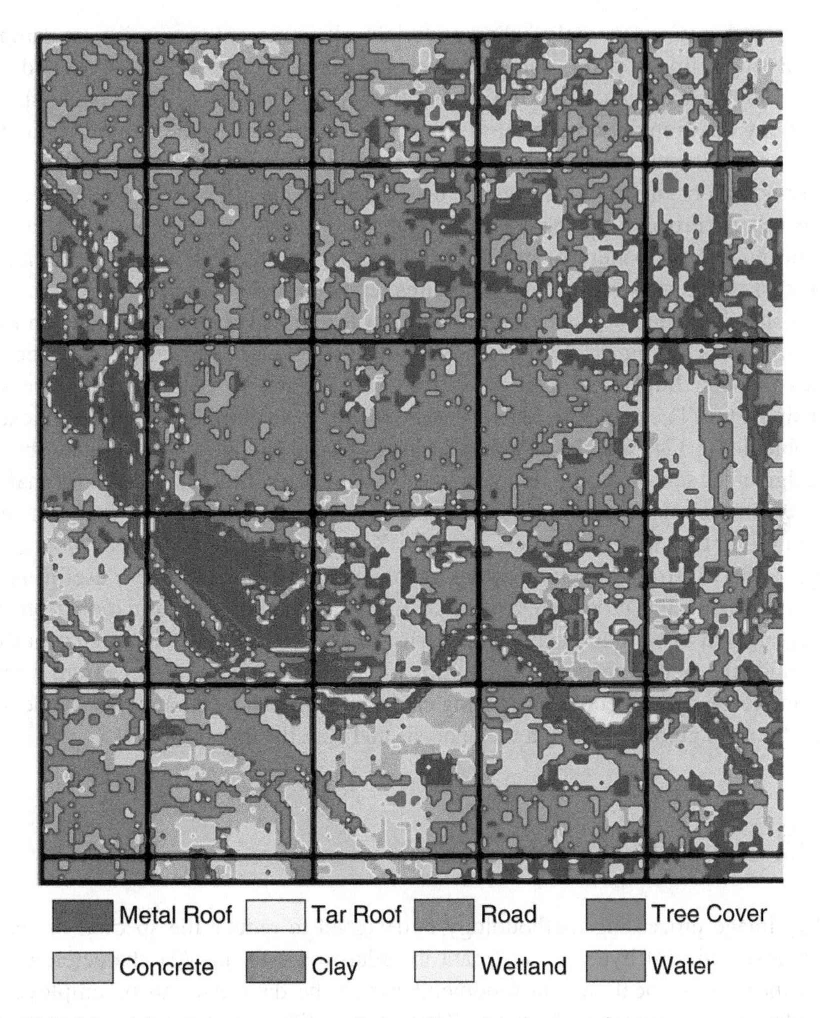

Fig. 10.2 Material cover representations derived from abundance surfaces

this logic, is typically defined by a measure of similarity between the unknown spectra and the spectra of the reference material. In simple terms the closer the unknown spectra is to the reference example as expressed by a "distance" metric, the closer is the match. A decision rule determines the threshold where an optimal match is achieved. The pixel is labeled accordingly based on its proximity or similarity to this threshold. Matching logic employs a range of approaches to est ablish matching thresholds. Common methods derive thresholds using distance-based, angle-based, or correlation-based measures (Drake et al. 1999; Goodenough et al. 2003; Kardi 2007).

Lastly, hyperspectral classification can be understood based on the logic of unmixing. The landscape we study is rarely composed of a single uniform material or condition. Consequently, when materials with different properties are sampled by a single image pixel, some degree of spectral mixing can be expected. We generally conceptualize mixing as the linear combination of reflectance spectra with the addition of some level of noise that contributes to the pattern in random ways. Classification based on the logic of unmixing focuses on reversing the mixing process that occurs and attempts to "take a part" in the pixel spectral. By reversing the mixing process, the goal is to isolate the various components (spectra) that contribute to the pattern of mixing explained by the pixel. Through the unmixing process the identity of spectral signatures (endmembers) hidden in the mixture are revealed and their corresponding proportions (fractional abundance) are quantified.

A variety of empirical procedures have been developed to guide hyperspectral classification. Implementing a specific procedure begins by selecting that processing chain that is (1) capable of addressing the requirements imposed by the application problem and (2) capable of compensating for the limitations introduced by the design of the sensor system that provides the hyperspectral data. End-to-end analysis of hyperspectral imagery can be approached following a sequence of recommended operations that have been described by Kruse et al. (2003) as the "hourglass" approach. Illustrative methods such as the "hourglass" model are a useful means of visualizing the procedures required to extract thematic information from our imagery and as well as a schematic device that serves as an organizing framework to guide the analyst through the classification problem. In the next section, we will examine the hyperspectral processing chain and employ the "hourglass" model as a road map to guide us through the analysis and classification process.

10.2 Hyperprocessing

Any image processing methodology is designed to reduce the spectral or spatial dimensionality of hyperspectral data in order to (1) locate, (2) characterize, and (3) indentify critical spectra (endmembers) in the data that can be employed to explain the remainder of the scene. The tasks specific to this general operation and the workflow required to produce the desired result is illustrated in Fig. 10.3. As with any remote sensing exercise, hyperspectral classification begins with the acquisition of data that complements the application problem. Although hyperspectral remote sensing has undeniable potential for advancing environmental analysis, hyperspectral imagery is not as readily available as other forms of remotely sensed data. At present there are very few operational satellite-based hyperspectral sensors and none provide the continuous repeat coverage common to multispectral sensors such as Landsat TM. The current situation suggests that satellite-based hyperspectral imaging is still in the developmental stages, therefore the majority of applications will rely on careful mission planning and data acquisition based on air-born systems for the foreseeable future. This near-term reality restricts the "off-the-shelf" availability of hyperspectral imagery which constrains the breath of applications to only those well-funded, carefully planned and clearly

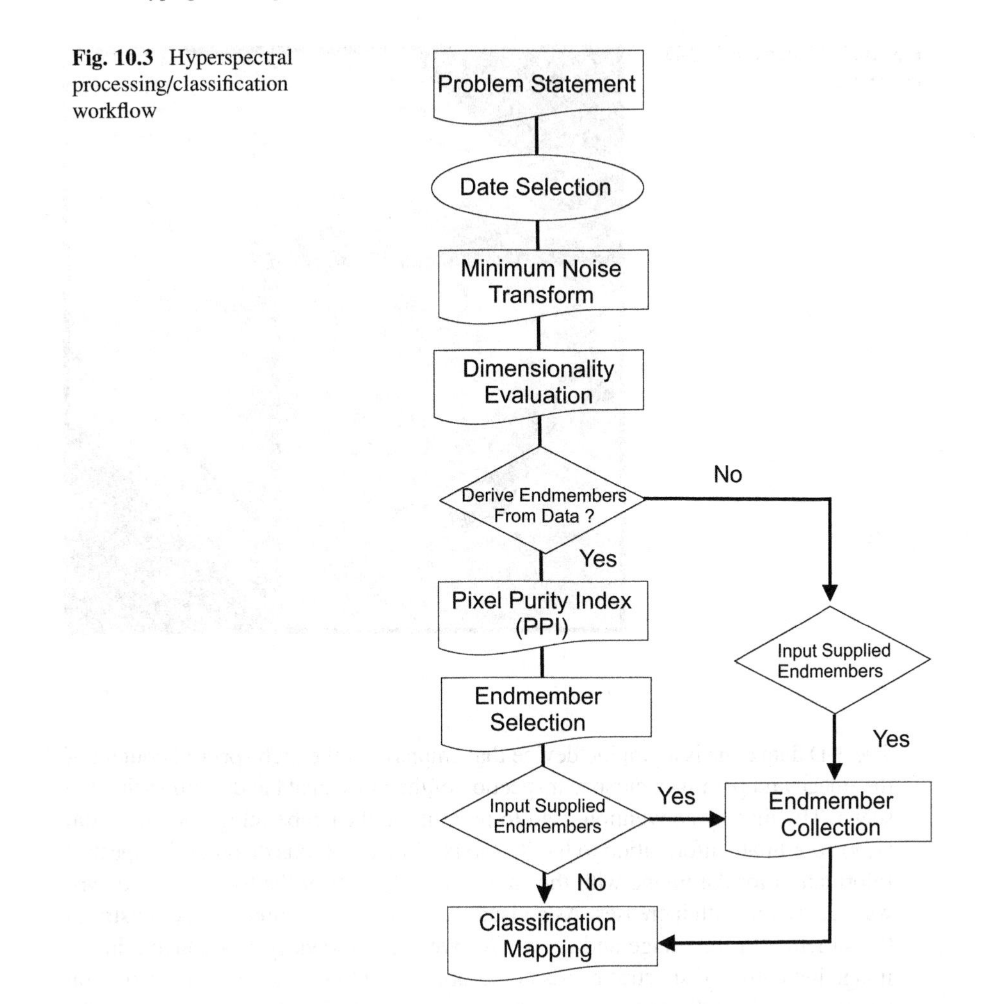

Fig. 10.3 Hyperspectral processing/classification workflow

targeted problems; the majority still somewhat experimental in nature. Despite this limitation hyperspectral data can be accessed from the EO-1 systems (EO-1 and Hyperion) and acquired through visualization servers such as GLOVIS supported by the United States Geological Survey (http://glovis.usgs.gov).

Once imagery can been acquired, either by means of a specific data collection mission or from Earth-orbiting platforms, formal processing can proceed. The stages involved include the following:

1. *Data preparation* – Data preparation begins the generation of a spatial subset of the hyperspectral image to focus attention on the study are germane to the application problem. Once the geographic extent of the study area has been identified, preparation continues with an initial assessment of image quality followed by radiometric and geographic preprocessing. One common method for displaying hyperspectral imagery for visual evaluation is the 3-D data cube (Fig. 10.4).

Fig. 10.4 Characteristic 3-D image cube

The 3-D data cube is a graphic device that emphasizes the high spectral content of the image and permits a cursory inspection of the individual bands comprising the scene. The display convention used to present the data cube relegates the spatial (x, y) coordinate information to the X–Y axis while the Z-axis displays the spectral information for the image with the shorter wavelengths at the top moving downward as wavelength increases. As a visualization device the image cube illustrates the structure of the image and is an effective way to identify black layers in the image indicative of spectral bands compromised by high atmospheric absorption where negligible reflectance reaches the sensor. An additional visualization tool for identifying absorption and noise in the data employs image animation. Animation of the scene with each band representing a frame in the sequence facilitates cycling through the wavelengths noting bands that have noticeable radiometric error and noise. Examples of image noise are provided in Fig. 10.5). Once bands plagued by error or noise have been identified they can be removed by spectral subsetting of the image. The result produces a relatively clean image that can be subjected to geometric correction and radiometric calibration. Geometric correction and orthorectification is a critical operation for imagery acquired from aircraft sensors in order to remove flight line distortions. Radiometric calibration attempts to remove any lingering striping and periodic cross-path swatch variations in brightness. Calibration also compensated for senor-related effects that modify the image and render the sensor-recorded values more proportional to actual at-sensor radiance.

Fig. 10.5 Image noise

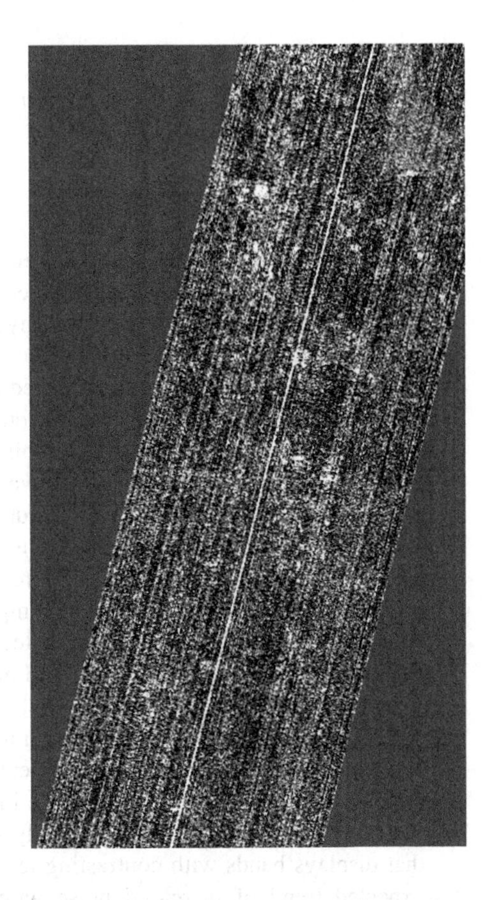

2. *Atmospheric correction* – Hyperspectral imaging sensors collect radiance data from either airborne or space-borne platforms that must be converted to apparent surface reflectance before classification operations can proceed. At this point, it is helpful to recall that radiance and reflectance are two distinct phenomena. Sensor radiance measures the amount of electromagnetic energy that reaches the sensor. Reflectance explains the percentage of electromagnetic radiation incident on a surface that is actually reflected by the material. Without converting radiance measures to reflectance values comparing image spectra with reference reflectance spectra (endmembers) will not yield meaningful results. Atmospheric correction techniques are called upon to remove spectral transmission of the atmosphere and scattered path radiance, producing an image expressed in units of reflectance in the process. Atmospheric correction can be undertaken using either:

   (a) *Simple atmospheric correction methods, such as*

      - *Flat field*
      - *IAR reflectance*
      - *Empirical line corrections*

or

(b) *Advanced atmospheric correction methods, such as*

- *ACORN*
- *FLAASH*
- *ATREM*

Under most circumstances model-based transformations using advanced correction methods are required for hyperspectral analysis.

3. *Data reduction* – Adjacent bands in a hyperspectral image are often visually and numerically similar. Reducing the dimensionality of the image data can speed up visual analysis and classification procedures. When this option is exercised, two of the most common data reduction techniques employed are principal components analysis (PCA) and the minimum noise fraction (MNF) transform. PCA is a familiar technique in multivariate statistics that seeks to produce an orthogonal linear transformation of data to form a new smaller number of uncorrelated variables. When applied to a hyperspectral image, PCA is a useful method of data reduction when there is sufficient evidence to believe that redundancy is an issue. Redundancy implies high levels of correlation between bands and suggests that these correlated bands are measuring the same condition. Pattern recognition in the high dimensionality defined by a hyperspectral image can be improved through a linear transformation to create a lower dimensional data set. The resulting transformed image presents a new set of bands ordered in a manner that presents important image information in the lower-ordered components with noise increasing as component (band) number increases. The MNF is mathematically similar to PCA; however, in an image that displays bands with contrasting levels of noise, PCA may not capture the expected trend of increased noise with increasing band number. The MNF transform addresses this potential limitation by normalizing each band in the image by its noise level. Normalization, in this context, acts to reduce the influence of noise by de-emphasizing problematic bands in the transformed image. The MNF algorithm projects the original image into a new space that is divided into two parts; one part associated with the signal while the second is dominated by noise. The MNF transform creates a set of images that contain weighted information regarding signal variance across all the bands that comprise the original scene. The transform retains specific band information since all of the original wavelengths contribute to this new data set. With either the PCA or MNF approach the data trend is generally explained by the first few bands in the new image. This feature requires the analyst to select the boundary separating signal (useful bands) from noise. In most instances selection is based on a careful evaluation of the eigenvalues visualized using a scree plot or by subjecting the new band combinations to review using image animation.
4. *Endmember selection* – As noted previously in our review of hyperspectral sensing, endmembers were characterized as spectra that can be used to classify pixels. Although they may represent materials of interest in the image, an endmember is

Table 10.1 Common methods for hyperspectral analysis and classification

Whole pixel	Subpixel
Spectral angle mapper (SAM)	Complete linear spectral unmixing
Spectral feature fitting	Matched filtering

neither the material nor the spectral signature of a surface type or class. Rather the endmember is a constituent part of a spectral mixture which may be pure or a composite of materials at the geographic scale represented by the image. For analysis to proceed, endmembers must be selected from the spectral library that was derived either directly from the image or by means of field or laboratory collection. The actual nature of the endmember and the material or substance it references depends on the application problem (Rogge et al. 2007). With the objectives of analysis clearly stated the appropriate library spectra can be gathered and made available for use during classification.

5. *Spectral classification* – Following satisfactory calibration and correction operations together with the selection of relevant endmembers hyperspectral processing enters the classification phase. A range of image analysis algorithms have been developed to extract the extensive information contained in hyperspectral imagery. The available methodologies fall in one of two categories: (1) whole-pixel approaches or (2) subpixel methods. Whole-pixel methods function to determine whether one or more material of interest are abundant within each pixel that forms the hyperspectral scene. The determination of abundances is accomplished on the basis of defining a measure of spectral similarity between the pixel and its reference spectra. Subpixel algorithms are used to calculate the quantity of a material in mixtures present in each image pixel (Borengasser et al. 2008). In both approaches the pixels in the scene are evaluated with reference to the spectra assembled into the reference library file. A listing of common processing algorithms is provided in Table 10.1 and each are reviewed later in this chapter. The classification logic applied in classification centers around two themes: (1) matching and (2) unmixing. As we recall from Chap. 9 matching asks the comparatively simple question, "can material A be found in the image"? The objective of matching is to find spectra of known materials in a reference set that match those found within the image. According to this approach, no assumptions are made regarding the materials on the ground or their spectral properties, although the number and type of reference spectra are usually limited based on the scope of the problem (Adams and Gillespie 2006). Through this example of spectral modeling physical interpretation, based on prior knowledge and matching logic, establishes a test to determine how well image spectra agree with a general hypothesis regarding what the image may contain. Unmixing can be conceptualized as the "picking apart" of the reflectance spectra. The objective of unmixing is to separate the spectrum addressing the question: "what does these spectra consist of"? Through unmixing the

membership of individual endmembers with reference to the source spectrum can be ascertained, creating a measureable result referred to as endmember abundance.

6. *Verification* – Spectral classification produces thematic maps that convey information about the nature and/or composition of the material surface. Since the map information are generated from material spectra they portray patterns of selected materials on the ground or labeled to describe more generalized types based on the reference endmembers. Since classification develops from the comparison of ground to reference spectra verification of the results is essential. Verification considers both the accuracy of a given label and how well materials at the surface were detected. The challenge using hyperspectral data is to establish the level of agreement which is made complicated by the spatial resolution of the imagery. At fine spatial scales pixel footprints can be sampled for representative locations. However, the pixel footprint requirement can be relaxed if aggregations of pixels produce a block pattern large enough to permit the clear identification of a material. From this point on, verification seeks to determine if the material is present in the sampling unit. This typically requires choosing an endmember with the highest abundance value as the label for the pixel to create a thematic representation that can be used to conduct a traditional accuracy assessment and produce standard accuracy statistics.

The stages outlined above suggest a general script that can be followed to guide the application of hyperspectral imagery in environmental analysis. The utility of the thematic information produced from the hyperspectral data, however, rests in how well the spectrum of measures has been exploited to generate a relevant information product. Converting the data in the image to information that solves a problem relies on how well a classification was executed relative to the strategy chosen to process the data. In the next section, the processing strategies alluded to above are examined in greater detail.

10.3 Hyperspectral Processing Strategies

Image classification is perhaps the most common method of extracting information from remotely sensed imagery. Although the product of hyperspectral classification may fit different objectives when compared to the thematic presentations created from multispectral analysis, an equally detailed body of techniques have been introduced to assist with hyperspectral classification that warrant careful review. A selection of the more widely applied methods is presented. These approaches can be separated into two dominating strategies: whole-pixel methods and subpixel techniques.

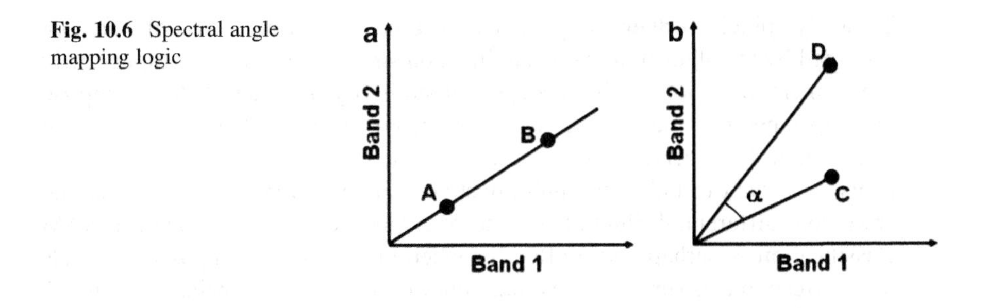

Fig. 10.6 Spectral angle mapping logic

10.3.1 Whole-Pixel Analysis

Whole-pixel processing methods are used primarily to ascertain the relative abundance of one or more target materials within each pixel in our hyperspectral image. The general processing convention employed by whole-pixel classifiers involves a similarity test that compares the target pixel with materials spectra assembled into a reference library. Common whole-pixel classifiers include the spectral angle mapper (SAM) and spectral feature fitting (SFF) algorithms each of which implement a specific spectral matching procedure.

- *Spectral angle mapper* – The SAM algorithm classifies hyperspectral data based on a set of reference spectra that define specific surface types or conditions of interest. The algorithm was introduced by Kruse et al. (1993) and is explained by the general relation:

$$\text{Angle} = \arccos\left(\frac{\text{Sum of } (T(i) \times R(i))}{|T| \times |R|}\right),$$

where, $T(i)$ is the test spectrum values ($i = 1$ to n), $R(i)$ is the reference spectrum values ($i = 1$ to n), $|T|$ is the square root of (sum of ($T(i) \times T(i)$)), and $|R|$ is the square root of (sum of ($R(i) \times R(i)$)).

According to this model, R explains the amplitude of the reference spectra at band (b) and (i) represents the amplitude of the spectra to be matched at band (b). Using this formula, the SAM algorithm computes a spectral angle between each pixel spectrum and each target spectrum. The smaller the spectral angle between the two conditions, the more similar the pixel is to the band vector or target (Fig. 10.6). The calculated spectral angle is measured in units of radians that are assigned to the corresponding pixel producing either a series of raster layers (one for each reference spectra in the library) or a single raster image that displays for each pixel in the image of the reference spectrum, which has the smallest angle and therefore represents the best match. Satisfying a match is a function of the minimum spectral angle threshold, where smaller angles define close matches even in situations where one spectrum is much brighter that another.

Typically, pixels further away than a specified similarity threshold are not classified by the algorithm. As a classification strategy, SAM is a comparatively robust method for identifying the spectral similarity of materials that comprise the hyperspectral image. The algorithm represses the influence of shading effects, thereby accentuating the reflectance characteristics of the target. SAM is also invariant to unknown multiplicative scaling and related deviations that arise from differential illumination and orientation conditions. However, SAM classification is perhaps most effective when the image is composed of pixels that correspond to surface types and conditions that are explained by "pure" pixels (single, spectrally distinct material classes) (Kruse et al. 1993; Hunter and Power 2002; Schwarz and Staenz 2001; Van Der Meer et al. 1997).

- *Spectral feature fitting* – SFF classification implements a matching algorithm based on the complete shape of the image spectra when compared to the reference spectra with a specified wavelength range. The SFF algorithm uses the absorption features of spectra to perform the comparison test and requires continuum removal of the spectra prior to processing. Drawing from the analyst's prior knowledge of specific materials, a range of spectra are selected to describe the spectral feature (absorption band). Next, continuum is removed from the image data and the reference spectra and the continuum-removed spectra is superimposed over the pixel spectra. A least-squares regression is calculated on a band-by-band basis where the total RMS (root mean square) error is used as a measure of goodness of fit. A spectral match is established based on two measures. The first measurement is the depth of the feature in the pixel compared to the depth of the features in the reference. The second is a measure of shape based on the correlation between the shape of the feature in the pixel and the shape of the feature in the reference (Clark et al. 2000; Gupta 2003). A scale image is produced for each endmember selected for analysis. This image is created by subtracting the continuum-removed spectra from one what inverts the spectra; rendering the continuum zero. A single multiplicative scaling factor is then determined that yields a reference spectrum match with the unknown spectrum. With selection of a reasonable spectral range, a large scaling factor is equivalent to a deep spectral feature and a small scaling factor describes a weak spectral feature. In some implementations of the SFF algorithm a ratio image of Scale/RMS is available that is used to explain how well the unknown spectrum matches the reference spectrum on a pixel-by-pixel basis.

10.3.2 Subpixel Analysis

Subpixel processing methods are robust detection algorithms that calculate the quantity or fraction of materials contained within the pixels that define our hyperspectral scene. The attraction to this family of image processing techniques

Fig. 10.7 Linear unmixing methodology

The steps involved in using linear spectral unmixing:

1. Convert the imagery to reflectance (preferably ground)
2. Endmember Collection
3. Chose your Constraints
4. Normalize the Fraction maps
5. Interpret your Results

rests in their capacity to detect quantities of a material or condition that fall below the spatial resolution of the sensor; features commonly referred to as subpixel phenomena. It has been demonstrated that in situations where the spectral contrast between a feature and its background is good, subpixel algorithms can indentify materials that cover as little as 13% of a pixel (Borengasser et al. 2008). Two of the more widely used subpixel analysis techniques in hyperspectral remote sensing are the spectral unmixing and matched filtering algorithms.

1. *Spectral unmixing* – Unmixing algorithms are based on the premise that the reflectance spectrum of a pixel is the product of a linear combination of the spectra of all image endmembers that fall within it. A linear combination in this context can be conceptualized as a weighted average of each endmember that contribute to the pixel's spectrum. The endmembers exert an influence that is assumed to be proportional to the area they occupy in the pixel. The unmixing procedure involves decomposing these mixtures of endmembers into their individual parts, expressing the resulting pattern as a set of corresponding fractions or abundances. The endmember fractions describe the proportion each explains pixel-by-pixel that complete the image (Keshava 2003; Keshava and Mustard 2002). The general unmixing algorithm solves for a set of n-liner equations for each pixel (Fig. 10.7). The unknown variables in these equations are the fractions of each endmember in the pixel. Therefore, to solve the linear set for an unknown mixed fraction, there must be more equations than unknowns. In practical application this suggests that unmixing can only proceed if there are more bands in the image than endmember spectra (materials). Endmember selection can also be problematic when conducting spectral unmixing, since endmembers are expected to represent "pure" pixels. The linear mixing model takes the basic form:

$$R_k = \sum r_{jk} F_j + V^k \quad (k = 1, \ldots p),$$

where R_k represents the mean spectral reflectance of a pixel in the k th spectral band, r_{jk} is the spectral reflectance of the j th component in the kth spectral band, f_j is the proportion of the jth component within the pixel, m is the number of components,

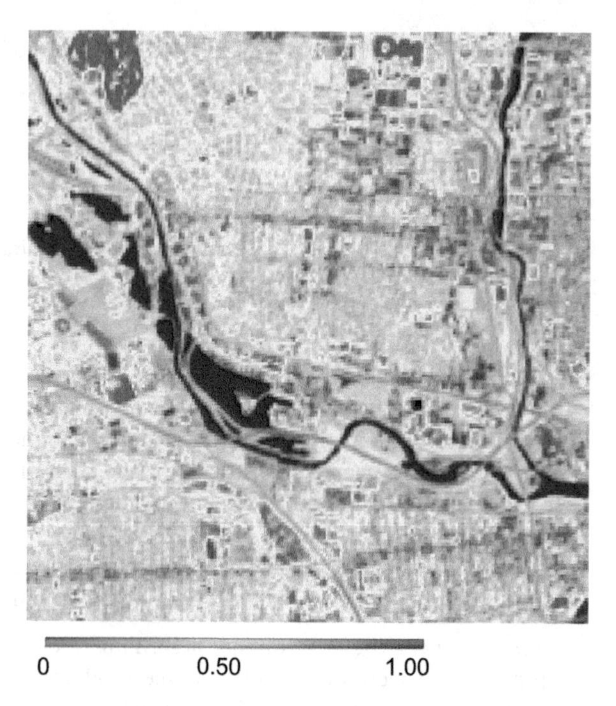

Fig. 10.8 Linear unmixing abundance layer

V^k is the residual, and p is the number of spectral bands for the kth spectral bands, respectively. Using this fundamental relationship with an image of k spectral bands, the ith endmember spectrum can be denoted as S_i and the abundance of the ith endmember as Ai. Taken together, the observed spectrum X for any pixel becomes

$$X = \mathrm{A}1S1 + A2S2 + \cdots + AmSm + V^k.$$

Linear models such as that expressed above assume no interaction between materials. There, if each photon on energy sees one material, the signals are additive. Multiple scattering involving several materials can be thought of as cascaded multiplications. When multiple scattering occurs nonlinear processes are at work. In the majority of examples nonlinear mixing tends to be a second-order effect. Although many surface materials mix in a nonlinear fashion, linear unmixing, while often producing an approximation of fractional patterns, can still provide satisfactory estimations. The product of the spectral linear unmixing algorithm are a series of fractional abundance images; one for each endmember, where the quantities assigned to each image pixel express the percentage of the endmember material it contains (Fig. 10.8).

(a) *Matched filtering* – Matched filtering (MF) is a derivative of the spectral unmixing algorithm where only a specific user-defined target is subject to analysis and thematic representation (Williams et al. 2002). Unlike the linear mixing model, the matched filtering does not depend on obtaining spectra of all endmembers characterizing the scene. This difference influences how an analysis is conducted and the type of application suited to the matching filtering approach. Because extensive endmember information is not required, matched filtering is often referred to as "partial unmixing" since the unmixing equations are only partially solved. As a mixed pixel classification strategy, MF suppressed background noise and estimates subpixel abundance of a single targeted material or condition based on a three phase procedure outlined by Mitchell and Glenn (2009) that begins with (1) a MNF of apparent reflection data, followed by, (2) a matched filtering of the selected material for abundance information, and culminating with, (3) mixture tuning to identify infeasible or false-positive pixels. Matched filtering operates under the assumption that the abundances of a target material in the hyperspectral scene is comparatively rare. Therefore, when this method is applied to move general thematic mapping situations, the interpretation of the results produced by the matching filtering algorithm can be problematic. Matched filtering scores are calculated for each pixel by projecting the MNF transformed data onto a matched filter vector. This vector is derived by transforming the target spectrum into MNF space; a process that requires projecting the vector onto the inverse covariance matrix of the MNF data and normalizing it to represent the magnitude of the target material's abundance. The product of this operation creates the match filter vector with unit length and corresponds to target components in the pixel that can range from 0 to 100%. Mathematically, the matched filter vector is expressed as follows:

$$V = [C_{\mathrm{mnf}}]^{-1} \times t^{*}_{\mathrm{mnf}} / (t^{*})^{T} \times [C_{\mathrm{mnf}}] - 1 \times t^{*}_{\mathrm{mnf}},$$

where V is the matched filter vector, $[C_{\mathrm{mnf}}]^{-1}$ is the inverse of the MNF covariance matrix, and t^* is the vector of the target spectra in MNF space. Values of the matched filter are calculated for each pixel to produce a target abundance image according to:

$$[\mathrm{MF}] = V \times [\mathrm{MNF}].$$

Because the abundance scores are normally distributed with zero mean, MF values of zero or less represent background clutter (no target) while pixels that scores greater than zero explain fractional target components proportional to the value of their MF score (Mundt et al. 2007). False-positives, however, are common with matched filtering and can be reduced by fine tuning the solution. The mixed tuning stage addresses the false-positive problem by assessing the probability of estimation error using the concept of mixed feasibility (Mundt et al. 2007). Calculations to determine the feasibility of the match involves the following:

- Determining the target component for a pixel
- Interpolating the variance eigenvalues respective to the target vector component

- Calculating the standard separation between a pixel and its ideal target vector

The defining equation used to compute the target vector component is given as follows:

$$C_i = \mathrm{MF}_i \, t^*_{\mathrm{mnf}},$$

where C_i defines the target vector component for pixel I, and MF_i represents the matched filter value for pixel i. The ideal pixel, according to this logic, will lie on the target vector; however, actual pixels will display some degree of variability due to noise and background mixing. The proximity of each pixel in the scene to its ideal location on the target vector served as a measure of infeasibility. This relationship describes the mixture tuning cone model and is calculated according to:

$$I_i = \frac{\|s_i - c_i\|}{\|e_i\|},$$

where I_i is the value of infeasibility for pixel I, s is the MNF spectra for pixel i, e is the interpolated vector of eigen values for pixel i and c is the target vector component. The product of mixed tuning is a suite of values that represent the statistical distance from the target mixing vector for each pixel. Pixels that fall within an infeasibility threshold match the target while those falling outside this threshold do not (Mundt et al. 2007). Incorporating the mixed tuning step in matched filtering reduces false-positives and improves the accuracy of matches above those obtained when using matching filtering alone since high infeasibility values are likely to be false-positives despite their matched filter score.

10.4 Interpretation and Explanation

When hyperspectral classification concludes, we are presented with an image set that portrays the spatial distribution of recorded pixel spectra that either satisfy the criteria of a "match" with a specific endmember material or explain an identifiable presence of endmembers in a scene confused by spectral complexity. Using the concept of "abundance" the unraveled pattern of spectra or the agreement between a known condition and its occurrence in the image can be expressed as a linguistic variable and quantified. For some environmental applications simply knowing the pattern of abundance of a given material is sufficient information to guide assessment or decision making. However, for detailed environmental studies we can consider the abundance patterns and matches derived via hyperspectra classification as intermediate data products that can be subject to more rigorous analytical treatment.

Treatment implies the desire to refine the patterns of fractional abundance into a thematic representation that conveys more decision-relevant information. Although the translation of the intermediate data product into a format that informs is problem-specific, there are several principles that can guide the process. Because hyperspectral imagery contains information regarding the material or chemical composition of the surface, the thematic information extracted from our imagery can be as varied as the reference spectra assembled for the given application. As demonstrated previously, spectra can be as diverse as worn concrete surface, oil-saturated soil, or hardwood trees infested by anthracnose (leaf spot disease). The possibilities that can be recovered using field or laboratory spectroscopy are limited only by the practicalities surround the collection of a material spectrum. When such reference information are available our attention can be directed at both the spatial identification of a material as well as the relationship this material may have with respect to other landscape characteristics, such as the pattern of contamination in relation to a sensitive receptor.

The abundance surfaces produced via hyperspectral classification, however, is somewhat abstract and may be inaccessible to a wider audience (Pu et al. 2008). Conveying abundance patterns in a meaningful way begins with interpretation. The spatial distribution of material abundance is expressed over the 0.0–1.0 continuum, much the same way we saw thematic information portrayed through soft-image classification. Like the soft-classification example, there is an element of subjective-technical judgment that is called upon in order to establish categorical boundaries that place pixels into distinct classes. Interpretation, albeit selective, is also accompanied by an explanation that is supported by the evidence (Heiden et al. 2007). Several questions inform this exercise:

- Is a pixel with a match filtering score of 0.40 a match?
- What is the criteria for a reasonable match given the nature of the material and the complexity of the surface?
- What does a pixel with a 10% soil mixture suggest?
- Should a pixel containing only 20% of an endmember for a contaminant be considered a contaminated pixel?

While these questions are hypothetical, they are not trivial. Rather, they underscore the issues that surround uncertainty with regard to image processing and analysis as well as the uncertainty inherent to a specific application problem. As noted by Adams and Gillespie (2006), image analysts are often disappointed when image units fail to coincide with units mapped on the ground. However, realizing the potential of hyperspectral sensing in environmental assessment begins by recognizing that the spectral properties of the surface are not the sole basis for thematic presentation. In actuality, thematic presentation is an interpretation and evaluation of data. The results attained via hyperspectral classification are, in essence, characterizations of the physical properties of the landscape deemed to be important by the analyst. The results of analysis also describe the culmination of a methodology designed to achieve specific objectives. Flaws in method wild yield flawed thematic representations as will logical inconsistencies that over-reach the limitations of hyperspectral technology.

Successful applications of imaging spectroscopy within the domain of environmental assessment are developed around four focusing principles:

1. *Definition* – Stressing the point that hyperspectral remote sensing moves analysis into the field of spectroscopy, problem definition centers on application problems that can benefit from the detailed examination of very accurate spectral data.
2. *Diagnostic potential* – Problems well suited for hyperspectral analysis are driven by an evidence-based paradigm where the properties of objects or features serve to support a given environmental condition.
3. *Spatial context* – Objects and features display a spatial context that hyperspectral imagery can exploit. These geographic relationships place measurement into a context whereby their occurrences are phenomena-driven and can be understood with reference to their spatial arrangement and juxtaposition.
4. *Process inference* – Ideas such as unmixing and matching contribute to problems that require the derivation of logical consequences from the circumstantial evidence as observed in the imagery. The resulting analysis engages a premise and provides confirmatory information for reaching a conclusion.

Over the next decade key developments in hyperspectral remote sensing are likely to propel this science forward and broaden its appeal. Among the likely technological developments will be the wider availability of hyperspectral data facilitated by new sensor systems designed with more sophisticated support infrastructure. The future is also likely to witness a new generation of hyperspectral sensors that employ on-board intelligent detection and classification algorithms capable of near-real time transmission of information to ground receiving stations. Such sensors will enjoy tighter coupling with emerging sensor Web technologies. As the decade unfolds hyperspectral sensing may give way to ultra-spectral instruments designed to recover surface reflectance over thousands of spectral bands with spatial resolutions below 15 m. With each of these potential advancements standardization and improved metadata conventions will evolve to assist the wider distribution of spectral libraries. Taken together these developments will encourage data sharing while facilitate the operational use of hypersensing in environmental analysis.

10.5 Summary

Building on the background introduced in the previous chapter, this chapter explored the methods and procedures that guide the hyperspectral image classification. The goal of hyperspectral classification is to achieve a successful identification of surface materials and conditions. Beginning with a discussion of processing fundamentals, this chapter details the methodologies involved in preprocessing as well and classification procedures following "whole-pixel" as well as "subpixel" strategies designed to define the material composition of surface materials. Whole-pixel analysis methods function to determine whether one or more target materials are abundant within each pixel in a hyperspectral image based on the spectral similarity between the pixel and target spectra. Subpixel analysis methods are used to calculate the quantity of target materials in each pixel of an image which can support the analysis and detection of quantities of a target that are much smaller than the pixel size itself.

References

Adams, J. and Gillespie, A. (2006) Remote Sensing of Landscapes with Spectral Images: A Physical Modeling Approach, Cambridge University Press, Cambridge, UK., 362p.

Borengasser, M., Hungate, W. and Watking, R. (2008) Hyperspectral Remote Sensing: Principles and Applications, CRC Press, Boca Raton, FL., 119p.

Clark et al., 2000; Clark, R. N, Swayze, G. A., Livo, K. E., Kokaly, R. F, Sutley S. J., Dalton, J. B., McDougal, R. R., and Gent, C. A., 2003a. Imaging spectroscopy: Earth and planetary remote sensing with the USGS Tetracorder and expert systems: Journal of Geophysical Research, 1080 (E12, 5131): 5–1–5-44, doi: 10.1029/2002JE001847

Drake, N. Mackin, S. and Settle, J. (1999) Mapping Vegetation, Soils, and Geology in Semiarid Shrublands Using Spectral Matching and Mixture Modeling of SWIR AVIRIS Imagery Remote Sensing of Environment, 68, 12–25

Goodenough, D., Dyk, A., Niemann, O., Pearlman, J., Chen, H., Han, T., Murdoch, M. and West, C. (2003) Processing Hyperion and ALI for Forest Classification, IEEE Transactions on Geoscience and Remote Sensing, 41, 1321–1331.

Gupta, R.P. (2003) Remote Sensing Geology. Springer-Verlag New York, 656p.

Heiden, U., Segl, K., Roessner, S. and Kaufmann, H. (2007) Determination of Robust Spectral Features for Identification of Urban Surface Materials in Hyperspectral Remote Sensing Data, Remote Sensing of Environment, 111, 537–552.

Hunter, E. and Power, C. (2002) An Assessment of two Classification Methods for Mapping Thames Estuary Intertidal Habitats Using CASI Data, International Journal of Remote Sensing, 23, 2989–3008.

Kardi, T. (2007) Remote Sensing of Urban Area: Linear Spectral Unmixing of Landsat Thematic Mapper Images Acquired Over Tartu (Estonia) Proceedings of the Estonian Academy of Biology and Ecology, 56, 19–32.

Keshava, N. (2003) A Survey of Spectral Unmixing Algorithms, Lincoln Laboratory Journal, 14, 55–78.

Keshava, N. and Mustard, J. (2002) Spectral Unmixing, IEEE Signal Processing Magazine, January, 44–57.

Kruse, F. A., Boardman, J. W. and Huntington, J. F. (2003). Comparison of airborne hyperspectral data and EO-1 Hyperion for mineral mapping. IEEE Transactions on Geoscience and Remote Sensing, 41(6), 1388–1400.

Kruse, F., Boardman, J., Lefkoff, A., Heidebrecht, K., Shapiro, A., Barloon, P. and Goetz, A. (1993) The Spectral Image Processing System (SIPS), Remote Sensing of Environmental, 44, 145–13.

Mundt, J., Streutker, D. and Glenn, N. (2007) Partial Unmixing of Hyperspectral Imagery, Theory and Methods, Proceedings of the American Society of Photogrammetry and Remote Sensing, Tampa, Florida, 46–57.

Mitchell, J. and Glenn, N. (2009) Subpixel Abundance Estimates' in Mixture-tuned Matched Filtering Classifications of leafy spurge (Euphorbia esula L.) International Journal of Remote Sensing, 30, 6099–6119.

Pu, R., Gong, P., Michishita, R. and Sasagawa, T. (2008) Spectral Mixture Analysis for mapping Abundance of Urban Surface Components from the Terra/ASTER Data, Remote Sensing of Environment, 112, 939–954.

Rogge, D., Rivard, B., Zhang, J., Sanchez, A., Harris, J. and Feng, J. (2007) Integration of spatial-spectral information for the improved extraction of endmembers Remote Sensing of Environment, 110, 287–303

Schwarz, I. and Staenz, K. (2001) Adaptive Threshold for Spectral Matching of Hyperspectral Data, Canadian Journal of Remote Sensing, 27, 216–224.

Van Der Meer, F. Vasquez-Torres, M. and Van Dyke, P. (1997) Spectral Characterization of Ophiolite Lithologies in the Troodos Ophiolite Complex of Cyprus and its Potential in Prospecting for Massive Sulphide Deposits, International Journal of Remote Sensing, 18, 1245–1257.

Van der Meer, F. (1998/1999) Imaging spectrometry for geological remote sensing - Geologie en Mijnbouw, 77, 137–151

Van der Meer, F. (2004) Analysis of spectral absorption features in hyperspectral imagery. JAG: International Journal of Applied Earth Observation and Geoinformation, 5(1), 55–68.

Van der Meer, F. (2006) The effectiveness of spectral similarity measures for the analysis of hyperspectral imagery International Journal of Applied Earth Observations and Geoinformation, 8, 3–17

Willams, P., Hunt, A. and Raymond, E. (2002) Estimation of leafy spurge cover from hyperspectral imagery using mixture tuned matched filtering Remote Sensing of Environment, 82, 446–456.

Chapter 11
Object-Based Analysis

In the preceding chapter, the techniques available to exploit the spectral content of our remotely sensed imagery were discussed. The methods and issues reviewed in Chap. 10 did much to demonstrate the unique and useful sources of information that can be gained from the expanded spectral domain offered by hyperspectral sensors. Although essential, spectral data alone does not provide a complete picture of the environmental system to fully understand process or guide environmental decision making. One under-utilized quality of an image that can add an additional dimension to the study of the environment is the explicit spatial arrangements, juxtapositions and patterns exhibited by our spectral measurements. These unique arrangements place spectral data in a geographic context where spatial descriptors such as shape, perimeter, and texture together with other geographic variables define objects in the image that add an element of knowledge into the classification problem. These image objects are derived exclusively from the spatial relationships found in the image and with this additional knowledge they present criteria beyond the spectral signature which enriches the image classification process. Object-based analysis offers new possibilities that may extend the role of remotely sensed data in complex mapping and assessment applications. In this chapter, we will examine the object-based paradigm and review the fundamental aspects of image classification based on the analysis of image objects.

11.1 Objects and Segments

When we examine an image display visually on a computer screen or as a "photographic" product printed on paper media, our eyes and brain strive to organize the image into definitive shapes and colors that enable us to decipher the scene and impart meaning to the arrangements we recognize. The geometries we observe begin to make sense to us and their juxtaposition together with our prior knowledge and experience enable us to add context to the image. Pattern together with context allow the image to become "known" to us and we can label the geometric elements

J.K. Lein, *Environmental Sensing: Analytical Techniques for Earth Observation*,
DOI 10.1007/978-1-4614-0143-8_11, © Springer Science+Business Media, LLC 2012

Table 11.1 Fundamental recognition elements of photo interpretation

Tone	Refers to the relative brightness or color of elements on a image and defines the most basic of the interpretive elements because without tonal differences none of the other elements could be understood
Size	The relative size of an object
Shape	The geometric shapes are usually indicators of surface use or form: cultural features – geometric, distinct boundaries whereas natural features – irregular shapes and boundaries
Texture	Impressions of "smoothness" or "roughness" of image features is caused by the frequency of change of tone in photographs where smoothness describes surface features of similar height and coarseness – irregular surfaces
Pattern	Characterizes spatial arrangement where the patterns formed by objects in a photo can be diagnostic Repeating simple geometric patterns tend to indicate cultural features More complex, curving patterns tend to indicate natural features
Shadow	Helps to determine the height of objects in aerial photographs
Site	Refers to topographic or geographic location and suggests relationships of a feature to its environment/location
Association	Objects are often found in association with other objects. This context surrounding an object can provide insight into what it is where the presence of one object is often correlated with the presence of another

that comprise the scene into information that is useful to us. Returning to the familiar visual cues that are fundamental to the visual interpretation of aerial photography, we are quickly reminded that there are important clues available in our images that get lost in the digital processing of spectral measurements that the computer cannot "see" (Table 11.1). The elements of photo interpretation listed in Table 11.1 take simple ideas like shape, texture and arrangement and communicate to us the content of a visual display. This inherently qualitative information can be interpreted through the lens of culture and assembled into thematic information. Qualities such as shape or texture define properties of the geometries our eyes see and our brains assemble into meaning (Fig. 11.1). From an exclusively spatial perspective these geometries can also be considered geographic primitives that describe the puzzle pieces that constitute the landscapes geographic form. For example, in a forested area we can anticipate distinctive separations between stands of trees, grassland areas, or barren fields where each suggested boundaries demarcated based on changes in shape, color, apparent texture, or other cues that are visually noticeable. A line drawn on the image enclosing area where common visual attributes suggest similar characteristics and imply a common theme. On the image the line encircling this visual commonality forms a polygon delineated on the basis of a set of shared attributes. The polygon extracted from the image represents an object, a geometric feature explained by a set of identifiable characteristics (Fig. 11.2). This object also becomes the principle unit of abstraction that drives the object-based paradigm (Lang 2008). The challenge is to move from this conceptual and qualitative definition to an entity that can be represented numerically and subject to digital computation. In the machine-processing environment of our remote sensing system, an object can be explained more precisely as an

Fig. 11.1 Simple spatial geometric shapes

Fig. 11.2 An image object

agglomeration of pixels that share similar spectral and spatial properties. Therefore unlike our visual analogy, image objects are recognized as contiguous regions of pixels that define uniform radiometric characteristics. Therefore an object in the image domain corresponds to an area of uniform shape, texture, topology,

Fig. 11.3 The pattern of an image following segmentation

heterogeneity, and spatial relationship. Such objects are formed by aggregating neighboring pixels with similar spectral or spatial characteristics following an object-building procedure known as segmentation (Navular 2007; Benz et al. 2004).

The segmentation process requires dividing a remotely sensed image into objects or areas that share specific properties in common (Fig. 11.3). A variety of approaches to segment an image into objects have been introduced (Pal and Pal 1993). Regardless of approach, the goal of segmentation is to decompose the image into spatially discrete, contiguous, nonintersecting and semantically meaningful regions. Image segmentation can be conducted following either a top-down, bottom-up, or combined strategy (Table 11.2). Three general forms of image segmentation have been described by Blaschke et al. (2006):

- *Pixel-based segmentation* – Pixel-based segmentation stratifies the image into pixels of two or more values by comparing pixel values to a predetermined threshold. Each pixel is examined individually to ascertain whether or not it should be assigned to the object relative to the mean digital value of all pixels already assigned to that object. The assignment test takes the general form of a Boolean decision rule, where

$$I(i,j) = \begin{bmatrix} 0 & \mathrm{BV}(i,j) < \text{threshold} \\ 1 & \mathrm{BV}(i,j) \geq \text{threshold} \end{bmatrix},$$

  with BV(*i,j*) representing the pixel brightness value at location (*i,j*) and *i* describing the segmented image (Gao 2009).
- *Edge-based segmentation* – Edges describe the boundaries created in image by abrupt changes in brightness values. Using edges as a delineating criteria, pixels

Table 11.2 Image segmentation strategies

Method	Description
Pixel-based	Stratifies an input image into pixels of two or more values by comparing pixel values with a predefined threshold. Pixels are examined in isolation to ascertain if they belong to a specific region based on their values
Edge-based	Edges describe boundaries between land covers where pixel values change abruptly. Using edge detection methods, pixels encompassed by edge pixels are organized into that homogeneous region
Region-based	Applies homogeneity criteria to identify regions in the image using either of two common seeding algorithms: (1) region growing; this approach to segmentation examines neighboring pixels of initial "seed points" and determines whether the pixel neighbors should be added to the region or (2) split and merge; where the input image is subdivided into squares of uniform size based on the initial seed pixels and adjacent regions that are similar are then merged together until a termination criteria is reached

that are enveloped within a polygon boundary formed by edges are assumed to be homogeneous and therefore can be defined as an object.

- *Region-based segmentation* – A region describes a geographic area delineated on the basis of one or more unifying criteria. As per segmentation logic, an image can be partitioned into regions by applying statistical criteria as a rationale for establishing uniformity (homogeneity). With these criteria explained segmentation proceeds following either a region-growing or split and merge decision rule. Region-growing segmentation employs seed pixels in a comparison test with neighboring pixels relative to the selected homogeneity criteria. Pixels are merged, in an iterative fashion, into larger regions until a homogeneity threshold is met. The process stops once regions grow to a point where no further pixels can be allocated and the entire image is segmented. The split and merge algorithm subdivides the image into uniform square based on initial pixel seeds. Adjacent regions that are statistically similar are merged until prespecified termination criteria are reached. The initial segmentation is refined to form more regular object boundaries using measures of texture, color, mean brightness value, size or shape. Refinement continues iteratively and ends once all the pixels in the image have been allocated to segments.

Other methods of image segmentation have been introduced including knowledge-based, multicriteria and multiresolution strategies. Each of these has been extensively reviewed by Gao (2009). Of the alternative approaches, the multiresolution method is worth noting due to its successful implementation in commercial remote sensing software systems. Multiresolution segmentation is a bottom-up region merging strategy where a sample of single seed pixels serve as initial image objects. As the algorithm proceeds, the individual regions are merged successively into fewer and larger objects. Merging builds a hierarchy based on a set of evaluative criteria such as object perimeter, area, and shape along with measures of compactness, texture, and color each of which are calculated from the segmented regions. Objects form with each iteration of the algorithm and are clustered based on the calculated variables which serve as indices that construct objects from the image. Optimization methods are also used to guide the merging process as objects form up through the hierarchy. As with other segmentation techniques, the multiresolution approach terminated when all pixels in the image have been assigned to regions, or when a predefined threshold has been reached. The overall success of this method is therefore dependent on the parameters supplied by the analyst which typically include a scale parameter and break-off value together with a set of single layer weights and homogeneity criteria that drive the algorithm (Baatz and Schape 2000; Gao 2009).

11.2 The Role of Homogeneity

Homogeneity is a critical variable in the segmentation process. Criteria selected to define homogeneity with respect to segments determine whether a seed pixel's neighbor belongs to the seed pixel's object (Guo et al. 2007). In many respects, the concept of homogeneity is comparable to the idea of similarity as applied in mutlivariate statistical operations such as cluster analysis. In the feature space of a remotely sensed image, two adjacent image objects (or initial seed pixels) $\Psi_{1\mathrm{d}}$ and $\Psi_{2\mathrm{d}}$ are considered similar if their features on the feature space of the image are close to one another. The shorter the distance, the more likely $\Psi_{1\mathrm{d}}$ and $\Psi_{2\mathrm{d}}$ belong to the same object. Closeness in this context can be expressed as the degree of fit (β) such that

$$\beta = \sqrt{\Sigma(\Psi_{1\mathrm{d}} - \Psi_{2\mathrm{d}})^2}.$$

The familiar Euclidean distance formula can then be normalized using the standard deviation to yield,

$$\beta = \frac{\sqrt{\Sigma(\Psi_{1\mathrm{d}} - \Psi_{2\mathrm{d}})^2}}{\sigma\Psi_{\mathrm{d}}}.$$

The value (β) direct the pair-wise merging of objects under the assumption that an image object should be merged with the neighboring segment where the reduction in homogeneity is the lowest.

In image segmentation object homogeneity can be defined in several ways:

1. *Color homogeneity* – Determined as the sum of the standard deviations of the spectral values of the object pixels over all the bands in the image
2. *Shape homogeneity* – Expressed as the relative deviation of the edge length from the most compact shape (i.e., a circle) which can be quantified using

(a) Edge criterion that approximate smoothness based on the ratio of object length (Λ) and the edge length of a bounding box (b) that explains the smallest rectangle that completely encloses the object

$$B = \Lambda / b.$$

(b) Area criterion that approximate the deviation of the object from an ideal shape expressed as the compactness ratio of object length (Λ) and the square root of object size in pixels (n):

$$B = \Lambda / \sqrt{n}.$$

Other examples of homogeneity criteria exist as well which include criteria that use measures such as

- Absolute spectral distance between the adjacent pixel and seed pixel expressed as:

$$B = \sum (\mathrm{BV}_{si} - \mathrm{BV}_{ni}),$$

where BV_{si} and BV_{ni} represent the digital value for a seed pixel and its adjacent pixel for the ith spectral band of the image.
- Absolute local spectral distance between two neighboring pixels based on the formula

$$B = \sum (\bar{u}_i - u_i)^2,$$

with $\bar{u}_i$ denoting the digital value of a known region pixel and μ_i is a neighboring pixel of $\bar{u}_i$ whose regional assignment has not been determined.
- Relative local difference between two neighboring pixels such that:

$$h = \frac{\left[\sum_{i=1}^{n} (x_i - \bar{x})(y_i - \bar{y}) \right]^2}{\sum_{i=1}^{n} (x_i - \bar{x})^2 \sum_{i=1}^{n} (y_i - \bar{y})^2}.$$

Supplied with homogeneity, criterion "best fit" optimization strategies guide the grouping of objects into similarity clusters. Here, optimization generally relies on the application of decision heuristics to merge segments using one of several options:

1. *Fitting* – Where object A is merged with any neighboring object B that fulfills the homogeneity criteria.
2. *Best fit* – Where object A is merged with object B that best fulfills the homogeneity criteria according to a minimum change constraint.

Table 11.3 Object properties produced from segmentation

Attribute	Explanation
Color	Mean "color" values for each object in each band
Shape	Squareness, Roundness, Length/Width Ratio, etc.
Size	How large or small/perimeter
Texture	Contrast, homogeneity, dissimilarity
Content	How does the object relate to its neighbors?
Relationship	How individual object characteristics interrelate to the entire image
	Relative location, subobjects, super objects

3. *Local mutual best fit* – Which finds for object A the neighboring object B that best meets the homogeneity criteria.
4. *Global mutual best fit* – This merges the pair of neighboring objects that best meet the homogeneity criteria across the entire image (scene).

The image segmentation procedure results in a surface abstracted as a collection of objects. This new object domain creates a data set that now contains more descriptive information (dimensions) of the image than simply the brightness value of reflectance or radiance as recorded for the pixel. With the image defined by a richer data set the image classification process can effectively draw from many more variables. The addition of these variables enhances categorization and facilitates a heightened level of precision. (An example of the data file created by image segmentation that defines the range of object properties as variables useful to the goals of classification is given in Table 11.3.) The object properties listed in Table 11.3 encapsulate the state of the object as a set of attributes and redefine the image according to the structure of an object-oriented file. Through this process the image is transformed into a database of objects that conforms to the representational schema of an object-based spatial database. Using this architecture enables the storage of more attribute information along with the objects as vectors which provide more efficient and flexible manipulation capabilities where compared to the raster/grid format.

11.3 The Object-Based Paradigm

The efficient storage of data for advanced computing purposes remains a central goal in computer science. The object-based paradigm is one of several methods used to represent and structure data for applications in a machine environment (Devereux et al. 2004; Blaschke 2010; Bock et al. 2005). The object database model emerged in the mid-1980s in response to certain limitations found with the relational database model that was widely touted at that time. One nagging problem with the relational mode was that relational databases did not allow users to define their own data types. The object-bases storage paradigm permits user-defined data types and enjoys other features such as:

- *Containers* – Means of representing the result of a query

- *Methods* – Which are pieces of computer code that can be associated with an object
- *Rules* – Describing computer code that is automatically activated when a specific event occurs.

  All of which give certain advantages to this database schema such as (1) a rich system type, (2) improved modeling of complex objects, and (3) enhanced performance. In a spatial context, the object-based data model *stores both the spatial and attribute data of spatial features in a single system.* This feature allows a *spatial feature* (object) to be *associated with* a set of *properties and methods.*

An object-based data model consists of seven essential ingredients:

1. Objects and object identifiers
2. Complex values and types
3. Classes
4. Methods
5. Subclass hierarchies
6. Inheritance and dynamic binding
7. Encapsulation

As a method for collecting real world entities in a logical manner, the object-based model attempts to structure data much the same way we conceptualize it. Extending this idea to the environment as abstracted via our remote sensing system, the object model is a convenient way to code the patterns and attributes generated by the segmentation process suggesting landscape entities that explain observable landscape features. Elements of the landscape abstracted as objects form entities that can be manipulated by our software such as a value, variable, or function. Object database methods are appropriate to this task because our data is complex as are the relationships that result from segmentation. This intuitive approach toward the regionalization of complex surface arrangements has contributed to a growing list of applications object-based classification (Gitas et al. 2006; Maxwell 2010; Jacquin et al. 2008; Bhaskaran, et al. 2010). The growing list of object-based approaches suggest that image analysis using this paradigm offers certain operational strengths when compared to pixel-based methods including

- An image portioned into objects in a manner that compliments how we conceptually organize the landscape
- A basic unit of representation that reduces computational effort while facilitating more complex analytical techniques
- The creation of objects that retain useful landscape descriptors such as texture, shape, and contextual relationships that cannot be provided at the pixel level
- The definition of spatial units that are less sensitive to the modifiable area unit problem by maintaining a structure that corresponds more closely to the phenomena under investigation
- A spatial representation that can be integrated into vector display and vector analysis using GIS

Given the challenges that confront environmental analysis, where concern is often directed at land units described as habitats, patches, or linear features defined within a complex matrix, the object-based paradigm decreases the level of detail, reduces image complexity and renders the content of our imagery more accessible. The image objects into which landscape entities form can be treated as individual elements of the landscape and the relationships that emerge between objects at different levels of representation can be subject to analysis. Thus, unlike the pixel, the object, once expressed according to a set of identifiers, can also be monitored over time. The only hurdle in this schema involves the question of "meaning". Segmentation is an ill-posed problem in that it produces no unique solution (Wang et al. 2010; Mueller et al. 2004). By simply changing a single parameter or heterogeneity measure can lead to different image segmentations. The resulting ambiguity is not unlike the example where to human photo interpreters delineate features on the same image in slightly different ways. In addition, there is often a lack of agreement regarding the conceptual foundations of the object-based approach; a problem that become readily apparent when considering the relationship between image objects (segments) and landscape objects (patches). The potential disparity underscores what has been termed the "orchard" problem originally introduced by Lang and Langanke (2006). The orchard problem speaks to the issue where geographic features exhibit conceptual boundaries rather than real one. Using the analogy of an orchard where the human eye sees the specific arrangement of trees in a matrix of grass and the brain applies gestalt rules (heuristics based on prior experience) to explain the context inherent to the object. In most instances there may be little evidence to support the belief that a segmentation-derived object is an understandable representation of a structural or functional unit of the landscape. Without a formal and accepted conceptual foundation there is no consistent means to establish if a given segmentation is appropriate or more correct than another (Jyothi et al. 2008).

Despite this potential flaw, object-based image analysis is attractive if only because it fosters a new way of thinking about remotely sensed data. Instead of the standard conceptualization of the image as a raster (grid) structure, the object paradigm redraws the image and casts it into a construct crafted using the principles of

- *Abstraction* – Denoting the essential characteristics of an object that distinguishes it from all other kinds of objects and provides crisply defined conceptual boundaries relative to the analyst's perspective
- *Encapsulation* – Explaining the process of compartmentalizing the elements of an abstraction that express its structure and behavior
- *Modularity* – Defining the process of decomposing a complex system (scene) into a smaller, self-contained collection of related classes of objects
- *Hierarchy* – Describing a ranking or ordering of abstractions, where hierarchies may be organized according to concepts such as categorization, aggregation, containment, partitions, or inheritance

The image database generated through the implementation of the object-based model demonstrates these principles and forms the basis for thematic classification.

11.4 Object-Based Classification

As suggested by the previous sections of this chapter, thematic extraction following the object-based model involves the classification of objects based on inquiries addressed to and exercised through the image transformed into a database. Image objects, defined on the basis of features connected with direct descriptions, logical conditions, or relationships between objects, can be assembled into thematic classes (Mallinis et al. 2008). However, landscape classification employs methods that sharply contrast to those described in earlier chapters. Object-based classification exploits the variables generated via segmentation together with spectral information to group neighboring pixels into meaningful regions that can be assigned with thematic labels (Conchedda et al. 2008; Duveiller et al. 2008; Hay et al. 2005). Through the inclusion of variables such as texture or shape, image classification expands beyond simple statistical pattern recognition and broadens the explanation of pattern by adding additional dimension into its expression. In theory with an expanded representation of pattern, improvements in thematic extraction should follow.

Object-based image classification centers around two main activities:

1. *Object identification* – A process is driven largely by the chosen segmentation procedure.
2. *Object labeling* – An activity involving the classification of image objects into thematic classes.

Object identification typically begins with a supervised image classification procedure. Supervised classification is used in the process to produce training data that serves as a spectral sample for the land surface types of interest. Samples may be general land cover classes or more specialized categorical designations depending on the application. Because segmentation is a form of automated processing, the results may or may not conform to the desired surface features. Supervised training creates a type of image knowledge that is called upon to support the classification of image objects. The training knowledge-base is referred to as a class hierarchy and it supplies the analyst with specific definitions of the categories we wish to extract from the scene. It should be understood, however, that there is no such thing as a perfect segmentation. Segmentation is a means to the end. Classification is the goal of object-based analysis not image segmentation. A general rule gathered from experience demonstrates that a single segmentation is typically never sufficient for image classification. Rather, repeated merging and segment growing tends to provide results that generate more meaningful categorizations (Verburg et al. 2009). In addition, image segmentation is highly scale dependent; therefore, some adjustment of the scaling parameters read by the segmentation algorithm is generally required.

Once a segmentation is produced, the next major phase in object-based analysis involves the construction of a class hierarchy. As noted above, all features of the land surface that are subject to our analysis must be declared in accordance to their

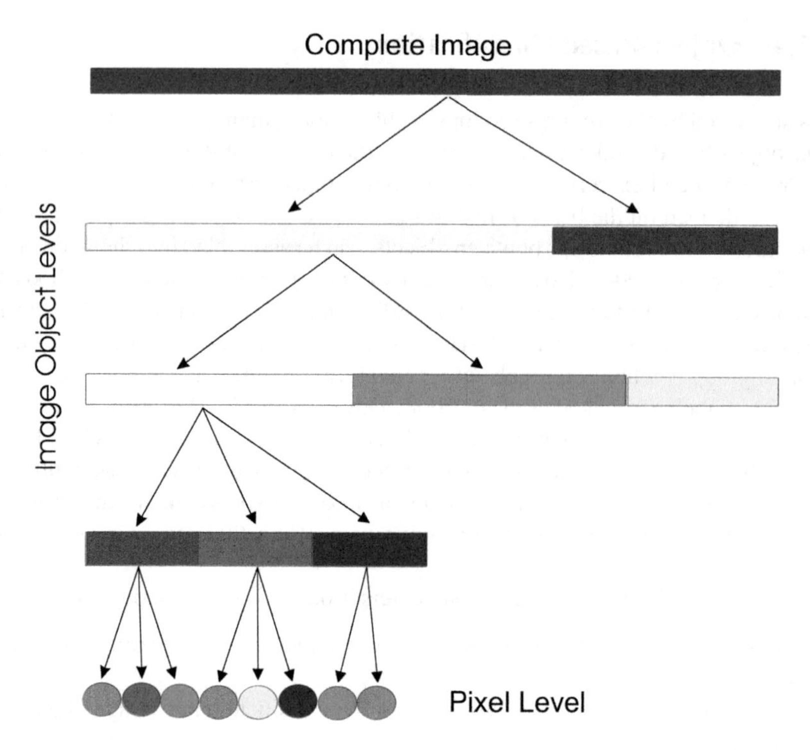

Fig. 11.4 A generalized schema of a class hierarchy

representative training samples. This selection procedure is identical to the sampling procedure followed when conducting a supervised classification and entails consideration of the same requirements. Samples should be representative of the theme and provide unambiguous spectral detail to reduce error and uncertainty. Depending on the classification system being used and the level of thematic discreteness available, the categories may be represented hierarchically to illustrate their complexity. This option requires the use of different spatial resolution in order to characterize a reasonable class structure (Fig. 11.4). A perfect class hierarchy may not be a reasonable or feasible solution given the nature of the problem or the types of surface features characteristic of the scene. At the conclusion of this stage in the classification process, the image is now abstracted and described by three elements: (1) the training data used to provide spectral signatures of the classes of interest, (2) a segmentation which defines a set of object primitives that further abstract the image, and (3) a class hierarchy that explains what the signatures mean with reference to the object primitives.

Depending on the degree of sophistication, the class hierarchy facilitates three possible object relationships that can be exploited during object labeling:

1. *Inheritance* – Describing the formation of parent and child classes where the parent class can pass on its feature descriptions to a child class (which can also have their own child classes) following the logic of the object-based data model.

2. *Groups* – Structures that are formed by semantic relationship among classes that combine to form larger units.
3. *Structure* – Identifying object classes that can be combined to form structural groups as a basis for classification-based segmentation.

In essence a classification hierarchy establishes the rules that will be used to determine which objects are assigned to which class. By supplying class descriptions with explicit rules, object assignment proceeds either in a deterministic fashion by means of thresholds, or through the supervised classification of sample sites. When sampling is employed, signatures for each class generate nearest neighbor classification rules that are used to perform object to class assignments. Classes in the hierarchy either contain or inherit these rules. Other rule-making strategies are possible including the use of membership functions derived using fuzzy set theory (Lizarazo and Barros 2010; Liu et al. 2008; Smits and Annoni 1999).

After the objects assembled from the image have been identified, they can be labeled. This step describes the actual classification of the segmented image using the decision rules that actively merge and name objects as specified in the hierarchy. Decision rules employed to conduct the classification of our imagery can be based on several different combinatorial/labeling strategies:

- Statistical
- Nearest neighbor
- Decision tree
- Semantic rules
- Neural networks
- Fuzzy logic

Implementing a decision rule directs the classification procedure to treat every image segment or object as a unit of analysis. Pixels comprising an object are assigned to an informational class by considering its detailed definition and how well the pixel corresponds to that definition. The comparison of pixels to definitions invokes a search strategy that employs sample objects that have been declared for each class. Search the functions to identify the closest sample object in feature space. The image object that is found to be closest to a given sample class is assumed to be representative of that class. That image object inherits the definition and is assigned (labeled) to that particular informational category.

Although the specific workflow followed when conducting an object-based classification will vary according to the software environment implementing the procedure, the general processing chain can be described according to seven central activities:

1. Image preprocessing
2. Selection of a classification scheme
3. Image segmentation
4. Training
5. Development of classification rules
6. Classification of segmented image
7. Accuracy assessment

11.5 Uncertainty Management

As with any image classification procedure error and uncertainty are always problematic issues that are difficult to resolve and object-based classification is no exception. The issues related to error when conducting an object-based analysis of remotely sensed imagery center around the segmentation process and introduce concerns regarding:

- Sample object size
- Sample object reliability
- Sample object density
- Spatial composition of objects

These four sources of uncertainty are compounded when the pragmatic implications of decision rule uncertainty, the selection of segmentation scaling parameters and the selection of threshold values are introduced. Considering the large array of segmentations algorithms available, deciding on the appropriate method often relies on the judgment of the analyst. In a review of the image segmentation problem Zhang (1996) identified three approaches to assess segmentation quality: (1) analytical comparison, (2) empirical goodness, and (3) empirical discrepancies. The method of analytical comparison considers the segmentation algorithm and how it performs by examining its underlying principles, requirements, and complexity. Quality assessment using empirical goodness methods evaluate the performance of the segmentation algorithm by judging the quality of the segmented image using "goodness" measures that evaluate:

- Intraregion uniformity
- Interregion contrast
- Region shape

A selection of these measures are presented in Table 11.4 based on Zhang (1996).Empirical discrepancy methods compare the segmented image to a reference image assumed to represent a correct or ideal segmentation. In some respects, this is analogous to "ground truthing" where discrepancies can be compared and quantified based on measures related to:

- The number of mis-segmented pixels
- The position of mis-segmented pixels
- The number of objects in the image
- The feature value of segmented objects

Although the use of empirical discrepancy indices is an involved procedure, they may prove to be more effective for remote sensing application focused on the production of thematic maps (Moller et al. 2007; Zhan et al. 2005). However, reliable reference data is not always available.

The lack of useful reference data encourages the use of global validation indices that examine the entire image in order to ascertain the appropriateness of the

Table 11.4 Evaluation methods for image segmentation

Empirical goodness methods
Goodness based on intraregion uniformity
Goodness based on interregion contrast
Goodness based on region shape
Empirical discrepancy methods
Discrepancy based on mis-segmented pixels
Discrepancy based on position of mis-segmented pixels
Discrepancy based on the number of objects
Discrepancy based on the feature values of segmented objects

Source: Zhang, Y. (1996) A Survey on Evaluation Methods for Image Segmentation, Pattern Recognition 29(8), pp 1335–1346

parameters selected to produce a segmentation. One widely used validation index is the normalized post-segmentation standard deviation (NPSS). This index is computed according to the formula:

$$\mathrm{NPSS} = \sqrt{\frac{\sum_{i=1}^{\mathrm{no}} n_i(\bar{x} - \bar{x}_i)^2}{\sum_1^N (\bar{x} - x)^2}},$$

where x is the object mean, no represents the number of objects, and N the number of pixels included in the ith object.

An additional measure of global validation is described by a map complexity metric termed the comparison index (CI) calculated from:

$$\mathrm{CI} = \frac{\sum_{i=1}^{n} C_i A_{C_i}}{n},$$

where C_i is the comparison class, A_{C_i} is the proportion of C_i within the reference space and n represents the total number of objects. A series of related complexity metrics have been evaluated by Stein and DeBeurs (2005). The role complexity metrics in the evaluation of error is to quantify the precision attained by a given method of image segmentation and help direct selection of the most feasible segmentation technique. While this is often a trial and error process, the appropriateness of an object's delineation and the correctness of boundary locations establishes the basis from which thematic accuracy can be evaluated and directly influences map precision. In practice, the implications are clear and suggest that accuracy when employing an object-based methodology is not only thematic but also a matter of semantic and geometric agreement.

11.6 Finding a Fit

As a method of analysis, object-based approaches offer unique image classification solutions that can benefit environmental assessment and decision-making activities. The attraction to object-based analysis stems from the inherently geographic nature of the object as both a focus of analysis (classification) and understanding (representation). Comparing the geometry of the image-object derived via segmentation to the inflexible sampling design imposed by the pixel, the representation of landscape elements evidences a morphology more closely matching natural form. This phenomenological agreement suggests that thematic maps developed from object-based classification techniques should characterize land surface information with a closer correspondence to natural boundaries. In studies involving habitat analysis, the delineation of environmentally sensitive areas or the simple description of land cover types, the potential exists to describe land units that conform to actual spatial arrangements. This facility of the object model is particularly valuable when these data are to be integrated into the database of a geographic information system. In addition, by taking advantage of a wider description of the landscape, features such as damaged environment, isolated wetlands, urban ecosystems become more visible. Incorporating a combination of variables from the spectral, spatial, morphological, and contextual relationships can be expressed that enhance the categorization of discrete surface associations. Illustrative examples of this potential include Frohn et al. (2009) documenting study where object-oriented analysis was used to detect isolated wetland patches achieving classification accuracies approaching 98%, Gitas et al. (2004) where object-based analysis was employed to map areas impacted by forest fire, Guo et al. (2007) whose study on the application of object-based classification demonstrated unique advantages in the identification of tree mortality, and Walker and Briggs (2007) study focusing on urban forest mapping.

Documentation and thematic mapping are only a fraction of the possibilities. Interest in object-based analysis has also been given to the question of environmental monitoring and change detection. Object-based change detection removes the pixel from direct analysis and refocuses the identification of change to the object. Object change in this context is both thematic and geometric which shifts the study of change and the problem of environmental monitoring to the per-parcel investigation of image differences. Analysis adopting this direction involves the comparison of corresponding objects expressed as a function of time. This implies that the expression of change is more than simple spectral in nature. Object change, based on a per-parcel investigation will necessarily include a spatial definition where the geographic extent of an object differs over time, a morphological change defined by variations in shape or texture and contextual differences corresponding to changed relationships to neighboring objects. This suggests that changing objects may not only change with respect to their thematic label, but also display changes that reflect geometric and topological properties as well. Introducing these properties into the study of environmental change facilitates detection of subtle deviation at the surface and the identification of small change features that are active in the landscape. This capacity has been well documented in the post-hazard consequence and damage assessment literature and can support on-going efforts to understand processes such as cumulative environmental impacts and the more general pattern of human disturbance in the environment (Aubrecht et al. 2009; Saura and Castro 2007). Conducting an object-based change detection follows the general procedures common to the per-pixel-based approaches, However, it is also possible to incorporate GIS modeling techniques in order to compare differences in the geometric patterns of surface arrangements (Bock 2003). The per-object strategy utilizes multitemporal segmentation and employs polygon overlay of the segmented geometry to distinguish boundary and shape contrasts. Examples of the object-based analysis of change can be found in Volker (2004), Im et al. (2008), Gamanya et al. (2009), and Hall and Hay (2003).

11.7 Limitations and Constraints

Object-based analysis in remote sensing has tremendous potential (Table 11.5); however, as with any image-processing method our enthusiasm must be tempered by a careful consideration of the limitations that surround this technique. Resting at the center of this image-processing paradigm is the assumption that objects derived through segmentation correspond with objects at the surface. Perfect one-to-one correspondence, however, may not hold true in all instances, particularly if the object is small, or if the resolution of the imagery is too coarse. In general, object-based analyses perform best using high or very high spatial resolution imagery or in situations where land surface objects are comparatively large. Consequently, object-based analysis and classification is scale dependent and the scale of observation is critical to a successful classification and image segmentation. This suggests that the scale parameters selected to guide the segmentation process can produce very different classification results. Since an optimal method for selecting an appropriate scale for segmentation, the process relies on experience and judgment. Although subjectivity cannot be avoided, not all surface objects occur at the same scale therefore a universal value may not be suitable for all applications. Scale influences can be a source of frustration when implementing the object-based approach since trial and error together with the looming presence of uncertainty can introduce time and cost constraints that undermine the effective use of this technique. Classification is also affected by the selection of training samples, which, as in the case of supervised image classification, direct the outcome and correctness of the process. Finally, reasonable concern exists regarding the boundary problem. Because object boundaries may have little relationship to actual

Table 11.5 Utility of object-based classification

Multiple scales – The spatial relationship information contained in image objects allow for more than one level of analysis
Spatial relationships – Objects can be classified using their spatial relationships with adjacent or nearby objects
Information filter – Object-based analysis is able to filter out meaningless information and assimilate other pieces of information into a single object
Fuzzy logic – Object-based analysis provides more meaningful information than pixel-based image analysis by allowing for less well-defined edges or borders between different classes

boundaries at the surface, boundary distinctiveness can be called into question. Boundary error can present itself particularly when natural phenomena exhibit gradual transitions and a "true" or crisp delineation does not exist. In these situations objects cannot be realized from a segmentation which can contribute to misclassifications and an inability to effectively represent a land surface condition thematically.

11.8 Summary

In recent years, exciting new techniques in remote sensing analysis have emerged utilizing object-oriented constructs. As long as pixel sizes remained typically coarser than, or similar in size to the objects of interest, emphasis on per-pixel analysis, or even subpixel analysis was a reasonable approach, but with increasing spatial resolutions alternative methods produce more useful results, particularly those aimed at deriving objects that are made up of several pixels. This chapter provides an overview of the development of object-based methods, which aim to delineate readily usable objects from imagery while at the same time combining image-processing capabilities to guide image classification. At the heart of object-oriented image analysis is the concept of automated image segmentation. In this chapter, the process of image segmentation was described leading to a detailed discussion of the object-based paradigm. From this discussion the chapter reviewed object-based classification and culminated with a treatment of uncertainty and the issues surrounding object-based accuracy; summarizing the limitations and constraints associated with object-based approaches.

References

Aubrecht, C.; Steinnocher, K.; Hollaus, M.; Wagner, W. (2009) Integrating earth observation and GIScience for high resolution spatial and functional modeling of urban land use Computers, Environment and Urban Systems, Vol. 33, Issue: 1, January, 2009. pp. 15–25.

Baatz, M. and Schape, A. (2000) Multi-resolution segmentation- an optimization approach for High Quality Multi-scale Segmentation in Strobl, J. (ed) Angewandte Geographische Informationsverarbeiung, XII, 12–23.

Benz, U. Hofmann, P., Willhauck, G., Lingerfelder, I., and Heynen, M. (2004) Multiresolution, Object-oriented Fuzzy Analysis of Remote Sensing Data for GIS-ready Information ISPRS Journal of Photogrammetry and Remote Sensing, 58, 239–258.

Berberoglu, S.; Akin, A. (2009) Assessing different remote sensing techniques to detect land use/cover changes in the eastern Mediterranean, International Journal of Applied Earth Observations and Geoinformation, Vol. 11, Issue: 1, February, 2009. pp. 46–53.

Bhaskaran, Sunil; Paramananda, Shanka; Ramnarayan, Maria (1993) Per-pixel and object-oriented classification methods for mapping urban features using Ikonos satellite data Applied Geography, Vol. 30, Issue: 4, December, 2010. pp. 650–665.

Blaschke, T. (2010) Object based image analysis for remote sensing, ISPRS Journal of Photogrammetry and Remote Sensing, Vol. 65, Issue: 1, January, 2010. pp. 2–16.

Blaschke, T. Burnett, C. and Pekkarinen, A. (2006) Image Segmentation Methods for Object-based Analysis and Classification, in de Jong, S. and van der Meer, F. (eds) Remote Sensing Image Analysis: Including the Spatial Domain, Springer, 211–236.

Bock, Michael (2003) Remote sensing and GIS-based techniques for the classification and monitoring of biotopes: Case examples for a wet grass- and moor land area in Northern Germany, Journal for Nature Conservation, Vol. 11, Issue: 3, 2003. pp. 145–155.

Bock, Michael; Xofis, Panteleimon; Mitchley, Jonathan; Rossner, Godela; Wissen, Michael (2005) Object-oriented methods for habitat mapping at multiple scales - Case studies from Northern Germany and Wye Downs, UK Journal for Nature Conservation, Vol. 13, Issue: 2–3, July 15, 2005. pp. 75–89.

Conchedda, Giulia; Durieux, Laurent; Mayaux, Philippe (2008) An object-based method for mapping and change analysis in mangrove ecosystems, ISPRS Journal of Photogrammetry and Remote Sensing, Vol. 63, Issue: 5, September, 2008. pp. 578–589.

Devereux, B.J.; Amable, G.S.; Posada, C. Costa (2004) An efficient image segmentation algorithm for landscape analysis, International Journal of Applied Earth Observation and Geoinformation, Vol. 6, Issue: 1, November, 2004. pp. 47–61.

Duveiller, G.; Defourny, P.; Desclee, B.; Mayaux, P. (2008) Deforestation in Central Africa: Estimates at regional, national and landscape levels by advanced processing of systematically-distributed Landsat extracts, Remote Sensing of Environment, Vol. 112, Issue: 5, May 15, 2008. pp. 1969–1981.

Frohn, R., Reif, M., Lane, C., Autr, B. (2009) Satellite Remote Sensing of Isolated Wetlands Using Object-Oriented Classification of Landsat-7 Data, Wetlands, 29, 931–941.

Gamanya, Ruvimbo; De Maeyer, Philippe; De Dapper, Morgan (2009) Object-oriented change detection for the city of Harare, Zimbabwe Expert Systems With Applications, Vol. 36, Issue: 1, January, 2009. pp. 571–588.

Gao, J. (2009) Digital Analysis of Remotely Sensed Imagery, McGraw-Hill, New York, 645p.

Gitas, I.Z.; Mitri, G.H.; Kazakis, G.; Ghosn, D.; Xanthopoulos, G. (2006) Fuel type mapping in Annapolis, Crete by employing QuickBird imagery and object-based classification, Forest Ecology and Management, Vol. 234, November 15, 2006. pp. S228.

Gitas, I., Mitri, G. and Ventura, G. (2004) Object-based Image Classification for Burned Area Mapping of Creus Cape Spain, Using NOAA-AVHRR Imagery, Remote Sensing of Environment, 92, 409–413.

Guo, Q., Kelly, M., Gog, P., Liu, D. (2007) An Object-based Classification Approach in Mapping Tree Mortality Using High Spatial Resolution Imagery, GIScience and Remote Sensing, 44, 24–47.

Hall, O. and Hay, G. (2003) A Multiscale Object-Specific Approach to Digital Change Detection, International Journal of Applied Earth Observation and Geoinformation, 4, 311–327.

Hay, G., Castilla, G., Wulder, M., Ruiz, J. (2005) An Automated Object-based Approach for the Multiscale Image Segmentation of Forest Scenes, International Journal of Applied Earth Observation, 7, 339–359.

Im, J., Jensen, J. and Tullis, J. (2008), Object-based Change Detection Using Correlation Image Analysis and Image Segmentation, International Journal of Remote Sensing, 29, 399–423.

Jacquin, Anne; Misakova, Lucie; Gay, Michel (2008) A hybrid object-based classification approach for mapping urban sprawl in periurban environment , Landscape and Urban Planning, Vol. 84, Issue: 2, February 6, 2008. pp. 152–165.

Jyothi, B. Babu, G., Krishna, I. (2008) Object Oriented and Multi-Scale Image Analysis: Strengths, Weaknesses, Opportunities and Treats – A Review, Journal of Computer Science, 4, 706–712.

Lang, S. (2008) Object-based Image Analysis for Remote Sensing Applications: Modeling Reality- Dealing with complexity, in Blaschke, T. Lang, S. and Hay, G. (eds) Object-based Image Analysis: Spatial Concepts for Knowledge-driven Remote Sensing Applications, Springer, 3–27.

Lang S. and Langanke, T. (2006) Object-based Mapping and Object relationships Modeling for Land Use Classes and Habitats, Photogrammetrie, Fernerkundung, Geoinformation 1, 5–18.

Liu, Yu; Guo, Qinghua; Kelly, Maggi (2008) A framework of region-based spatial relations for non-overlapping features and its application in object based image analysis ISPRS Journal of Photogrammetry and Remote Sensing, Vol. 63, Issue: 4, July, 2008. pp. 461–475.

Lizarazo, I., and Barros, J. (2010) Fuzzy Image Segmentation for Land Cover Classification, Photogrammetric Engineering and Remote Sensing, 76, 151–162.

Mallinis, Georgios; Koutsias, Nikos; Tsakiri-Strati, Maria; Karteris, Michael (2008) Object-based classification using Quickbird imagery for delineating forest vegetation polygons in a Mediterranean test site, ISPRS Journal of Photogrammetry and Remote Sensing, Vol. 63, Issue: 2, March, 2008. pp. 237–250.

Maxwell, Susan K. (2010) Generating land cover boundaries from remotely sensed data using object-based image analysis: Overview and epidemiological application Spatial and Spatio-temporal Epidemiology, Vol. 1, Issue: 4, December, 2010. pp. 231–237.

Moller, M., Lymburner, L. and Volk, M. (2007) The Comparison Index: A Tool for Assessing the Accuracy of Image Segmentation, International Journal of Applied Earth Observation and Geoinformation, 9, 311–321.

Mueller, Marina; Segl, Karl; Kaufmann, Hermann (2004) Edge- and region-based segmentation technique for the extraction of large, man-made objects in high-resolution satellite imagery, Pattern Recognition, Vol. 37, Issue: 8, August, 2004. pp. 1619–1628.

Navular, K. (2007) Multispectral Image Analysis Using the Object-Oriented Paradigm, CRC Press, Boca Raton, FL, 163p.

Pal, N and Pal, S. (1993) A Review on Image Segmentation Techniques, Pattern Recognition, 26, 1277–1294.

Saura, Santiago; Castro, Sandra (2007) Scaling functions for landscape pattern metrics derived from remotely sensed data: Are their subpixel estimates really accurate? ISPRS Journal of Photogrammetry and Remote Sensing, Vol. 62, Issue: 3, August, 2007. pp. 201–216.

Smits, P.C.; Annoni, A. (1999) Towards operational knowledge-based remote-sensing image analysis, Pattern Recognition Letters, Vol. 20, Issue: 11–13, November, 1999. pp. 1415–1422.

Stein, A. and DeBeurs, K. (2005) Complexity Metrics to Quantify semantic Accuracy in Segmented Landsat Images, International Journal of Remote sensing 26, 2937–2951.

Volker, W. (2004) Object-based classification of remote sensing data for change detection, ISPRS Journal of Photogrammetry and Remote Sensing, Vol. 58, Issue: 3–4, January, 2004. pp. 225–238.

Verburg, Peter H.; van de Steeg, Jeannette; Veldkamp, A.; Willemen, Louise (2009) From land cover change to land function dynamics: A major challenge to improve land characterization, Journal of Environmental Management, Vol. 90, Issue: 3, March, 2009. pp. 1327–1335.

Walker, J. and Briggs, J. (2007) An Object-oriented Approach to Urban Forest Mapping in Phoenix, Photogrammetric Engineering and Remote Sensing, 73, 577–583.

Wang, Zhongwu; Jensen, John R.; Im, Jungho (2010) An automatic region-based image segmentation algorithm for remote sensing applications Environmental Modelling and Software, Vol. 25, Issue: 10, October, 2010. pp. 1149–1165.

Zhang, Y. (1996) A survey on Evaluation Method for Image Segmentation, Pattern Recognition, 29, 1335–1346.

Zhan, Q., Molenar, M., Tempfli, K., Shi, W. (2005) Quality Assessment for Geo-spatial Objects Derived from Remotely Sensed Data, International Journal of Remote sensing, 26, 2953–2974.

Chapter 12
Forensic Remote Sensing

The application rich areas of remote sensing devoted to environmental analysis and assessment successfully demonstrate how this technology can be employed to address important questions concerning resource management, environmental impact analysis, and environmental planning. Whether focused on understanding and documenting natural processes or studies that fall under the broad umbrella of human impact, the capabilities of image processing techniques to provide thematic information that illuminates critical land surface patterns or confirms underlying process-response hypotheses is irrefutable. Given the wealth of accumulated knowledge together with the agglomeration of methods designed to extract information from raw imagery, environmental remote sensing is poised to embark on a new agenda that moves remote sensing technology beyond its predilection for application research and into a more directed operational setting. Borrowing from the Latin word "forensic," meaning public, we can call this new agenda forensic remote sensing.

Forensic remote sensing considers the investigative use of image processing technology to support policy decisions regarding the environment and the regulation of human activities that interact with environmental process and amenities. In this chapter we will examine the forensic use of remote sensing and discuss its relevance in the context of human-directed alterations within the environmental system. The alterations induced by human actions on environmental processes are the nonrandom consequence of human decisions made to satisfy myriad of social and economic desires. It has been argued that this decision process is often flawed to the degree that policy directives that embody these decisions invite environmental damage without a sound mechanism to identify and evaluate their potential to damage environmental functioning (Purdy 2006, 2009). Forensic remote sensing is envisioned as one solution to address the gap between policy and action which places remote sensing in the role of a practicable decision support technology.

J.K. Lein, *Environmental Sensing: Analytical Techniques for Earth Observation*,
DOI 10.1007/978-1-4614-0143-8_12,

12.1 Remote "Fingerprinting"

The term "fingerprinting" is a concept taken from the field of environmental forensics. Within this subject area, fingerprinting is a general term that includes methods developed to identify specific associations between patterns of a material sample and sources of pollutants in the environment that may be responsible for the observed contamination. Fingerprinting is a integral part of environmental forensics; a field that involves the systematic examination of environmental information in order to ascertain the source of chemical contamination, the timing of pollutant releases to the environment, the spatial distribution of contamination and the identification of entities that may be responsible for the observed situation. Central to the definition of fingerprinting is the critical role played by investigative methodologies that strive to establish patterns of transport and fate together with explanations or origins and impacts. Frequently, fingerprinting methodologies require the ability to

1. Distinguish an entity's contribution to the nature and extent of a contamination problem
2. Determine the degree of an entity's involvement in the activities that generated the pattern of contamination
3. Assess the level of care exercised by the potentially responsible entity

Adapting these ideas to the larger question of human impact introduces an important spatial component to environmental forensics that can be addressed using remote sensing techniques (Brilis et al. 2000).

In a forensic application, satellite imagery highlights the spatial footprint of human actors in very real and compelling ways (Kalacska et al. 2009). Given the continuous nature of satellite-based data collection, the spatial relationships between natural land surface patterns and observed environmental or ecological damage are presented as "time-stamped" documentation that aptly describes the extent of human activities (Fig. 12.1). As documenting evidence, imagery can focus the scope of a forensic investigation. In addition, owing to the highly visual nature of our imagery and the thematic produces that can be derived via processing, the spatial representation of documenting evidence is a direct and persuasive way to identify patterns of damage and potentially connect it to a responsible human action. The documentation of occurrence patterns, present-day or historical, can be placed immediately into a geographic context and examined in relation to locational, physiographic, or other environmental qualities pertaining to the site and situational characteristics of the aberrant case. To illustrate this capability, consider for example a documented decline on a riparian wetland. Utilizing remotely sensed data, activities can be defined to have occurred upstream from or adjacent to this environmental damage area. Those offending actions can be tracked and the spatial changes in the extent of damage they produce can be quantified. This information can be employed to assess the scope of damage, characterize the continuing threat to the wetland area and guide specific remediation and

Fig. 12.1 The documented footprint of human damage

enforcement directives designed to protect this environmentally sensitive feature. By means of land cover analysis the extent and pace of decline can be measured and the geographic distribution of land uses known to promote damage can be "fingerprinted" in the image. In this example and in situations fraught with "unknowns," image analysis readily supports investigative assessments to indentify the cause or pinpoint the source of environmental damage. (Xiao and Ji 2007).

The capacity to employ remote sensing in this role is an important departure in the design of remote sensing applications (Howard et al. 2002). Forensic remote sensing capitalizes on the generation of information to investigate a specific event of condition not to provide broad thematic explanation. In a forensic analysis there is less concern for pure research but rather immediate problem resolution. How well satellite image analysis contributes to investigative applications rests on a clear understanding of:

- The objectives surrounding the specific case
- The compelling enforcement functions or activities involved
- The necessary analytical procedures required
- The explicit requirements that define quality assurance

Remote sensing technology can be introduced into different phases of an environmental investigation (Middleton 2002). During the targeting of an prioritization phase, remotely sensed data at low to high spatial resolutions can provide needed information regarding an area's population, natural systems, and obvious environmental impacts. During the planning phase, remote sensing can support change detection activities that will provide insight into active processes or operational changes occurring within the specific environmental setting. Change analysis can document possible activities that fail to comply with legal requirements and situations where compliances can be called into question. As an investigation progresses into the field inspection phase, remotely sensed data at medium to high spatial resolution supports a unique form of virtual field reconnaissance by giving the analyst an "over-the-fence" view of critical facilities and related land uses. This perspective allows the site and situational characteristics of the area in question to be carefully examined and comprehensively explored. During the inspection phase high spatial resolution imagery serves as a base map to guide geophysical sampling, ground verification, and related field assessment activities. In addition, the systematic collection and archiving of imagery facilitates site monitoring programs that can greatly improve long-term efforts at environmental amelioration and future environmental appraisals (Cartalis et al. 2000).

The systematic application of satellite data coupled with standardized procedures for information extraction increases the opportunities for a wide range of environmental oversight activities. Oversight implies a guided review, which in concept has been an integral element of environmental legislation including diverse instruments and directives such as

- Post-EIA auditing
- Compliance assessment and enforcement
- Policy review and program assessment
- Zoning and regional plan monitoring
- Land evaluation planning
- Verification studies
- Environmental site assessment

12.2 Baselines and Benchmarks

Remote sensing technology has long been used to study the Earth's surface. As a research tool the value of satellite imagery is well understood; however, when placed into a forensic role remote sensing must contend with policy and operational constraints that do not introduce themselves in a research setting. These constrains include factors such as

- Financial limitations
- Institutional and organizational barriers
- Absence of expertise within the jurisdictions involved
- Poor capacity to implement the technology

These constraints can greatly influence the "life-cycle" costs associated withadopting remote sensing technology and need to be understood. However, communicating the investigative value of satellite imagery processing begins by developing cost-effective information products and engaging activities that extend the effectiveness of environmental policies and programs. In this context, satellite remote sensing is a transformative technology with respect to environmental management and enforcement activities. The challenge is to separate the technical aspects of remote sensing from the pragmatic policy "tasks" common to environmental protection programs and establishing the connection that link environmental remote sensing methods to direct and real policy needs. Using the examples listed previously in this chapter, we can selectively examine where remote sensing can fill the void and provide a useful and feasible alternative.

1. *Environmental impact assessment and post-EIA auditing* – Environmental impact assessment (EIA) describes a systematic evaluation of the potential effects of a major human action on the environmental system. The EIA process involves a detailed study of the action under review and how this action may induce alterations in the physical and human landscape where it is being proposed. The purpose of this study is to document environmental change before it occurs. In this regard, the EIA serves as a form of pro-active decision making that informs policymakers, stakeholders, and the public on the consequences of a proposal that could significantly affect environmental quality. As a methodology, EIA embodies a form of "future-casting" where a human action is placed into an environmental setting and projected forward into an unspecified future. Although EIA methods strive to maintain a rigorous scientific footing, uncertainty is an unavoidable aspect of any assessment. Because EIA can be considered an exercise in forecasting, the obvious question stemming from this process is whether the changes forecast by an EIA actually occur. To address this question Post-EIA auditing has been a recommended addition to the EIA process. Post-EIA audits are a means of evaluating the post-action status of the affected environment to ascertain the accuracy of EIA projections and to identify impacts not discovered during the initial assessment (Dipper 1998). In practice, post-EIA auditing is a field-based activity. The audit team selects the appropriate field techniques, field sites and develops a detailed study plan (Wilson 1998). Often included in the field protocol are activities such as site reconnaissance, field measurements, and the visual interpretation of aerial photography. The prospects for conducting Post-EIA audits by means of remote sensing depend on the type of action involved. Several options for remote auditing include
   (a) Developing land use/land cover baseline inventories
   (b) Examining pre- and post-action conditions of the local environment
   (c) Examining changes in selected landscape indicators that define critical aspects of environmental quality

   By reducing dependence on field investigation, remote sensing facilitates wider post-action review that can be useful for detecting unanticipated changes and tracking development trends that contribute to cumulative environmental effects (Lein 2002). Developing a remote sensing solution for post-EIA auditing can be conducted in either of two ways. Other approaches rely on the final environmental impact statement (EIS) to provide the information that will focus the audit. Drawing from the discussion outlining the adverse impacts noted in the

IS, those that exhibit a spatial expression that can be resolved using remotely sensed data can be placed into a change detection strategy for evaluation using a pre-action and post-action framework. Because environmental changes resulting from an action may occur differentially over time, satellite data supports the systematic revisit of the site which can be helpful when attempting to establish trends and document change trajectories. The environmental impacts associated with an action may be manifested as direct alterations that can be comparatively easy to identify such as the decrease in habitat or the increase in impervious surface. More challenging to target are the indirect and secondary changes attributed to an action; however, with carefully selected indicators surrogate measures can be obtained that illuminate their presence and consequence. The alternative approach to post-EIA auditing directs attention to anomaly detection techniques and their capacity for stopping outlier cases without specific reference to pre-action conditions. Used in this manner, anomaly detection is not based on a preselected set of impacts, but rather concentrates observation on the action, noting the presence and status of deviations in existing condition. Conditional deviation in the patterns of descriptive landscape attributes serve as indicators of changes that are potentially novel, unexpected, or unanticipated. Remote surveillance of this type can help document the actual pattern of impacts that result from the action and may provide essential information regarding the success and effectiveness of impact mitigation strategies. In the broad setting of human development actions, adopting a satellite-assisted approach to environmental assessment offers a source of "early warning" information that encourages regular post-action review or projects, particularly those that may be controversial in nature. The results taken from the remote audit signals changing environmental conditions that can help prioritize mitigation measures. Furthermore, the synoptic scale of the satellite imagery improves the use of post-EIA auditing as a means of oversight that integrates monitoring as an active component of long-term environmental review. Overtime, an information resource is created that enhances our understanding of human-induced environmental change that can assist future attempts at EIA prediction (Cashmore 2004).

2. *Environmental compliance assessment* – Human activities prone to promote environmental damage or threaten environmental quality are often the target of policy instruments designed to regulate their scope and mitigate the consequences associated with these activities (Berry and Dennison 2000). A critical element of many environment policy directives and protection strategies are the enforcement mechanisms introduced to insure regulatory compliance. The enforcement and systematic review of actions known to damage the environmental system is fraught with complications and constraints that limit serious attempts to engage in comprehensive assessments. Satellite remote sensing offers courses of information that can support environmental compliance operations (Purdy 2006; Lein 2009). Environmental compliance may be defined as the state of being in accordance with a set of guidelines, specifications, or legislative mandates designed to protect or manage environmental resources

or amenities (Heyes 2000; Vincoli 1993; Winter and May 2001). As a means of moderating human impact on the environment, identifying activities that fail to agree with established control standards remains on solution society relies on despite its uneven success (Lein 2009). Integrating remote sensing into the legal setting where environmental compliance takes place requires an understanding of the governing laws that provide opportunities for the application of this technology. The physical inspection regimes that typically define how compliance with regulatory statutes is enforced are generally directed toward industrial, extractive, or agricultural activities whose operation must conform to a specific set of performance standards. The site-specific nature of compliance assessment therefore implies that

(a) Evaluation of a given activity is an on-going process that extends over it operational life-cycle
(b) Evidence exists (in the form of complaints) suggesting that the activity may not be functioning (or operating) in accordance with one or more governing statutes or regulations

These defining assumptions supply the needed context that directs integration of remote sensing into environmental compliance operations. Three analytical support functions that remote sensing facilitates can be identified:

(a) *Systematic damage detection* – Although the role of satellite remote sensing in monitoring environmental processes was discussed previously in Chap. 5, damage detection carries a more rigorous expectation when included as part of a compliance inspection regime. Damage characterizes the situation where an aberrant deviation in the status or behavior of a landscape feature can be identified in the scene. When observed or detected on the image damage assumes a pattern that is irregular, abnormal, and difficult to explain in context to the surrounding and expected land cover relationships. An example of this form of discovery on an image might include events such as the failure of hazardous waste containment or an illegal discharge signaled by an anomalous pattern of vegetation decline down slope or downstream from the site. Similarly, events may trigger thermal patterns the deviate from the expected as well. The remote sensing strategy supporting this form of discovery and detection would depend on (1) a systematic revisit of the site to define its normal or "ambient" status and (2) a method of detection where the anomaly emerges from the baseline norm. Success of this type of analysis is predicated in the selection of an optimal spatial scale to anchor the analysis, the acquisition of data sufficiently close in time to the event, and use of an appropriate indicator that highlights the offending pattern.
(b) *Remote inspection* – Current compliance practice relies heavily on field-based inspections as the primary means of detecting violations and evaluating performance. The synoptic view of satellite-based sensors that enables observations of a site were (1) access is restricted by physical or institutional barriers, (2) activities are geographically dispersed over the

region of interest, or (3) resources to support field inspection are limited. Remote inspection facilitates routine review of facilities in relation to a set of performance criteria; particularly those criteria that assume a geographic expression that can be resolved by the sensor (Fig. 12.2). If the location of features are known, sites falling into the same descriptive category can be scrutinized based on factors such that the physical placement of storage and containment facilities, on-site material and waste piles, the location and flow of discharges into receiving bodies of water, the design, placement, and characteristics of roads, fencing, buildings, and treatment works. Remote surveillance directed at a specific class of facility or over a specific geographic area can produce intermediate thematic products that can be employed to prioritize field inspections. In this role satellite data serves as a screening device that can eliminate certain area that satisfy an initial "remote" evaluation.

(c) *Change over time assessment* – Perhaps one of the more vexing aspects of environmental policy solutions aimed at improving environmental quality is the factor of time. Policy decision making takes place within the institutional framework of governments and is inherently captive to human conceptualizations of time. Policies are drafted, laws passed, but the attainment of specific objectives are left for the future to resolve. The environmental system also exhibits distinct temporal behaviors, but often at time scales that are cross-generational. Efforts to reclaim or restore, re-establish, or repair damaged environments are not likely to evidence significant success at the time frames implied by the majority of environmental policies. Similarly, human actions also require the passage of time before their consequences can be fully realized in a given environmental setting. Change over time assessment directs environmental analysis toward the tracking of the "long-term" progression of land use or land cover trend linked directly to the goals and objective articulated in environmental policy. Policy goals may be couched in terms that characterize management, remediation or preservation activities but the effectiveness of this application depends on linking policy of specific metrics. Ideally, such metrics can be derived from satellite imagery and implemented in a systematic monitoring program. In the USA for instance, legislation such as the Surface Mining and Reclamation Act, the National Historic Preservation Act, National Environmental Policy Act, and the Coastal Zone Management Act all carry explicit requirements to reduce human-induced environmental change; however, none of these mandates indentify remote sensing as a support technology. Utilizing specific indicators linked to policy objectives provides crucial information on the success of these and other similar programs and can offer early warning feedback of failures that can be used to manage direct management resources and remediation strategies (Backhaus and Beule 2005). A useful example demonstrating the use of satellite data to evaluate habitat conservation plans pursuant to the US Endangered Species Act was reported by Schweik and Thomas (2002). Institutional performance was

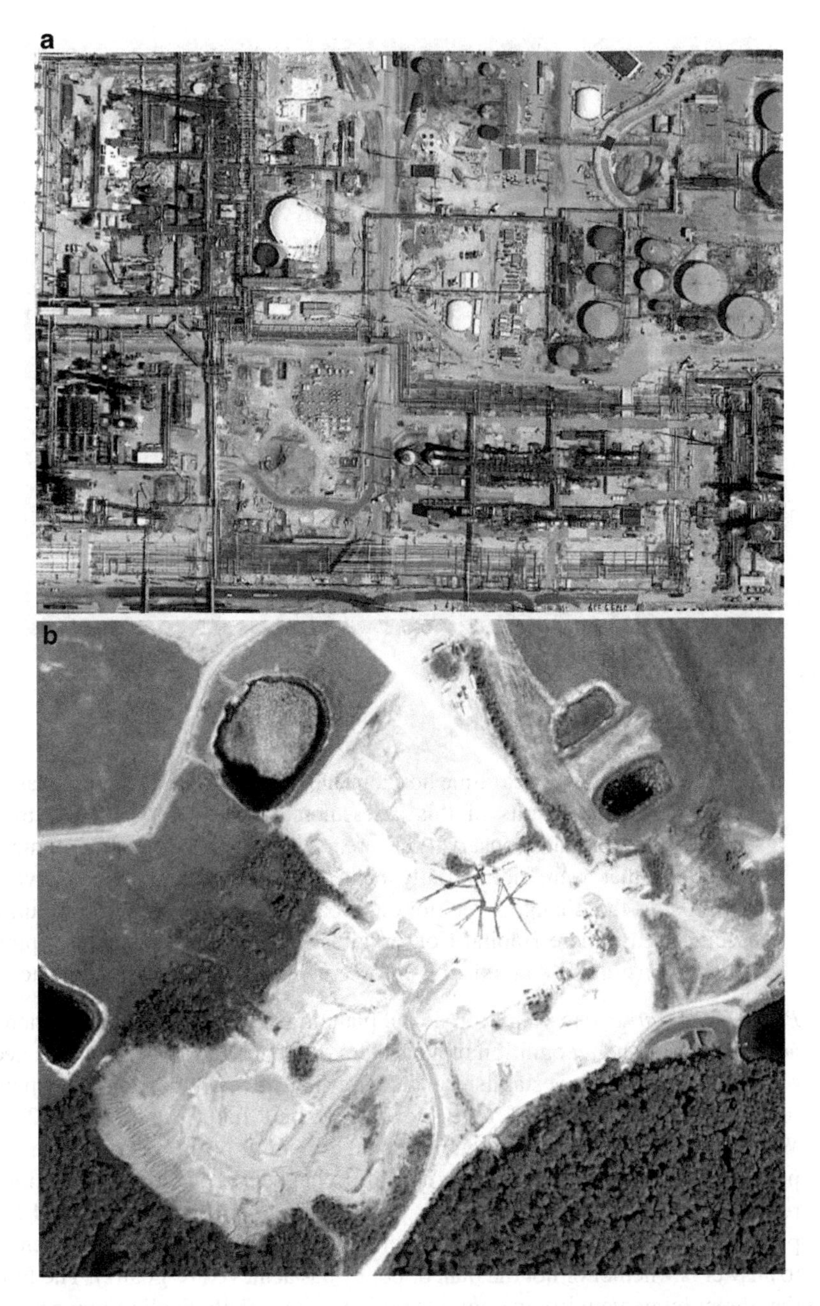

Fig. 12.2 Elements of the compliance landscape: (**a**) Industrial facility, (**b**) material processing/ impoundment, and (**c**) extractive activities

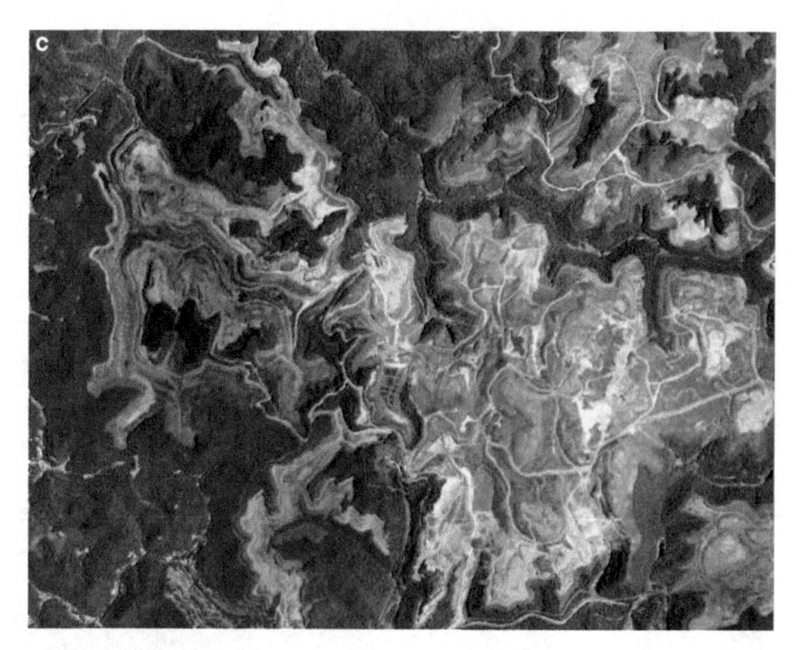

Fig. 12.2 (continued)

evaluated across a 12 year time horizon using data acquired from the Landsat TM system. The results of this assessment showed that land use trends beyond the habitat conservation area were modifying environmental conditions that were not initially considered in the habitat conservation plan. Using the change over time approach, it was possible to document specific areas where Habitat Conservation Plan "rule breaking" was prevalent, demonstrating that existing enforcement regimes were inadequate.

3. *Policy and plan review* – Environmental planning and policy are future-oriented activities. With the recognition of a problem or identification of a societal need, planning sets out to establish goals and objectives together with policy recommendations designed to achieve a desired future state (Lein 2003). Whether addressing land use, environmental quality, resource conservation or public safety concerns, the production of a plan sets out a course of action to resolve the compelling issues, many of which have a discernable spatial footprint. Because planning targets the future, one question that can be challenging to answer is whether or not the plan or policy has achieved its goal(s). The forces that direct human/environmental interaction in a spatially explicit manner must consider the interacting influences of socio-economic drivers, policy decisions and the environment that anchors the plan. Using remote sensing to guide

Fig. 12.3 The synoptic plan view in the context of agricultural land valuation

planning capitalizes on the unique characteristics of remotely sensed data (Miller and Small 2003):

(a) The self-consistent synoptic nature of the imagery
(b) The capacity for routine periodic and unobtrusive updating and comparison
(c) The capacity for the description, classification, and measurement of critical physical properties

Examples of these capabilities include the use of Thematic Mapper imagery to evaluate permit reviews under the section 404 program of the Clean Water Act (Swenson and Ambrose 2007), the use of Thematic Mapper data to examine the impact of land use zoning policy on habitat fragmentation (Munroe et al. 2005), the application of Landsat ETM+ data to evaluate the appropriateness of development density policies with respect to environmental quality objectives (Wilson et al. 2003) and the application of Landsat TM imagery to assess the effectiveness of goals guiding watershed restoration strategies (Basnyat et al. 2000). In each of these case studies the actual plan or a product derived from the plan served as the basis for comparison. Remotely sensed imagery was employed largely to provide indicators whose relationship to planning goals could be examined based on predetermined benchmarks. The implications for plan review are instructive, showing that remote sensing places planning squarely into a spatial context where specific planning outcomes can be observed (Fig. 12.3). In addition, this selective review underscore the forensic and diagnostic value of remote sensing where the products generated from the raw satellite data become useful instruments of environmental assessment.

12.3 Sensing Beyond the Obvious

The wealth of application research within the realm of environmental analysis supports the use of remote sensing as a means of establishing evidentiary data concerning the disposition of human activities relative to existing environmental controls. Research has also moved remote sensing technology into more diagnostic roles with forensic value. Selections of methods with this type of operational relevance include the topical areas of environmental health, environmental pollution, and environmental hazards.

1. *Environmental health* – The continued population pressures exerted on planetary ecosystems coupled with the probable shifts in environmental conditions resulting from climatic and ecological change introduce broad global and community health concerns. Satellite-based sensors with improved spectral, spatial, and temporal resolution have been shown to provide useful data products for the assessment of health risks (Beck et al. 2000; Herbreteau et al. 2007; Leblond et al. 2007). Applications focus largely on converting satellite observations into vegetation land surface temperature, atmospheric moisture and rainfall indices to produce models of critical environmental conditions that promote disease transmission, vector production, the emergence and maintenance of disease foci and risk factors influencing human–vector contact. The derivative products generated from the satellite imagery typically characterize habitat conditions, vector sources and human settlement relationships that can be extracted from the data (Goetz et al. 2000). Although specific methodologies will vary, the common theme centers around the identification of environmental conditions that are suitable for the specific disease vector to establish itself. The list of environmental factors that can effect suitability include

   (a) *Vegetation/NDVI*
   (b) *Land use type(s)*
   (c) *Soil type*
   (d) *Moisture*
   (e) *Temperature*

   The relationship between known cases of disease and the satellite measured environmental factors are used to construct a statistical model of the general form:

   $$P = b\alpha + bX_1 + bX_2 + \cdots + b_nX_n,$$

   where the presence (incidence of the disease vector (P) is functionally related to a set of independent environmental variables (X). The statistical procedures selected to derive a model range from multiple regression discriminant analysis, principal components analysis or the focus on developing specific disease incidence indices (Tran et al. 2002; Arboleda et al. 2009). The statistical surface generated by these models describe a prediction that is then compared to human population and

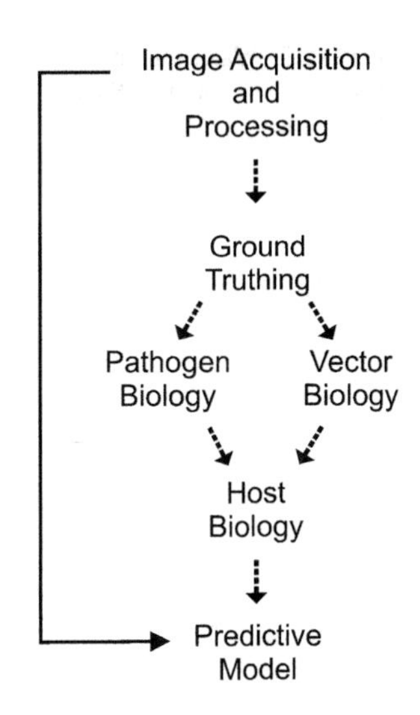

Fig. 12.4 Satellite-based health-exposure modeling

activity pattern data in order to assess environmental exposure patterns. The salient features of this approach have been described by Ford et al. (2009) (Fig. 12.4). The majority of modeling efforts, however, are retrospective in design. This is due largely to geographic and seasonal influences that modify the distribution of many infectious diseases in relation to the environmental conditions to which they are inherently linked. Establishing effective correlates between satellite measurements and disease incidence rate can yield predictive indicators that can be employed in early warning systems and support studies tracing outbreaks back to their probable geographic origins.

2. *Environmental pollution* – The forensic analysis of pollution using space-borne sensors exploits reflectance contrasts in the environmental media that can then be related to approximate concentrations of specific contaminants. In the atmosphere this could be particulate matter or a specific gas or compound. In water, contaminants can range from suspended sediment, algae, or various forms of organic matter. In either instance, remote sensing methods focus on measuring those substances or conditions that become recognized as a change in the optical or thermal properties of the environmental media. Substances that do not directly alter optical or thermal characteristics can only be inferred by measuring surrogates, such as vegetation, which may display a changed response to an input of a chemical or biological contaminant. Although not a panacea, the value of remote sensing lies in its ability to provide both a spatial and temporal view of selected environmental quality parameters that may not be available using field sampling techniques. A selection of methods concentrating on water quality and air pollution analysis demonstrates remote sensing's salient contributions.

   (a) *Water quality assessment* – The basic logic applied to remote estimation of water quality involves separating the radiance of the parameter of interest

Fig. 12.5 Water Quality Issues in the Maumee Basin Lake Erie

Table 12.1 Factors impacting water quality

Microbial contamination
Mineral content
Heavy metal contamination
Chemical contaminants

from all other radiance components recorded by the sensor. Symbolically this can be explained simply as the relation:

$$R_1 = R_t - (R_p + R_s + R_b),$$

where R_1 is the radiance of the parameter of interest, R_t is the total radiance recorded by the sensor, R_p is the atmospheric attenuation, R_s is the surface reflectance, and R_b is the bottom reflectance.

As various substances are introduced into a body of water they produce a change in its backscattering characteristics. A sample of the factors that affect water quality are listed in Table 12.1. The changes produced when a substance is introduced can be measured as a spectral response pattern as a function of the water body in relation to the modification the substance creates in backscattering (Fig. 12.5). Recognizing that the optimal wavelength used to measure water quality parameters will depend on the substance involved, its concentration and the resolution characteristics of the sensor, it becomes possible to relate measured contrasts in backscatter to a sampled concentration of a given water

Table 12.2 Typical water qualities indices

Typical water qualities indices
Dissolved oxygen
Fecal coliform
pH
BOD (Biochemical Oxygen Demand)
Temperature
Total phosphate
Nitrates
Turbidity
Total solids

quality parameter using empirical or analytical models (Ritchie et al. 2003). Empirical solutions take the general form

$$Y = a + bX \text{ or}$$
$$Y = ab^x,$$

where Y defines the satellite measured radiance, X explains the selected water quality parameter and a and b express empirically derived coefficients. Analytic approaches replace statistical modeling with algorithm based on the optical properties of water and water quality parameters. Such approaches often employ unique band ratios together with parameter-specific indices (Table 12.2). Simple classification of radiance differences can also be developed using unsupervised or supervised classification procedures. The main objective following this strategy is to discover the presence of contaminant in a water body without reference or fit to a known concentration level. Generally, wavelengths between 0.45–0.52 μm and 0.63–0.69 μm are useful for these purposes as are the thermal bands. Overall, remote assessment of water quality tends to focus on measuring suspended sediments, chlorophylls, and temperature. Suspended sediments have the distinction of being among the most typical pollutant in terms of volume and weight in an aquatic system. Suspended sediment is also a useful surrogate contaminant for other pollutants such as insecticides, metal, and other inorganic substances that adhere to sediment particles and contribute to ecological exposure. In the visible and near infrared portions of the spectrum, suspended sediments increase reflectance of surface waters; a pattern clearly discernable using wavelengths between 0.7 and 0.8 μm. Thermal discharges and thermal enrichments define pollution sources emanating from sites where water is used as a cooling agent. Utilizing thermal brightness temperature, illegal discharges as well as improper cooling impoundment facilities can be retrieved from satellite imagery. Estimates of chlorophyll content can also be re-retrieved from satellite data. Because the concentration of chlorophyll serves as an indicator of eutrophication, developing associations between radiance in narrow bands or band ratios can guide the management on impacted watersheds and receiving bodies of water. Several algorithms have been introduced to estimate chlorophyll-a content (Gohin et al. 2002, 2008). Examples include band ration algorithms and statistical models developed with the assistance of field sampling.

(b) *Air quality* – Satellite remote sensing of air quality has developed rapidly over the last decade (Hoff and Christopher 2009). Much of the focus has been directed toward (1) the development of processing methods to infer aerosol optical (thickness) depth from space and (2) producing estimates of air quality that compare with ground-based measurements. Aerosol optical thickness defines the degree to which aerosols in the atmosphere prevent the transmission of light. It can be expressed more formally as the integrated extinction coefficient over a column or unit cross section of the atmosphere. Aerosol optical thickness is a measure widely used to monitor sources and sinks of suspended particles and gases such as haze, smoke, and a range of air pollutants. Satellite observations of aerosols offer a source of information with broad scale spatial coverage that facilitates:

- Emission characterization for the source and transport of contaminants such as NO_2, NO_x, and particulate matter (PM)
- Estimation of biogenic VOC through the process of satellite-derived formaldehyde signals
- Pin-pointing the location and source strength of wildfires and dust plumes
- Accountability analysis through the assessment of air quality management programs

Useful reviews and case studies detailing satellite-based air quality analysis can be found in Hidy et al. (2009), Martin (2008), Liu et al. (2005), and Gupta et al. (2006). The success of these applications hinges on the sophistication of the aerosol optical thickness (AOT) retrieval algorithm (Retalis and Sifakis 2010; Hoyningen-Huene et al. 2003). A simplifying solution to the algorithm issue involves the use of processed data such as the MODIS MOD04 Aerosol Product (Fig. 12.6). The MODIS product monitors the ambient optical thickness and can be sued to evaluate the sources and sinks of aerosols over both the land and ocean surface (Remer et al. 2005).

3. *Environmental hazards* – Environmental hazards describe a range of events that can lead to harm. Although not a concept that is simple to explain, attention is general directed toward geophysical processes that can harm and disrupt human populations (Gillespie et al. 2007). Also important in this explanation is the manner by which social systems interact with natural processes that are often extreme. It should be understood, however, that environmental hazards also can include

(a) *Chemical*
(b) *Physical*
(c) *Mechanical*
(d) *Biological*
(e) *Psychological*

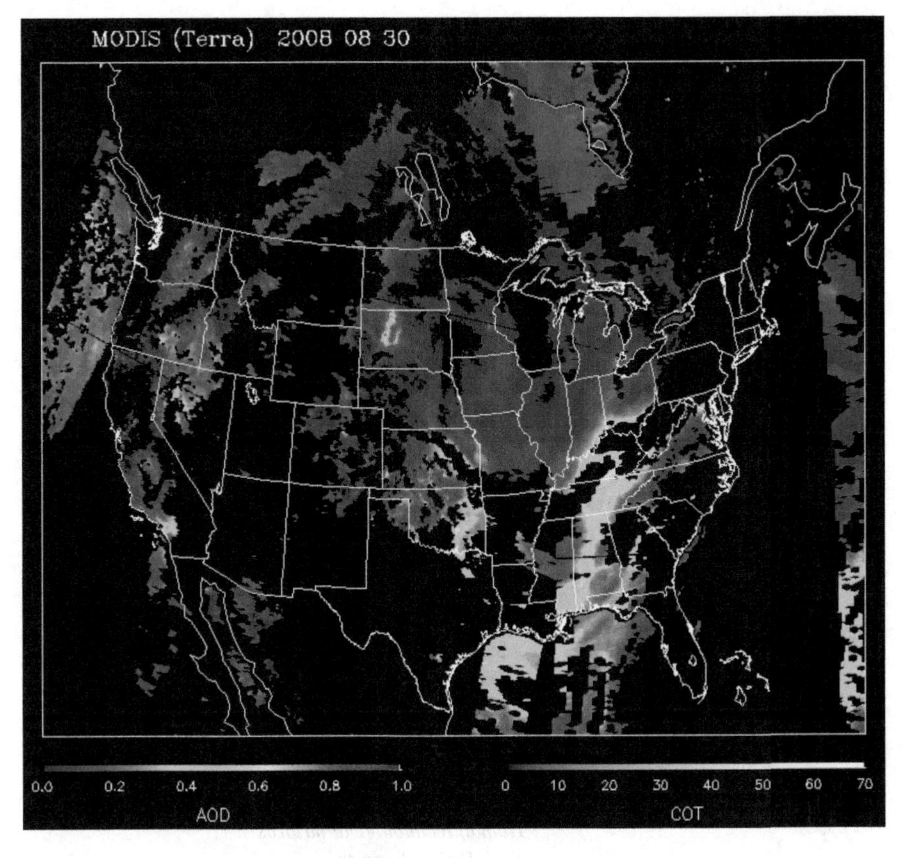

Fig. 12.6 Satellite-derived aerosol product

Actors that present "threats" to society that require some type of response or adjustment. Under this expanded definition we can characterize environmental hazards as either natural events such as floods, earthquakes, tornados, or elements of human creation which are toxic or harmful to some degree (Table 12.3). The hazard exists because human populations are exposed to their behavior in an otherwise neutral environment. A pivotal factor in the explanation of an environmental hazard is the notion of risk; a concept that is applied in close association to the definition of a hazard. Risk explains the actual occurrence of an event and is typically expressed as the probability of a given occurrence and its consequence. The relationship between a hazard and a risk is complex and it is not always possible to distinguish the effects of an event from their causes or to determine their significance. Attempts to assess environmental hazards build from the recognition that as an inherent element of the landscape, events that maybe classed as hazards describe a spatiotemporal, physical, and human dimension which implies that a set of common elements exist that can be understood using remote sensing technologies':

Table 12.3 Environmental hazards

Natural hazards
Geological hazards
Avalanche
Earthquake
Lahar
Landslides and mudflows
Sinkholes
Volcanic eruption
Hydrological hazards
Flood
Limnic eruption
Whirlpool
Maelstrom
Seiche
Tsunami
Climatic and atmospheric hazards
Blizzard
Drought
Hailstorm
Heat wave
Cyclonic storms
Ice storm
Tornado
Climate change
Geomagnetic storm
Wildfire hazard
Disease
Human/technological hazards
Sociological hazards
Crime
Civil disorder
Terrorism
War
Technological hazards
Industrial hazards
Structural collapse
Power outage
Fire
Hazardous materials
Transportation

(a) Areal extent
(b) Intensity of impact
(c) Duration of impact
(d) Rate and pattern of onset
(e) Spatial dispersion or spread

With these characteristics in mind, the assessment of risk becomes a data-driven procedure that involves

(a) Identification and description of the hazard, its geographic distribution and potential effects
(b) Assessment of vulnerabilities
(c) Identification and measurement of exposure patterns of population and infrastructure
(d) Identification of major public problems associated with the event
(e) Identification of the costs and features of mitigation measures
(f) Identification and description of the public policies that enable hazard response, reduction, and mitigation

These elements of classic risk assessment draw from both qualitative and quantitative information sources. Qualitative assessment is informed by gathering knowledge of the hazard and patterns of exposure to characterize risk. Quantitatively, assessment strives to collect measures of critical attributes of risk. Remote sensing provides important information to support both activities in key aspects of the assessment problem including

(a) *Hazard characterization and mapping* – Relying on fine to moderate resolution sensors, landforms and land use types can be delineated to assist with the identification of flood plains, detection of thermal hot spots, infrastructure review, mitigation assessment and the mapping of natural hazard patterns such as landslide zones and volcanic features (Tralli et al. 2005; Sanyal and Lu 2004).
(b) *Vulnerability assessment* – Estimating populations and utilizing these estimates to characterize exposure patterns and risk (Harvey 2002; Wu and Murray 2007).
(c) *Event detection and early warning* – Involving the development or event detection systems and the tracking of events using satellite-based data (Zschau and Kuppers 2003).
(d) *Post-event analysis* – Employing the synoptic scale nature of the imagery to examine damage patterns to provide profiles of infrastructure failures and population displacement (Adams and Huyck 2006).
(e) *Disaster recovery* – Utilizing fine to moderate resolution data to track the pace of recovery and the effectiveness of emergency response (Hall et al. 2006).

12.4 Creative Problem Solving

Although the forensic use of remote sensing is not a panacea, important information can be gathered by recognizing what the technology does well and where its constraints limit its effectiveness. Remote sensing can be an effective method of data collection for

- Comparatively large areas where time constraints limit other approaches
- Inaccessible regions or dangerous locations

The data acquired from our sensors facilitate machine processing techniques that can be difficult to produce using manual methods and with the addition of creative thinking novel solutions can obtained that address unique problems. This highlights the experimental nature of the technology. Since forensic investigation centers on deriving clues that suggest a pattern or point to an anomalous situation, remote sensing accommodate ill-defined problems. Given the repeat coverage aspects of the data, investigations can span time scales that vary with respect to duration and frequency that filed analysis cannot achieve. This forensic investigation can effectively review a previous state of the system of interest and continue over weeks, months or beyond, paced to the cyclic nature of an event of the intermittent behavior of human activities. Satellite data, however, are not direct samples of phenomena, they require calibrations against a reality and this calibration is never exact. Nevertheless, remote sensing technology provides the opportunity to densely characterize the near surface where success is realized by:

1. Achieving a sound theoretical understanding of the sensor
2. Comprehending the significance of measurement uncertainty
3. Possessing detailed knowledge of the phenomena under considerations

12.5 Summary

Exploring the role of remote sensing beyond land cover mapping formed the focus for this chapter. Using the concept of "forensic" analysis, this chapter explored the use of remote sensing in an investigative capacity. Although not necessarily directed on specific legal issues, the notion of forensic remote sensing suggests that data analysis can be performed to derive information products that hold specific analytical value in areas that are concerned with environmental compliance and enforcement issues, as well as those related to impact analysis, policy review, and topics concerning environmental health and quality. Beginning with a discussion of baselines and benchmarks, the review of methods and applications suggested a value-added quality of satellite-based image processing that can contribute to a wider array of environmental questions. Although the forensic use of remote sensing is not a panacea, important information can be gathered by recognizing what the technology does well and where its constraints limit its effectiveness.

Remote sensing can be an effective method of data collection for comparatively large areas where time constraints limit other approaches or for inaccessible regions or dangerous locations.

References

Arboleda, S., Gorla, D., Porcasi, X., Saldana, A., Calzada, J., Jaramillio-O, N. (2009) Development of Geographical Distribution Model of Rhodnius pallescens Barber, 1932 Using Environmental Data Recorded b Remote Sensing, Infection, Genetics and Evolution, 9, 441–448.

Adams, B. and Huyck, C. (2006), The Emerging Role of Remote Sensing Technology in Emergency Management, Infrastructure Risk Management Processes – Natural, Accidental, and Deliberate Hazards, Monograph 1, ASCE, Reston, Virginia.

Backhaus R. and Beule, B. (2005) Efficiency Evaluation of Satellite Data Products in Environmental Policy, Space Policy, 21, 173–183.

Basnyat, P., Teeter, L. , Lockaby, B. and Flynn, K. (2000) The use of remote sensing and GIS in watershed level analyses of non-point source pollution problems Forest Ecology and Management, 128, 65–73.

Beck, L., Lobitz, B. and Wood, B. (2000) Remote Sensing and Human Health: New Sensors and New Opportunities, Emerging Infectious Diseases, 6, 217–226.

Berry, J and Dennison, M. (2000) The Environmental Law and Compliance Handbook, McGraw-Hill, New York, 807p.

Brilis,G., Gerlach, C., van Waasbergen, R. (2000) Remote Sensing Tools Assist in Environmental Forensics. Part I: Traditional Methods, Journal of Environmental Forensics, 1, 63–67.

Cartalis, C., Feidas, H., Glezakou, M., Proedrou, M., and Chrysoulakis, N. (2000) Use of Earth Observation in Support of Environmental Impact Assessments: Prospect and Tends, Environmental Science and Policy, 3, 287–294.

Cashmore, M. (2004) The Role of Science in Environmental Impact Assessment: Process and Procedure versus Purpose in the development of Theory, Environmental Impact Assessment Review, 24, 403–426.

Dipper, B. (1998) Monitoring and Post-auditing in Environmental Impact Assessment: A Review, Journal of Environmental Planning and Management, 41, 731–747.

Ford, T., Colwell, R., Rose, J., Morse, S., Rogers, D., Yates, T. (2009) using Satellite Images of Environmental Changes to Predict Infectious Disease Outbreaks, Emerging Infectious Diseases, 15, 1341–1346.

Gillespie, T., Chu, J., Frankenberg, E., Thomas, D. (2007) Assessment and Prediction of Natural Hazards from Satellite Imagery, Progress in Physical Geography, 31, 459–470.

Goetz, S., Prince, S. and Small, J. (2000) Advances in Satellite Remote Sensing of Environmental Variables for Epidemiological Applications, Advances in Parasitology, 47, 289–307.

Gohin, F., Lampert, L., Druon, J. (2002) A five channel chlorophyll concentration algorithm applied to SeaWiFS data processed by SeaDAS in coastal waters, International Journal of Remote Sensing 23, 1639–1661.

Gohin, F., Saulquin, B., Oger-Jeanneret, H., Lozac'h, L., Lampert, L., Lefebvre, A., Riouand, P., and Bruchon, F. (2008) Towards a better assessment of the ecological status of coastal waters using satellite-derived chlorophyll-a concentrations Remote Sensing of Environment,112, 3329–3340.

Gupta, P., Christopher, S., Wang, J., Gehrig, R., Lee, Y., Kumar, N., (2006)Satellite remote sensing of particulate matter and air quality assessment over global cities. Atmospheric Environment 40, 5880–5892.

Hall, S., Carolla, M., Deason, J. (2006) The Use of Imagery in Environmental Disaster preparedness and Response, Federal Facilities Environmental Journal, Winter, 65–72.

Harvey, J. (2002) Population Estimation Models Based on Individual TM Pixels, Photogrammetric Engineering and Remote Sensing, 68, 1181–1192.

Herbreteau, V., Salem, G., Souris, M., Hugot, J. and Gonzalez, J. (2007) Thirty Years of Use and Improvement of Remote Sensing applied to Epidemiology: From Early Promise to Lasting Frustration, Health and Place, 13, 400–403.

Heyes, A. (2000) Implementing Environmental Regulation: Enforcement and Compliance, Journal of Regulatory Economics, 17, 107–129.

Hidy G., Brook J., Chow J., Green M. Husar R., Lee C, Scheffe R., Swanson A. Watson J., (2009) Remote sensing of particulate pollution from space: have we reached the promised land? Journal of the Air Waste Management Assoc., 59, 1130–9.

Hoff, R. and Christopher, S. (2009) Remote Sensing of Particulate Pollution From Space: Have We reached the Promised Land?, Journal of Air and Waste Management Association, 59, 645–675.

Hoyningen-Huene, S, Freitag, M., and Burrows, J. (2003) Retrieval of aerosol optical thickness over land surfaces from top-of-atmosphere radiance JOURNAL OF GEOPHYSICAL RESEARCH, 108, AAC 2-1 – 2-20.

Howard, H., Pacifici, K., Pacifici, J. (2002) The Evolution of Remote Sensing, Environmental Protection, 13, 28–34.

Kalacska, Margaret E.; Bell, L., Arturo Sanchez-Azofeifa, G., and Caelli, T. (2009) The Application of Remote Sensing for Detecting Mass Graves: An Experimental Animal Case Study from Costa Rica, Journal of Forensic Sciences, 54, 159–166.

Leblond, A., Sandoz, A., Lefebvre, G., Zeller, H. and Bicout, D. (2007) Remote Sensing Based Identification of Environmental Risk Factors Associated with West Nile Disease in Horses in Camargue, France, Preventative Veterinary Medicine, 79, 20–31.

Lein, J. (2002) New Directions in the Use of Remotely Sensed Data for Cumulative Impact Assessment, Environmental Practice, 4, 220–235.

Lein, J. (2003) Integrated Environmental Planning, Wiley- Blackwell, New York, 240p.

Lein, J. (2009) Implementing Remote Sensing Strategies to Support Environmental Compliance Assessment: A Neural Network Applications, Environmental Science and Policy, 12, 948–958.

Liu, Y., Sarnat, J.A., Kilaru, V., Jacob, D.J., Koutrakis, P. (2005) Estimating ground-level PM2.5 in the eastern United States using satellite remote sensing. Environmental Science & Technology 39,3269–3278.

Martin, R. (2008) Satellite Remote Sensing of Surface Air Quality, Atmospheric Environment, 42, 7823–7843.

Middleton, C. (2002) Imagery Use in Environmental Enforcement, Proceedings: Pecora 15 Conference, Land Satellite Information, Denver, CO.

Miller, R. and Small, C. (2003) Cities from Space: Potential Applications of Remote Sensing in Urban Environmental Research and Policy, Environmental Science and Policy, 6, 129–137.

Munroe, D., Croissant, C., York, A. (2005) Land Use Policy and Landscape Fragmentation in an Urbanizing Region: Assessing the Impact of Zoning, Applied Geography, 25, 121–141.

Purdy, R. (2006) Satellites: A New Era for Environmental Compliance, Journal of European Environment and Planning Law, 3, 406–413.

Purdy, R. (2009) Using Earth Observation Technology for Better Regulatory Compliance and Enforcement of Environmental Laws, Journal of Environmental Law, 22, 59–87.

Remer, L., Kaufman, Y., Tanre, D., Mattoo, S., Chu, D., Martins, J., Li, R., Ichoku, C., Levy, R., Kleidman, R., Eck, T., Vermote, E., Holben, B. (2005) The MODIS aerosol algorithm, products, and validation. Journal of the Atmospheric Sciences, 62(4), 947–973.

Retalis, A., Sifakis, N. (2010) Urban aerosol mapping over Athens using the differential textural analysis (DTA) algorithm on MERIS-ENVISAT data, International Journal of Photogrammetry and Remote Sensing, 65, 17–25.

Ritchie, J., Zimba, P. and Everitt, J. (2003) Remote Sensing Techniques to Assess Water Quality, Photogrammetric Engineering & Remote Sensing, 69, 695–704.

Sanyal, J and Lu, X (2004) Application of Remote Sensing in Flood Management with Special Reference to Monsoon Asia: A Review NATURAL HAZARDS 33, 283–301.

Schweik, C. and Thomas, C. (2002) Using Remote Sensing to Evaluate Environmental Institutional Designs: A Habitat Conservation Planning Example, Social Science Quarterly, 83, 244–262.

Swenson, D. and Ambrose, R. (2007) A Spatial Analysis of Cumulative Habitat Loss in Southern California Under the Clean Water Act 404 Program, landscape and Urban Planning, 82, 41–55.

Tralli, D., Blom, R. Zlotnicki, V., Donnellan, A., Evans, D. (2005) Satellite Remote Sensing of Earthquake, Volcano, Flood, Landslide and Coastal Inundation Hazards, ISPRS Journal of Photogrammetry and Remote Sensing, 59, 185–198.

Tran, A., Gardon, J., Weber, A., Polidori, L. (2002) Mapping Disease Incidence in Suburban Areas Using Remotely Sensed Data, American Journal of Epidemiology, 156, 662–668.

Vincoli, J. (1993) Basic Guide to Environmental Compliance: Wiley and Sons, New York, 272p.

Wilson, J., Clay, M., Martin, E., Stuckey, D., Vedder-Risch, K. (2003) Evaluating Environmental Influences of Zoning in Urban Ecosystems with Remote Sensing, Remote Sensing of Environment, 86, 303–321.

Wilson, L. (1998) A practical Method for Environmental Impact Assessment Audits, Environmental Impact assessment Review, 18, 59–71.

Winter, S. and May, P., (2001) Motivation for Compliance with Environmental Regulations, Journal of Policy Analysis and Management, 20, 675–698.

Wu, C. and Murray, A. (2007) Population Estimation Using Landsat Enhanced Thematic Mapper Imagery, Geographical Analysis, 39, 26–43.

Xiao, H. and Ji, W. (2007) Relating Landscape Characteristics to Non-point Source Pollution in Mine Waste-located Watersheds using Geospatial Techniques, Journal of Environmental Management, 82, 111–119.

Zschau, J., and Kuppers, A. N., (Editors), 2003. "Early Warning Systems for Natural Disaster Reduction", Springer, Berlin, 834 pp.

Chapter 13
Integrative Sensing

An exercise in remote sensing commonly results in the production of a thematic map that depicts characteristics of the land surface germane to the motivating application. The cartographic convention of the map effectively communicates information that documents the spatial disposition of the desired elements, landscape elements and aids decision making as the patterns revealed on the map are interpreted and appraised. However, the outcome of a remote sensing investigation also serves as data for additional analytical treatment and wider forms of inquiry. When working in concern with complimentary geospatial technologies, remote sensing embodies a core methodology within the expanding field of geoinformatics; a discipline which integrates the acquisition, modeling, analysis, and management of spatially referenced data. Understanding remote sensing's contribution to this larger endeavor not only strengthens its role as an information technology, but further demonstrates the uniqueness of the remote sensing solution. As innovation continues to propel geospatial technologies forward, integrative sensing with remote sensing as the center-piece can be anticipated. Realizing this future depends on how and when remote sensing can be marshaled to support extended geomatic studies and where an appropriate technological fit can be achieved. In this chapter, we will examine the concept of integrative sensing. From this discussion we can explore the connection between geo-imaging as realized via satellite remote sensing and the field of geoinformatics as evidenced in the defining technologies of geographical information systems (GIS), environmental sensor networks, and geosimulation.

13.1 Integrative Sensing: Geomatics in Action

The multifacetted nature and complexity of present-day environmental concerns reflect high-variety and ill-structured problems that demand a synthesis of information and a means of analysis that can elucidate critical processes and define causation (Dar et al. 2010). Although the thematic map may be the intended product of image

J.K. Lein, *Environmental Sensing: Analytical Techniques for Earth Observation*,

DOI 10.1007/978-1-4614-0143-8_13,

Table 13.1 Fundamental geographic questions in directing GIS analysis

Where?	→	Location
What is there?	→	Basic inventory
What spatial patterns exist?	→	Patterns
Why there?	→	Cause and effect
What has changed since...?	→	Trends
What if...?	→	Modeling

processing methods, increasingly environmental analysis seeks to address more detailed question whose answers reside just below the thematic representation of the pixel (Merchant and Narumalani 2009). Exploiting the products generated from image processing and analysis for these purposed concentrates on the analytical manipulation of the surface representations created. In this context, remote sensing provides input to a higher-order geographical interrogation assisted by a set of specialized operations standardized within the software environment of a GIS (Weng 2010).

The combined use of remote sensing and GIS is well established (Hinton 1996; Estes 1992; He et al. 1998; Wilkinson 1996). Through the integration of these technologies a "value-added" dimension is available and new information products are possible that amplify and advance critical aspect of the problem under investigation. In many cases these products reveal hidden characteristics of the problem and in other situations they describe sources of derived data that assists in the management of ill-defined and ill-structured problems. Developing integrative solutions depends exclusively on how GIS is employed given the questions asked of the data. Because our data and processing environments are inherently spatial our questions assume a geographic disposition focused on considerations of

Location and extent
Distribution, pattern, or shape
Spatial association
Spatial interaction
Spatial change

The GIS helps to form, generate, and define geographic questions and support methodologies that can be developed to provide answers to them (Table 13.1). When the results of a sensing application are imported into the GIS environment, analysis is informed by these questions in order to determine (1) what and where features of interest are, and (2) how and why they are there. By asking the right question via GIS answers unfold as spatial realization that impart meaning to the observed arrangements that help unravel the spatial associations evidenced in the scene. A compliment of analytical functions exist in GIS that assist in crafting solutions to the questions asked of the data. These functions are assembled to form specific methodologies within a given problem-domain.

Although the analytical engine of a GIS is constantly evolving, there is a set of fundamental operations that can be drawn from to enrich a remote sensing investigation. These operations fall into two broad categories: (1) spatial analysis and (2) raster modeling. Under the spatial analysis heading three widely applicable operations include (1) selection and classification, (2) distance and proximity analysis, and (3) overlay. Typically, these spatial operations are used sequentially to address a problem forming chain operations which may end with a single map output or as "new" data to incorporate into other GIS operations. Raster modeling is based on the principles of map algebra; a set of operations performed on raster data using Boolean logic, Bayesian probability or approximate reasoning methods such as fuzzy set theory or Dempster–Shafer Theory of Evidence (Malpica et al. 2007). Raster analysis is generally performed using overlay operations to implement local functions such as

Mathematical operations – Addition, subtraction, exponentiation
Logical operations – Designed to reduce, simplify or combine data based on Boolean AND, OR, and NOT
Neighborhood operations – Employing "moving window" operations to modify or calculate attributes populating the raster surface

Taken in total, these GIS capabilities create new representations of the original data at local (cell), neighborhood (window), or global (surface) levels.

13.2 GIS Primitives

The analytical manipulation of spatially referenced data in a GIS proceeds according to the level of functionality descriptive of the software environment. Functionality reduces to a selection of task-specific algorithms that act on the spatial data to model a desired quality of characteristic. Algorithm function centers around three activities or actions:

Selection and classification – Because the primary data model of a remotely sensed image is a two-dimension grid or raster, selection and classification actions involve manipulating the values represented by the pixels comprising our scene. Selection, within this representational schema explains the process of identifying a feature (label) in the data that satisfies a condition or criteria, preserving that selected feature onto a new map representation and eliminating all other irrelevant features. In this example, classification entails recoding the desired value or label according to a simple "binary" logic where the desired condition is preserved by recoding the appropriate cells with the value (symbol) 1 and relegating all other cells in the raster to a value of zero (Fig. 13.1). As Fig. 13.1 demonstrates, the thermal surface characterizes surface temperatures across the scene; however in this example, analysis would like to select out only those areas that have temperatures about 20°C. The resulting selection displays those areas as a new image layer. A second type of selection uses a classification operation to impose order or structure the data into a meaningful categorization. In the example of continuous data such as the

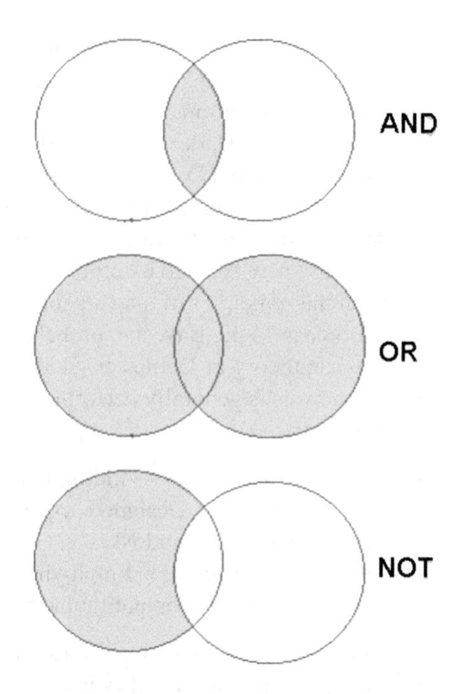

Fig. 13.1 Simple Boolean logic

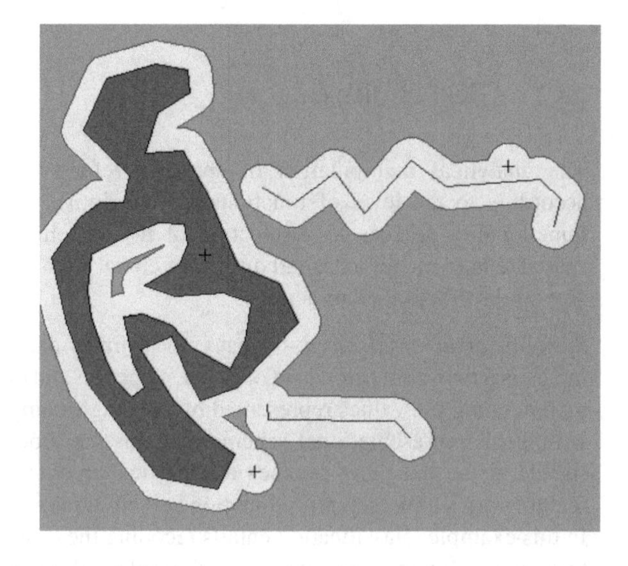

Fig. 13.2 GIS produced buffer generation

thermal surface presented in Fig. 13.1, the range of temperatures may be grouped into a pre-determined number of classes or based on mathematical properties or arbitrary system chosen by the analyst.

Proximity analysis – Proximity operations implement distance metrics to modify existing features on the image to form a new feature relationship. Distance as an operand produces a buffer geometry that encompasses the feature of interest (Fig. 13.2). Buffer generation entails calculating the separation outward from a

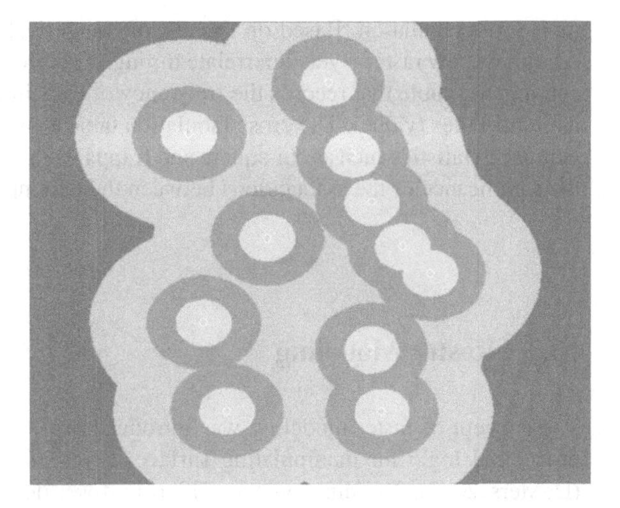

Fig. 13.3 A simple proximity surface

source pixel to all other pixels centers. The resulting geometry defines a geographic area that meets a specific distance requirement. For example, a remote sensing investigation using hyperspectral analysis may have identified land areas that match the spectra of an industrial gray-field. Environmental officials would now like to know if any of these sites are within 1 km of a ephemeral stream. Generating this information uses distance to calculate an outward spread from the target (Gray-field) pixels which is then followed by a binary classification that simplifies distance into a two-valued categorization; areas inside the threshold and areas greater than the threshold (Fig. 13.3).

Raster overlay – Overlay operations are perhaps the most recognized feature of GIS analysis. With respect to raster data, overlay describes the pixel-by-pixel combination of two surfaces. Assuming that the input raster surfaces are compatible, data from one layer at a given pixel location (row, column) is combined with the corresponding pixel in the second layer. Combination implies that a logic is in use that guides how the two layers are joined in to produce a new representation. The guiding logic is a pseudo-arithmetic manipulation of the attribute values within the raster. The arithmetic operation yields a result that satisfies the desired combination. For example, in the previous illustration where thermal data was subject to reclassification, suppose instead we wanted to know what land cover types correspond to areas that display the highest temperature. We could use addition to simply add the value of one surface to those of another. In this case given six land cover classes the possible combinations created from the addition operation would be confusing to interpret. However, if we multiplied the highest temperature surface (1) by land cover class only those areas where both 1 (highest temperature) and the land cover class existed together would be preserved. All other cells would result in zero or null values in the product image. A second type of map surface combination does not involve direct overlay but can be extremely useful when attempting to establish the statistical association between two raster surfaces. This is the procedure referred to as cross-tabulation. Based on the use of categorical data, cross-tabulation help illuminate how raster layers interrelate through the calculation of a two-dimensional contingency table that records the frequency of cells in a cross-wise comparison of their attributes (value). The cross-tabulation table provides a wealth of information including statistics such as chi-square and Kappa that can be interpreted to determine the significance of the associations between the two input surfaces.

13.3 Raster Modeling

The concept of raster modeling was introduced by Tomlin (1990) and explains a procedural logic for manipulating surface representations in a GIS environment (DeMers 2002). Building on the GIS primitives described previously, rater (or cartographic) modeling organizes these basic operations into a logical sequence of analysis designed to develop a specific information product from the GIS data. As a logic, raster modeling can be conceptualized as a collection of raster surfaces registered to a common cartographic reference. The individual layers are linked by the basic GIS operations which perform on the data in a manner analogous to how one would solve a set of algebraic equation to determine unknowns (Fig. 13.4). Depending on the nature of the problem, raster modeling can require a detailed series of steps that involve numerous geoprocesing operations and several raster data layers, or it can be relatively simple in design. The value of this technique is that it demands considerable fore-thought in crafting the solution. Fore-thought is often communicated in the form of a flow chart that outlines the relationship between data inputs, the GIS functions needed to transform them, and the anticipated output results. Raster modeling unlocks the power of GIS and facilitates its role in descriptive and prescriptive analysis (Lein 1997).

Deriving meaningful answers to the problems addressed using raster modeling places emphasis on two critical aspects of a modeling application:

- *Knowledge of the problem* – Because environmental processes are often complex, prior understanding of the variables required to both define the problem and explain causality is critical. With this fundamental understanding, the next challenge is to appreciate what tasks are needed to move the data toward the solution. As with any algebraic equation, to obtain an answer, one needs to understand the steps needed to solve it.
- *Knowledge of GIS primitives* – Seeing the solution also focuses attention on knowing which geoprocessing functions are needed to transform the data into the desired information. Since raster modeling algorithms are application specific, the conceptual framework outlining the approach has to be formalized ultimately into a set of geoprocessing operations that when executed in the proper sequence craft the solution.

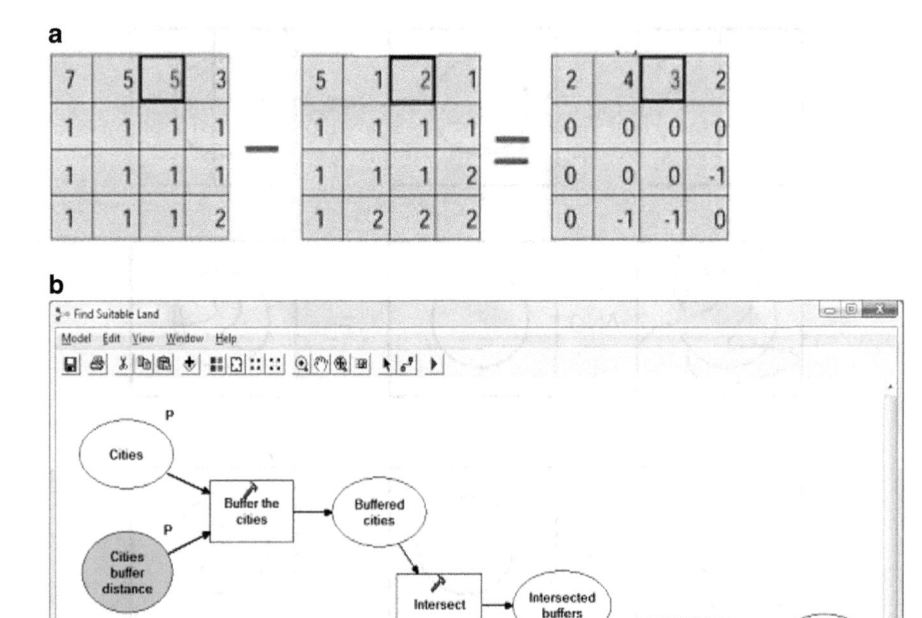

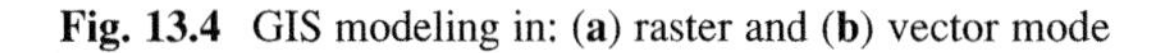
Fig. 13.4 GIS modeling in: (**a**) raster and (**b**) vector mode

Raster modeling operations are frequently described according to four main categories that explain how our data are being modeled with respect to spatial juxtaposition:

- *Local operations* – Operations of this type define GIS procedures that create an output surface on which the value of each cell is a function of the corresponding cell location on the input surface. Examples of local operations include mathematical functions, Boolean operations, reclassification, and overlay.
- *Focal operations* – Working within "neighborhoods" this class of operations implements a moving window algorithm to modify cell values in the input layer.
- *Global operations* – Global operations explain functions applied to an extended neighborhood to produce a new raster surface.Examples of global operations include various statistical operations, proximity analysis, geometric transformations, buffering and surface modeling such as viewshed analysis and cost surface mapping.

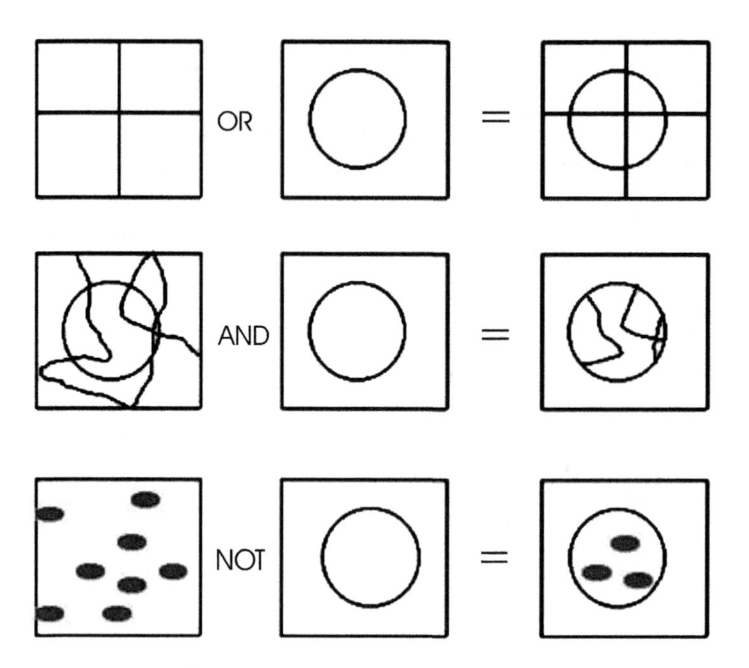

Fig. 13.5 The concept of GIS overlay

- *Zonal operations* – These functions are applied to define regions or zones in the input surface. Zones explain collections of cells that exhibit similar attributes which can be created using (1) reclassification calculations based on area, shape or perimeter or (2) categorical overlay developed from binary "cookie cutters" to extract cell values from a raster layer (Fig. 13.5).

13.4 Image-Based Modeling

Environmental modeling and simulation are integral activities in both the study of environmental process and the analysis of human impact on the environmental systems (Ford 1999; Shenk and Franklin 2001; Wainwright and Mulligan 2004). The activity of modeling and the production of models that capture the behavior of complex environmental processes support a societal need to consider the consequences of human decisions on the environment and to develop a better understanding of the interconnectedness and interactions that shape the landscape system. When examined from this perspective modeling becomes a useful way to characterize those patterns of the environmental system we are interested in knowing more about and although models are abstraction and simplified analogies of real systems, they have a value as a means of (1) prediction and comparison, and (2) as an aid to thought, communication and as a device for experimentation. An additional quality some models posses is that the process they explain are often spatial in nature.

Models that are inherently spatial represent the real world setting in a geographic context that preserves locational attributes and describes behaviors in a spatially explicit manner. In most cases models of this type abstract the landscape using a grid or polygonal arrangement. Using this geometry, calculations are performed to mimic critical aspects of key processes we wish to understand, calculations drawn from the physical laws, measured values and assumed relationships. Through the simulation process the ensuing results are assembled onto the abstracted landscape and the spatial patterns of the modeled response variables are displayed.

Spatial models, of course, are fed data where the driving input variables and parameters are themselves, spatial. Increasingly, remote sensing technology has been looked upon as an important source of data to support spatial modeling (Brimicombe 2003). Remote sensing serves other central aspects of the modeling problem as well including:

- *Data input* – The information products derived from image analysis and classification, from land cover, NDVI to temperature, represent land surface characteristics in a grid-based format that easily accommodates modeling efforts. In addition, objects at the surface, such as building footprints, roads and related features can be extracted from imagery and used as input to drive or parameterize the appropriate model.
- *Model validation* – Validation and verification are critical steps in the modeling and simulation process designed to establish the level of confidence in the model and the inference drawn from the simulation. Since models are essentially theories describing the structure and interrelationships of an observed phenomena, how well the model "fits" or compares against the real process determines whether or not the insights gained from the model are reasonable. Remotely sensed data can be employed to validate select models by providing an observed condition against which modeled results can be compared. In some instances, it may also be desirable to assemble a long-sequence time series to capture the actual progression of a system and its change over time with intervals set to the time step used to model the process.

A range of methodologies have been introduced to engage the environmental modeling problem. Rather than attempt to review each method in detail we can focus on a selection of two common modeling recipes that are readily adaptable to the use of remotely sensed data. The recipes described are basic representational schemas for expressing causality and approximating the spatial behavior of elements that form the landscape system.

1. *Markov processes* – This modeling recipe is based on the assumption that the future state (behavior) of an active process is influenced largely by its previous state of condition. Characterizing the environment following this logic explains system behavior in terms of probabilities where the likelihood of a process under investigation being in a given state at a particular point in time is determined from its previous state (Harbaugh and Bonham-Carter 1981). The progression of a system under the Markov design can be conceptualized as a chain of discrete

states in time or space where the probability of moving from one state to any other possible state in the sequence depends on the previous condition exhibited by the system. Therefore, a Markov process describes a series of transitions between different states defined by probability values associated with each transition exhibited by the system. Consequently, Markov processes in their general form contain a finite number of states. The probabilities associated with the transitions from state to state explain process for only a single step at a time; a property reference to as a first-order Markov chain. Perhaps the most significant aspect of Markov chains is that they exhibit a dependence on the probabilities associated with each transition of the immediately preceding state. In order to apply this modeling recipe the phenomena we wish to model must possess this property. In most instances, the thematic information we extract from remotely sensed data satisfies this condition which has supported the use of Markov models to explore a range of dynamic landscape processes (Riitters et al. 2009; Weng 2002; Wu et al. 2006; Guo et al. 2009).

2. *Cellular automata* – A cellular automata is a model that can be employed to explain how elements of a system interact. Models based on this design define a simulation environment represented by a grid space. On this grid space a set of rules are used to determine the attributes of a cell taking into account the attribute condition of its neighbors. The cell is not only an element in this raster representation of a geographic area, but also a type of memory element that stores states that represent characteristics of the system being modeled. In a cellular automaton all cells behave identically, have the same connectivity, and are directed based on three characteristics properties:

   (a) *States* – The number of distinct states or conditions a cell can be in
   (b) *Neighborhood* – The description of how cells are connected to each other
   (c) *Transition rules* – The decision o how a cell's state should change based on the state of its neighbors

   Change is introduced based on the rules added to the model which direct the system to its ext time step. Defining these rules rests at the core of the model since they (1) establish the logical relationships of the process subject to simulation and (2) determine their spatial evolution. All interactions in a cellular automaton are local, with the next state of a cell determined as a function of the current state of that cell and its neighbors. The rules for transition from one cell to another explain a general condition of either growth or decline, such as a change from land cover type x to land cover type y. This direction of change is a function of behaviors taking place in the neighborhood surrounding the cell, a neighborhood region that can assume several different geometries (Fig. 13.6). At the most basic level, we can reduce the concept of a transition rule to a single IF–THEN relationship that can take the form:

   IF – something acts on the neighborhood of a cell
   THEN – some related behavior alters the cell

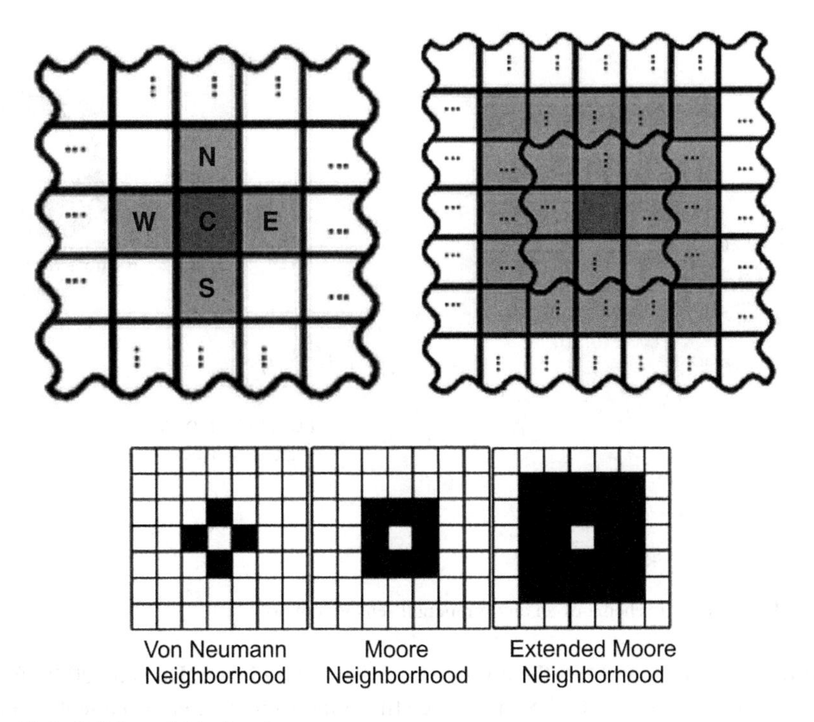

Fig. 13.6 Cellular neighborhoods

The example above describes a type of decision rule that moves the simulation forward. Although cellular automata are based on transformations produced locally (at the cell level) they generate a global pattern. The "cell" framework permits encoding the desired attributes of the landscape into the simulation model such that the state of cell can be made to represent land use, land cover or other conditions at the surface. The rules created to drive the model capture the dynamics of change and can be devised to mimic how phenomena in the real world behave. Interesting implementations of cellular automata using satellite-derived data inputs include Fan et al. (2008), Han et al. (2009), Bone et al. (2007), and He et al. (2005).

13.5 Environmental Sensor Nets and Webs

Although the definition of remote sensing implies that in situ data collection can be replaced by the imaging technologies embedded in the space-borne sensors, field data collection remains an important and often indispensible part of an environmental remote sensing investigation (McCoy 2005). Recently, sensor networks are related technologies have emerged that moves field data collection in a very different direction and presents new opportunities for gathering land surface

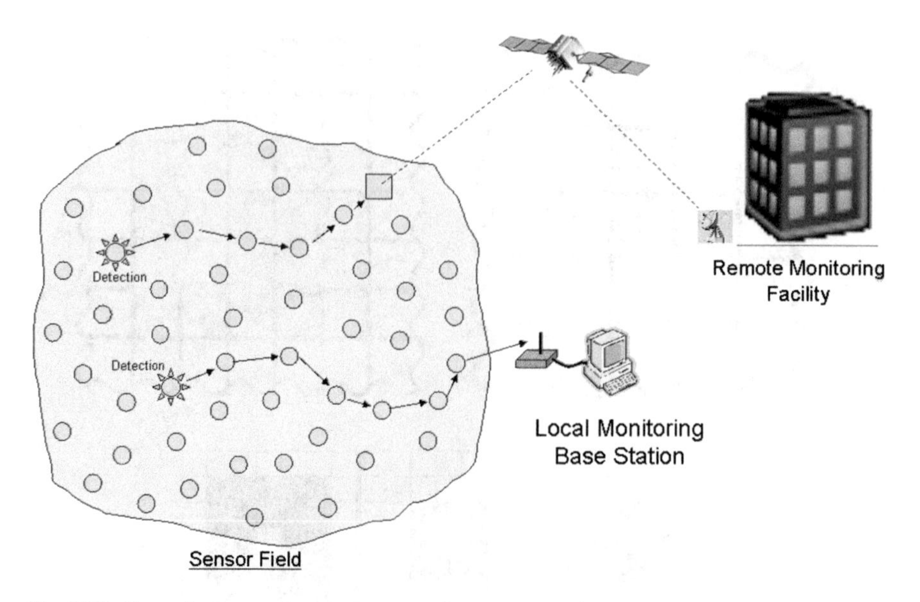

Fig. 13.7 General schema of an environmental sensor network

information that amplifies the utility of remotely sensed data (Ho et al. 2005; Porter et al. 2005; Kussul et al. 2009). While still in their infancy, environmental sensor networks and their assimilation into sophisticated sensor webs describe powerful combinations of distributed sensing capacity, real-time data, geographically anchored analysis, and coupled integration with adjacent networks and remote sensing data streams (Rundel et al. 2009). As this technology evolves, environmental analysis will be supplied with an enabling capacity to measure, map, monitor, and model earth environments in detailed and dynamic ways.

An environmental sensor network can be defined broadly as an array of sensor nodes distributed in the field together with a communication system that allows data they collect to reach a server (Hart and Martinez 2006). A general configuration illustrating the basic structure of an environmental sensor network is shown in Fig. 13.7. As Fig. 13.7 suggests, the sensor nodes deployed in the network collect data autonomously and this data stream is transmitted to one or more base stations which then forward the data stream to a computer functioning as a network server. Ideally, communication in the network is managed using wireless communications systems. The wireless connectivity allows sensor networks to be placed into remote and sensitive environments. Also, by employing wireless protocols, environmental sensor networks are capable of capturing local and broadly dispersed information simultaneously. In addition, sensor networks posses the capacity to respond to changes across the network of sensor elements by triggering observations selectively. Sensor networks therefore are a robust and dynamic approach to environmental observation (van Zyl et al. 2009).

The sensor nodes configured into a network can collect data on a wide range of environmental variables. The actual sensor elements selected depends largely on the goals established for the network. Typical environmental variables include temperature, humidity, pressure, soil moisture, and solar radiation; however, more specialized sensors can be introduced including GPS, air quality, and water quality indicators (Rundel et al. 2009). Ideally, sensor nodes can store data and make decisions about what data to transmit and when condition are appropriate (or optimal) to actually take measurements. The sensor network is therefore not only an observation platform, but also a communication platform where the communication abilities of the sensor nodes enable remote deployment. To be effective and efficient observation devices serving as sensor nodes should share a set of common traits including:

- Low cost
- Low power need
- Automated
- Robust
- Scalable
- Wireless
- And plug and play capable

Since sensor nodes are likely to be deployed in uncertain environmental conditions they should be designed to operate un-obtrusively in rugged outdoor setting.

As an information technology, environmental sensor networks are a type of ad hoc network that does not rely on a pre-existing infrastructure such as router or access points common to most managed computer network systems. Because of their ad hoc nature, sensor networks function well where central nodes cannot be easily configured. Their minimal design requirements and quick deployment features when coupled with dynamic and adaptive routing protocols enable ad hoc networks to form rapidly. However with ireless ad hoc networks, nodes compete to access the shared wireless medium which often results in data packet collisions. This problem is addressed by using some form of cooperative wireless communication such as a wireless mess topology (Akyildiz and Wang 2005). A wireless mesh network consists of radio nodes (sensors) organized in a mesh design (Fig. 13.8). Networks following this topology have three main elements: (1) mesh clients, (2) mesh routers, and (3) gateways. Clients in a mesh network can range from laptop computers to cell phones, but any wireless device such a sensor node can be incorporated into the network. Mesh routers forward information to and from gateways which can be connected to the internet. The coverage area for a mesh is called the mess cloud and access to the cloud depends on the nodes working in concert to form the network. An important feature of this design is its redundancy which makes this topology suitable for a range of environmental applications. With redundancy, when one node fails the remaining nodes can still communicate with directly or through one or more of the intermediate nodes that complete the network.

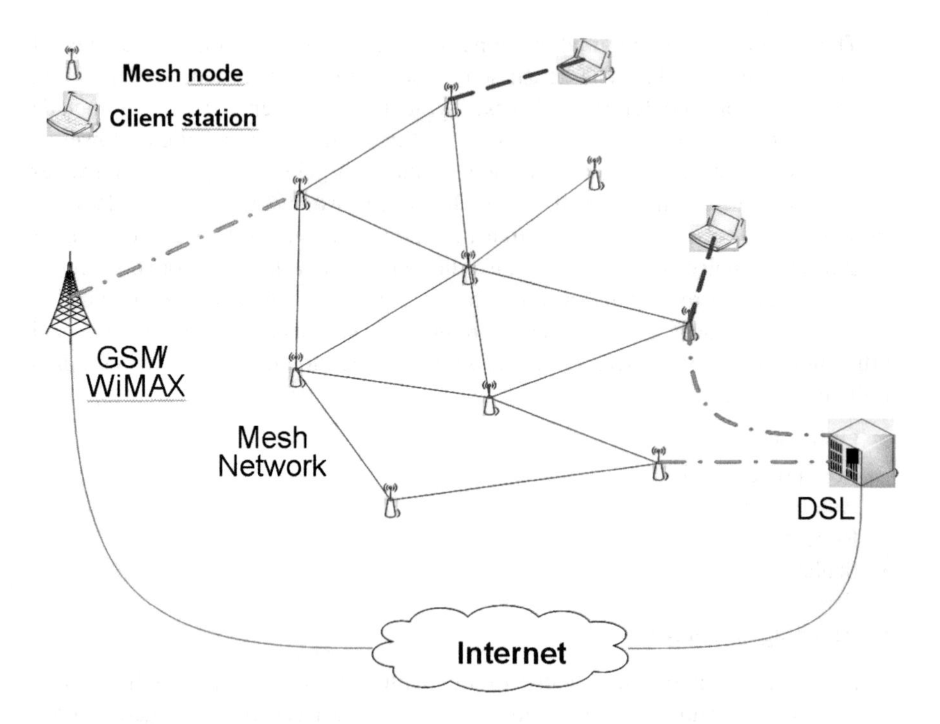

Fig. 13.8 Characteristic design of a wireless mesh network

Since most networked elements relay their data using radio links, communications are a critical feature of sensor networks. Network communication may be hampered by the environmental setting into which the sensor network is placed. Radio transmission is often constrained by power requirements; however, in complex terrain high power levels can be impractical. Generally, sensor networks transmitting on high frequencies are more depended on line-of-sight between stations. Unfortunately, current radio frequencies used by the majority of sensor nodes developed transmit over comparatively short distances. These factors conspire to restrict the current generation of sensor networks to areas not exceeding one square kilometer (Hart and Martinez 2006).

The challenges encountered in the design and deployment of environmental sensor networks has been summarized by Martinez et al. (2004) and includes consideration of factors such as

Miniaturization – Elements comprising the network must be unobtrusive, consequently size is a consideration that must be carefully balanced against the power requirements of each sensor node.

Power management – Long-term operation is critical to a successful deployment which demands the integration of systems to manage power and switch power supplies.

Scalability – Scalability explains the ability to add sensors to the network. Ideally sensor networks should be able to grow and needs dictate.

Remote management – Environmental sensor networks deployed in remote locations require a means to enable remote access.

Usability – Components of a network should be comparatively easy to install and maintain particularly if "off-the-shelf" turn-key systems are involved.

Standardization – Standardization implies compatibility with network components, operating systems, and data exchanges.

Security – Security issues are critical at all levels of the network to protect against physical and data interference.

An attractive feature of an environmental sensor network is that it enables the integration of information obtained from multiple sensors into a larger world-view not detectable by any single sensor alone (Rundel et al. 2009). Moving beyond the sensor network takes this concept of a distributed sensing into the realm of the sensor web. The sensor web concept has been described in detail by Delin (2002). In its most general form, a sensor web explains a "macro-instrument" comprised of spatially distributed sensor platforms that for temporally synchronous, geographically amorphous networks that form an embedded monitoring presence. This evolving vision of a "network or networks" provides an infrastructure that supports the integration of sensor systems that allow access to a range of instruments, sensor networks and their corresponding observational data sets (van Zyl et al. 2009). The central theme of this vision is that the sensor web instrument facilitates spatio-temporal understanding of an environment through the coordinated measurement of a multiple number and type of sensing platforms including both orbital and terrestrial, fixed and mobile (Delin 2002; Delin et al. 2005). As an open-complex adaptive system the sensor web can be defined by four unique characteristics (van Zyl et al. 2009):

- Heterogeneous components arranged into a self-organizing network structure that is robust and dynamic
- Open sensor resources that interact either directly or by means of a mediator or proxy
- A network of sensor networks that highlight internet communication protocols
- External access to the constituent components that form the sensor web that serve as sensor resources defined as data or metadata

By design the sensor web spreads collected data and processed information throughout its entire network. The general schema of this design is illustrated in Fig. 13.9. However, unlike a typical wireless network, there is no design criterion for routing a sensor web. Information is spread everywhere and the communication protocol is structured for omni- and bi-directional flow. Because there is no specific routing of information, all sensor elements share everything with each other. These qualities together with the schema illustrated in Fig. 13.9 suggest that sensor webs are comprised of three fundamental layers; (1) an information layer, (2)

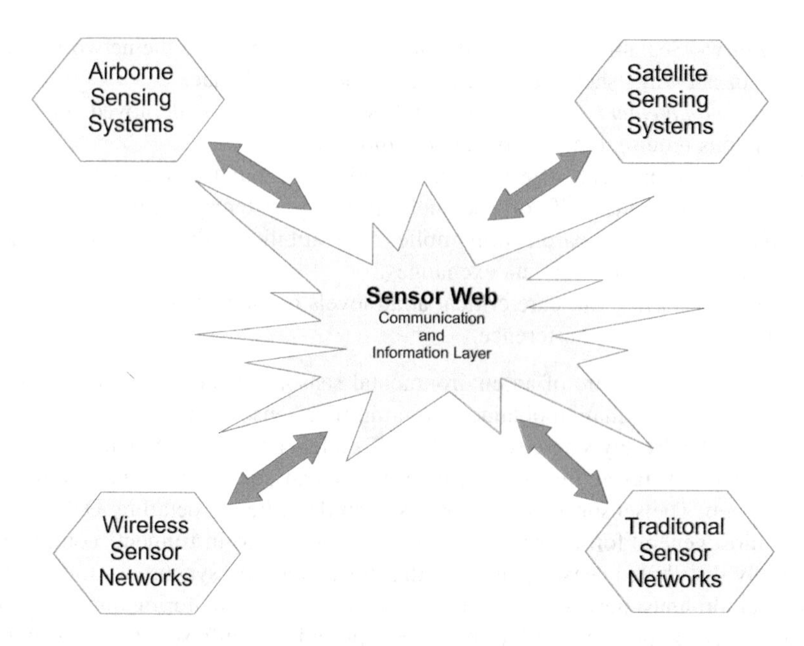

Fig. 13.9 Generalized pattern of an environmental sensor web

a communication layer, and (3) a sensor logic (Liang et al. 2005). These layers can be described accordingly:

Information layer – This is the component of the sensor web where the sensing resources are stored, disseminated, exchanged displayed and analyzed. Sensing resources include the various sensors deployed in the web, the sensor locations, their real-time, near-real time or archived measurements, control elements and other information pertinent to sensor web users. This layer displays significant variety of data transport and access demands, data uses and data users where interoperability is essential. The information, given these demands should support data from different sources and facilitate their combination and integration.

Communications layer – This layer controls how data and commands are transmitted within the sensor web. Included at this level are the media, protocols and topologies that organized the web. Typically this takes form as an internet, satellite, cell-phone, or radio-based network. The specific configuration of this layer depends exclusively on the environmental setting and the requirements and constraints imposed by the purpose the web was developed to support.

Sensor layer – The sensors identify devices that provide usable output in response to a specific physical quantity, property or condition which is the subject of measurement. Advances in sensor technology have produced devices capable of measuring physical, chemical and biological attributes. Many of these sensors have become smaller, more reliable, less expensive, more efficient, and intelligent and capable of operating in situ or remotely. In situ sensors tend to be less costly per unit and enjoy higher accuracies and temporal resolutions. Remote sensors have the advantage of a much wider spatial extent. However, sensor webs employ both in situ and remote sensors to achieve truly integrated sensing capabilities.

Examples of integrated sensing using sensor web technologies coupled with satellite-based platforms have been discussed by Chien et al. (2005) and Mandl (2004). Although the systems described are largely experimental, these prototypes demonstrate the value of sensor web implementations and how Earth observing satellites such as the Advanced Land Imager (ALI), Hyperion, and EO-1 together with the MODIS and ASTER instruments can be assimilated in a collaborative manner to address high-value environmental applications.

13.6 Sensing on the Horizon

The nature of environmental problem-solving and the need to provide timely information to guide decision making will encourage continued research in the use of remotely sensed data. One of the more promising and interesting developments that has the potential to advance remote sensing science and its role in environmental analysis in the near-future is the deployment of small-satellite systems (Xue et al. 2008; Kramer and Cracknell 2008). Small satellites, although common in the early days of space exploration, have witnessed resurgence due to their comparative cost advantage, short development cycle, and simplicity (Fig. 13.10).

The term small satellite is typically applied to systems that have an on-orbit mass ranging from 500 kg to as a little as 1 kg (Table 13.2). Aside from the smaller,

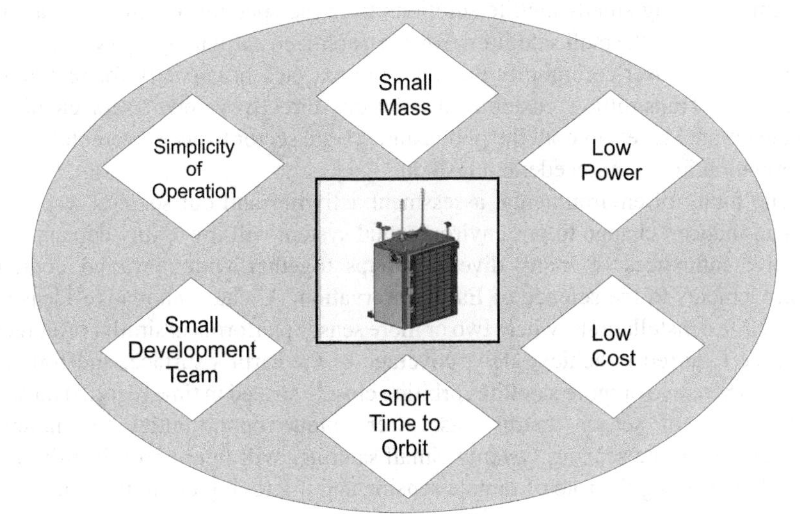

Fig. 13.10 Qualities of small-satellite systems

Table 13.2 Characteristics of small satellites

Parameter	Low-end buses (w/o options)	High-end buses (w/o options)
Design life (years)	1–3	≫5
Reliability (at design life)	0.8–0.9	0.8–0.9
Avionics redundancy	Limited	Extensive to full
Bus mass (kg)	150–300	425–650
Payload mass (kg)	100–300	300–500
Payload power (orbital average, W)	60–125	100–500
Propulsion authority (kg hydrazine)	0–25	33–75
Pointing accuracy (deg 3-sig)	0.02[a]–0.25	0.01[a]–0.03[a]
Pointing knowledge (deg 3-sig)	0.001[a]–0.2	0.003[a]–0.008[a]
Data storage (Gbit)	2–64	12–200
Downlink (Mbps)	2–4 at S-band; 100 at X-band available on SA200S	2 at S-band, 320 at X-band

Note: The low-end buses are the Spectrum Astro SA200S, Swales, and the three-axis TRW STEP; the high-end buses are the Ball RS2000, Lockheed Martin LM900, and TRW SSTI-500
Source: Rapid Spacecraft Development Office (RSDO). 1999. Available online at http://rsdo.gsfc.nasa.gov
[a]With star trackers

cheaper, faster philosophy propelling development of small satellites, one attractive feature of this technology is their potential for customization. Given this low cost over traditional systems, it becomes feasible to deploy small satellites as remote sensing platforms that can meet very specific and specially tailored information needs. Here, the design capabilities of the satellites can be matched exclusively to the requirements of a well-defined and narrow application problem. An example of this potential is described by Guelman and Ortenberg (2009) were the requirements for implementing small-satellite solutions for hyperspectral imaging is examined. As experience with small satellites grows, customized earth imaging systems can be presented to a user community that may present lack of capacity for remote data collection. Transmitting customized solutions directly to end users enables all organizational levels in both the public and private sector to access remotely sensed information in an unprecedented fashion.

The future of environmental assessment activities and our societal response to human-induced change to our environmental system will invariably depend on the creative influences of many diverse groups together with new and continued improvements to the science of Earth observation. Already innovative ideas such as satellite constellations, where two or more sensor platforms in similar orbit reduce the time required to achieve daily coverage of the earth's surface, and formation flying, where two or more satellites orbiting closely spaced in time record data across a full suite of sensor instruments, offer unique opportunities to implement distributed remote sensing systems. Small satellites will figure prominently in this model, expanding the role of remote sensing and improving our ability to:

- Access any area on the earth within 24 h
- Obtain coordinated measurements over a range of sensor instruments
- Collect critical data regarding the disposition of our environment

When these innovations are brought together the future of environmental sensing promised to be dynamic and exciting.

13.7 Summary

Satellite remote sensing can be considered one facet of an expanding list of geospatial technologies. In this chapter, the integration of remote sensing into this larger information technology framework was described examining first how the results of a remote sensing investigation can fit seamlessly into the analytic engine of GIS. Here, by exploring a set of GIS primitives, the information content of a remote sensing product can be maximized and the raster format of the remote sensing product can flow directly into a range of raster modeling applications. Image-based modeling concepts were examined using Markov processes and cellular automata has examples. Geographic information systems, however, are only one integrative technology. Also considered in this chapter was the potential for deploying environmental sensor networks and the expansion of environmental sensing opportunities through the design of sensor webs. From this discussion the future of remote data collection was examined through the introduction of small-satellites and their role in environmental analysis and assessment.

References

Akyildiz, I and Wang, X. (2005), "A Survey on Wireless Mesh Networks," IEEE Communications Magazine, 43, s23–s30.

Bone, C., Draagicevic, S., Roberts, A. (2007) Evaluating Forest Management Practices using GIS-based Cellular Automata Modeling Approach with Multispectral Imagery. Environmental Modeling and Assessment, 12, 105–118.

Brimicombe, A. (2003) GIS, Environmental Modelling and Engineering, Taylor and Francis, London, 312p.

Chien, S., Cichy, B., Davies, A., Tran, D. (2005) An Autonomous Earth-Observing Sensorweb, IEEE Intelligent Stem, 16–24.

Dar, I., Sankar, K Mithas, D. (2010) Remote sensing technology and geographic information system modeling: An integrated approach towards the mapping of groundwater potential zones in Hardrock terrain, Mamundiyar basin *Journal of Hydrology*, Vol. 394, Issue: 3-4, November 26, 2010. pp. 285–295.

Delin, K. (2002) The Sensor Web: A Macro-Instrument for Coordinated Sensing, Sensors, 2, 275–286.

Delin, K, Jackson, S., Johnson, D. Burleigh,S. (2005) Environmental Studies with the Sensor Web: Principles and Practice, Sensors, 5, 103–117.

DeMers, M. (2002) GIS Modeling in Raster, Wiley and Sons, New York, 203p.

Estes, J. (1992) Remote sensing and GIS integration - Research needs, status and trends , ITC Journal , no. 1, pp. 2–10. 1992.
Fan, F., Wang, Y., Wang, Z. (2008) Temporal and Spatial Change Detecting (1998-2003) and Predicting of Land Use and land Cover in Core Corridor of Pearl River Delta (China) by Using TM and ETM + Images, Environmental Monitoring and Assessment, 137, 127–147.
Ford, A. (1999) Modeling the Environment: An Introduction to System Dynamics Modeling of Environmental Systems, Island Press, Washington, DC., 401p.
Guelman, M., Ortenberg, F. (2009) Small Satellite's Role in Future Hyperspectral Earth Observation Missions, Acta Astronautica, 64, 1252–1263.
Guo, Zhongyang; Dai, Xiaoyan; Wu, Jianping (2009) Study on land use/land cover change in Jintai and Weibing districts of Baoji city in Western China based on remote sensing technology and Markov method, Journal of Applied Remote Sensing, Volume 3, 332–347.
Han, J., Hayashi, Y., Cao, X., Imura, H. (2009) An Application of and Integrated System Dynamics and Cellular Automata Model for Urban Growth Assessment: A Case Study of Shanghai, China, Landscape and Urban Planning, 91, 133–141.
Harbaugh, J and Bonham-Carter, G. (1981) Computer Simulation in Geology, Robert E. Krieger Publishing, Malabar, FL, 575p.
Hart, J., Martinez, K. (2006) Environmental Sensor Network: A revolution in the Earth Systems Science?, Earth-Science Review, 78, 177–191.
He, C., Zhang, Q., Li, Y., Li, X., Shi, P. (2005) Zoning grassland Protection Area Using Remote Sensing and Cellular Automata Modeling- A Case Study in Xilingol, Steppe Grassland in Northern China, Journal of Arid Environments, 63, 814–826.
He, Hong S., David J. Mladenoff, Volker C. Radeloff, and Thomas R. Crow. 1998. INTEGRATION OF GIS DATA AND CLASSIFIED SATELLITE IMAGERY FOR REGIONAL FOREST ASSESSMENT. Ecological Applications 8:1072–1083. [doi:10.1890/1051–0761(1998) 008[1072:IOGDAC]2.0.CO;2].
Hinton, J.C. (1996) GIS and remote sensing integration for environmental applications. Int. J. GIS 10: 877–890.
Ho, C., Robinson, A., Millerm D. and Davis, M. (2005) Overview of sensors and needs for environmental monitoring, Sensors, 5, 4–37.
Kramer, H. and Cracknell, A. (2008) An Overview of Small satellites in remote Sensing, International journal of remote Sensing, 29, 4285–4337.
Kussul, N., Shelestov, A. and Skakun, S. (2009) Grid and sensor web technologies for environmental monitoring Earth Science Informatics. Vol. 2, no. 1–2, pp. 37–51. June 2009.
Lein, J. (1997) Environmental Decision Making: An Information Technology Approach, Blackwell Science, Malden, MA., 213p.
Liang, S., Croitoru, A., Tao, V. (2005) A Distributed Geospatial Infrastructure for Sensor Web, Computers and Geosciences, 31, 221–231.
Malpica, J A | Alonso, M C | Sanz, M A Dempster-Shafer Theory in geographic information systems: A survey Expert Systems with Applications. Vol. 32, no. 1, pp. 47–55. Jan. 2007.
Mandl, D. (2004) Experimenting with Sensor Webs Using Earth Observing 1, IEEE Aerospace Conference Proceedings, 176–183.
Martinez, K., Hart, J., Ong, R. (2004) Environmental Sensor Networks, Computer, 37, 50–56.
McCoy, R. (2005) Field Methods in Remote Sensing, The Guilford Press, New York, 158p.
Merchant, J.W.; Narumalani, S. Integrating remote sensing and geographic information systems. In *The SAGE Handbook of Remote Sensing*; Warner, T.A., Nellis, M.D., Foody, G.M., Eds.; SAGE Publications Ltd: London, UK, 2009; pp. 257–268.
Porter, J., Arzberger, P., Braun, H., Brynat, P., Gage, S., Hansen, T., Lin, C., Lin, F., Kratz, T., Michener, W., Shapiro, S. and Williams, T. (2005) Wireless Sensor Networks for Ecolog, BioScience, 55, 561–572.
Riitters, K., Wickham, J., Wade, T. (2009) An indicator of forest dynamics using a shifting landscape mosaic, Ecological Indicators, 9, 107–117.

Rundel, P., Grham, E., Allen, M., Fisher, J., and Harmon, T. (2009) Environmental sensor networks in ecological research, New Phytologist, Volume 182, Issue 3, pages 589–607.

Shenk, T and Franklin, A. (2001) Modeling in Natural Resource Management, Island Press, Washington, DC., 223p.

Tomlin, C. (1990) Geographic Information Systems and Cartographic Modeling, Prentice-Hal, Englewood Cliffs, NJ., 249p.

van Zyl, T., Simonis, I. and McFerren, G. (2009) The Sensor Web: Systems of Sensor Systems, International Journal of Digital Earth, 2, 16–30.

Wainwright, J. and Mulligan, M. (2004) Environmental Modeling: Finding Simplicity in Complexity, Wiley and Sons, New York, 408p.

Weng, Q. (2002) Land Use Change Analysis in the Zhujiang Delta of China using Satellite , GIS and Stochastic Modeling, Journal of Environmental Management, 64, 273–284.

Weng, Q. (2010) Remote Sensing and GIS Integration: Theories, Methods and Applications, McGraw-Hill, 416p.

Wilkinson, G.G. (1996) A review of current issues in the integration of GIS and remote sensing data. *International Journal of Geographical Information Systems*, 10(1), pp. 85–101. ISSN (print) 0296–3798.

Wu, Q., Li, H., Wang, R, Paulussen, J, He, Y., Wang, M, Wang, B., Wang, Z. (2006) Monitoring and Predicting Land Use Change in Beijing Using Remote Sensing and GIS, landscape and Urban Planning, 78, 322–333.

Xue, Y., Li, Y., Guang, J., Zhang, X., Guo, J. (2008) small Satellite remote Sensing Applications – History, Current and Future, International Journal of Remote Sensing, 29, 4339–4372.

Index

J.K. Lein, *Environmental Sensing: Analytical Techniques for Earth Observation*, DOI 10.1007/978-1-4614-0143-8, © Springer Science+Business Media, LLC 2012

I

9781461401421